AF430629

THE PHYSICS OF CANCER

Research Advances

THE PHYSICS OF CANCER
Research Advances

Editor

Bernard S. Gerstman, Ph.D.
Florida International University, USA

NEW JERSEY · LONDON · SINGAPORE · BEIJING · SHANGHAI · HONG KONG · TAIPEI · CHENNAI · TOKYO

Published by

World Scientific Publishing Co. Pte. Ltd.

5 Toh Tuck Link, Singapore 596224

USA office: 27 Warren Street, Suite 401-402, Hackensack, NJ 07601

UK office: 57 Shelton Street, Covent Garden, London WC2H 9HE

Library of Congress Control Number: 2020947402

British Library Cataloguing-in-Publication Data
A catalogue record for this book is available from the British Library.

THE PHYSICS OF CANCER
Research Advances

Copyright © 2021 by World Scientific Publishing Co. Pte. Ltd.

All rights reserved. This book, or parts thereof, may not be reproduced in any form or by any means, electronic or mechanical, including photocopying, recording or any information storage and retrieval system now known or to be invented, without written permission from the publisher.

For photocopying of material in this volume, please pay a copying fee through the Copyright Clearance Center, Inc., 222 Rosewood Drive, Danvers, MA 01923, USA. In this case permission to photocopy is not required from the publisher.

ISBN 978-981-122-348-8 (hardcover)
ISBN 978-981-122-349-5 (ebook for institutions)
ISBN 978-981-122-350-1 (ebook for individuals)

For any available supplementary material, please visit
https://www.worldscientific.com/worldscibooks/10.1142/11915#t=suppl

Typeset by Stallion Press
Email: enquiries@stallionpress.com

Desk Editor: Xiao Ling

CONTENTS

Preface

Bernard S. Gerstman

Department of Physics
Florida International University
Miami, Florida, USA

Cancer deaths per capita have decreased in recent years, but the improvement is attributed to prevention, not treatment. The difficulty in treating cancer may be due to its "complexity", in the mathematical physics sense of the word. Tumors evolve and spread in response to internal and external factors that involve feedback mechanisms and non-linear behavior. Investigations of the properties of the individual units, biological cells, are necessary but not sufficient. In addition, the non-linear interactions among cells, and between cells and their environment require investigations that are equally crucial for developing a sufficiently detailed understanding of the system's emergent phenomenology to be able to eventually control the behavior. In the case of cancer, controlling the system's behavior will mean the ability to treat and cure the disease. Physicists have been studying various complex, non-linear systems for many years using theoretical, mathematical, computational, and experimental approaches. These investigations have provided insights that allow physicists to make unique contributions towards the treatment of cancer.

A large part of cancer treatments involves attacking and killing cancer cells using modalities such as chemotherapy and radiation. These methods of treatment are valuable, but it is not enough for two reasons. Cancer cells by nature proliferate at an enormous rate. The amount of treatment necessary to attack the proliferating and invading army of cancer cells usually proves to be toxic to healthy tissues and causes dreadful side-effects. There are treatments that attempt to minimize the terrible side-effects by targeting very specific aspects of cancer cells. However, the high proliferation rate naturally leads to cancer cell heterogeneity that results in some cells becoming resistant to the specialized molecular therapy while still maintaining their cancerous properties. A second reason for the necessity of a wider paradigm in cancer treatment is the horrifying ability of cancer to migrate to other regions and invade heathy tissues. Therefore, it is crucial to develop an understanding of the mechanisms of how cancer cells communicate and interact with each other, and with other cells in the environment. This understanding may produce new treatments that complement the modalities that attack individual cells.

This interdisciplinary book presents recent advancements in physicists' research on cancer. The work presented in this volume uses a variety of physical, biochemical, mathematical, and computational techniques to gain a deeper molecular and cellular understanding of the horrific disease that is cancer. Some of these chapters report on work that aims to better understand the properties of individual cancer cells, while many of these chapters report on research that investigates the dynamics of cancer cells' interactions with each other, and with their environment.

The first chapter in the book by Sara Hamis, Stanislav Stratiev, and Gibin G Powathil, entitled, "Uncertainty and Sensitivity Analyses Methods for Agent-Based Mathematical Models: An Introductory Review" provides an excellent gateway into multiscale, agent-based mathematical modeling of biological systems. Uncertainties and strong sensitivities to choices of parameter values are unavoidable. In light of this, they recommend a consensus approach using multiple analysis platforms. They discuss three different platforms and very helpfully describe how to implement these methods.

The non-linear, complex cellular dynamics of cancer produces a system that intrinsically can be investigated using the mathematics of game theory. To do this well requires using the formalism of relevant physical and mathematical methodologies along with biochemical input that allow a game theoretic approach to adapt to the specific aspects that make cancer so challenging. Yinan Zheng, Yusha Sun, Gonzalo Torga, Kenneth Pienta, and Robert Austin apply mean-field evolutionary game theory in an astute manner in the chapter, "Game Theory Cancer Models of Cancer Cell–Stromal Cell Dynamics Using Interacting Particle Systems".

The mathematical techniques of fractal analysis have been used extensively by physicists to investigate a variety of systems. The chapter entitled, "Occupancy and Fractal Dimension Analyses of the Spatial Distribution of Cytotoxic ($CD8^+$) T Cells Infiltrating the Tumor Microenvironment in Triple Negative Breast Cancer" by Juliana C. Wortman, Ting-Fang He, Clare Yu, and others use fractal analysis to better understand how to increase favorable clinical outcomes in breast cancer. By using novel applications of fractal dimensionality, they can better understand how specific variations in the spatial density of tumor infiltrating lymphocytes result in better clinical results.

Tumor cell heterogeneity is a major problem for developing efficacious treatments that are robust against drug resistance. Xin Li and Devarajan (Dave) Thirumalai investigate the critically important issue of cancer cell heterogeneity in their chapter, "Cooperation Among Tumor Cell Subpopulations Leads to Intratumor Heterogeneity". Using an evolutionary model, they develop quantitative theoretical models for explaining data from a variety of *in vitro* experiments involving pancreatic cancer, as well as *in vivo* data from glioblastoma multiforme. These physical and mathematical models are an important complement to experimental studies in developing effective therapies.

Cell migration and invasion is, unfortunately, a hallmark of the progression of cancer. The switch between cell states, such as epithelial, mesenchymal or amoeboid states, seems to be mainly based on epigenetic changes and environmental cues that induce the reversible transition of cells toward another state and thereby promote a specific migration mode. In her chapter, "Cell Mechanics Drives Migration Modes", Claudia Mierke provides an excellent review of how mechanical features of cells and their matrix environment regulate and subsequently determine individual migration modes.

The immense influence of the physical properties of the extracellular matrix (ECM) on tumor progression and malignancy is further explored by Hanna Engelke in her chapter, "Physics of the Extracellular Matrix and Biology of Tumors — A Close Relationship". In addition to physically impeding or promoting tumor progression, the ECM can also indirectly modulate tumors via interactions with cellular signaling pathways, such as integrin or YAP/TAZ signaling. Conversely, tumors remodel and re-build the extracellular matrix. This leads to feedback loops and cascades of mutual interaction. Therefore, combinations of therapies that target the tumor's interaction with the extracellular matrix may improve the success of cancer therapy.

Computational models are promising complementary tools for guiding future immuno-oncology research, because they are fast, powerful and cost-efficient. In their chapter, "Computational Modelling and Experimental Investigations to Enhance the Successful Response of Anti-PD-1 Cancer Immunotherapies", Damijan Valentinuzzi, Robert Jeraj and coworkers describe a computational model that they use to explore the tumor characteristics associated with resistance to anti-PD-1 immunotherapy. The model is designed to account for the heterogeneity of tumor cells in MHC class I expression and PD-1 ligand expression. Importantly, they validate their computational model with *in vitro* and *in vivo* experiments and conduct sensitivity analyses to explore the impact of the measured tumor characteristics on the observed tumor response to anti-PD-1 immunotherapy.

The chapter by Matthew Zanotelli, Cynthia Reinhart-King and coworkers, "The Physical Microenvironment of Tumors: Characterization and Clinical Impact", also focuses on the important role played by the tumor microenvironment in tumorigenesis and metastasis. Extracellular matrix homeostasis is lost during tumor progression, and altered physical properties of the tumor microenvironment alter cancer cell behavior, limit delivery and efficacy of therapies, and correlate with tumorigenesis and patient prognosis. They discuss significant mechanical and structural properties of the tumor microenvironment, and provide numerical values and clinical impact across cancer type and grade. This leads to a better understanding of how aberrant extracellular matrix dynamics contribute to cancer progression and will be crucial in effectively detecting, monitoring, and treating cancer.

Yu Ling Huang, Mingming Wu and coworkers discuss the link between tumor architecture and tumor invasion in their chapter, "The Architecture of Co-Culture Spheroids Regulates Tumor Invasion Within a 3D Extracellular Matrix". Most current 3D tumor invasion assays consist of single tumor cells embedded within an extracellular matrix but do not account for cell–cell adhesion within the tumor. They developed a micrometer scale 3D co-culture spheroid invasion assay. Real-time microscopic imaging of tumor spheroid invasion revealed that the spatial distribution of two cell types critically regulated tumor invasion. This work highlights the importance of biophysical cues within the bulk of the tumor for tumor invasion.

Physical techniques are crucially important in understanding the molecular mechanisms of specific cancer treatments. This is demonstrated in the chapter by D. Bilge, N. Civelek and Z. Özçelik Çetinel, "Investigations of Interactions Between Altretamine and Model Membranes: Spectroscopic and Calorimetric Analysis". They studied at a molecular level, the interactions of the drug altretamine with model membranes. It is an alkylating, antineoplastic drug utilized for the therapy of progressive ovarian cancer. This type of physical investigation of the interactions of chemotherapeutic compounds with biological membranes provides valuable information about the therapeutic process and the drug's toxicity.

The relationship between cancer and the immune system is critically important but not fully understood. Activation of oncogenic pathways in tumor cells is known to reduce induction of a local antitumor immune response, and the tumor growth becomes faster. Vladimir Zhdanov discusses this topic in his chapter, "Kinetic Aspects of the Interplay of Cancer and the Immune System". He proposes a kinetic model describing the role of the immune system in the lifetime risk of cancer. His analysis shows how specific details can determine the size of tumors.

Charles Lineweaver and Paul Davies provide a thoughtful discussion in their chapter, "Comparison of the Atavistic Model of Cancer to Somatic Mutation Theory: Phylostratigraphic Analyses Support the Atavistic Model". The atavistic theory postulates that many of the hallmarks of cancer on a cellular level such as cell proliferation, are from a cellular reversion to deeply embedded, pre-existing functions that were originally advantageous. They distinguish this atavistic idea from the somatic mutation theory that proposes that cancer cells evolve new functionalities. By identifying differences between the two ideas, they make specific predictions of the atavistic model that can be compared with phylostratigraphic information.

I would like to thank Xiao Ling in the Editorial Department at World Scientific Publishing. She has been wonderfully helpful in shepherding this project to completion. I would also like to thank my wife, Noelle Gerstman. Her biologist's insight has been extremely helpful throughout this process. Finally, I would like to dedicate this book to my mother, Frances Shapiro Gerstman, who died many years ago from cancer. She passed away much too early for both her and me.

About the Editor

Professor Bernard S. Gerstman, Ph.D., is Professor of Physics at Florida International University. His research speciality is in theoretical and computational molecular and cellular biophysics. After receiving his Ph.D. from the Physics Department at Princeton University, Prof. Gerstman moved to a post-doctoral position at The University of Virginia. He then accepted a faculty position at Florida International University. He has recently served two terms as Chair of the Physics Department at FIU. He has also been an Executive Editor for the American Institute of Physics, and Editor-In-Chief for the Springer-Nature book series, *Biological and Medical Physics, Biomedical Engineering*.

Chapter 1

Uncertainty and Sensitivity Analyses Methods for Agent-Based Mathematical Models: An Introductory Review

Sara Hamis[1,2], Stanislav Stratiev[3] and Gibin G. Powathil[2,*]

[1]School of Mathematics and Statistics
University of St Andrews
St Andrews, KY16 9SS, Scotland, United Kingdom

[2]Department of Mathematics
College of Science, Swansea University
Swansea, SA1 8EN, United Kingdom

[3]Department of Physics
College of Science, Swansea University
Swansea, SA2 8PP, United Kingdom
**g.g.powathil@swansea.ac.uk*

Multiscale, agent-based mathematical models of biological systems are often associated with model uncertainty and sensitivity to parameter perturbations. Here, three uncertainty and sensitivity analyses methods, that are suitable to use when working with agent-based models, are discussed. These methods are namely Consistency Analysis, Robustness Analysis and Latin Hypercube Analysis. This introductory review discusses origins, conventions, implementation and result interpretation of the aforementioned methods. Information on how to implement the discussed methods in MATLAB is included.

Keywords: Sensitivity analysis; agent-based models; cancer model; consistency analysis; robustness analysis; latin hypercube analysis.

1. Introduction

Mathematical models of biological systems are abstractions of highly complex reality. It follows that parameters used in such models often are associated with some degree of uncertainty, where the uncertainty can be derived from various origins. Epistemic uncertainty refers to uncertainty resulting from limited knowledge about the biological system at hand, whilst aleatory uncertainty stems from naturally occurring stochasticity, intrinsic to biological systems.[1,15,23] Model parameters may thus be naturally stochastic, theoretically unknown, and unfeasible or impossible to measure precisely (or at all). Further magnifying the contributions of uncertainty in mathematical models of biological systems, in particular, is the fact that *one* parameter in the mathematical model may correspond to *a multitude* of underlying biological mechanisms and factors in the real, biological system. This is especially true for minimal parameter models, *i.e.* mathematical models that aspire to be as non-complex as possible whilst still capturing all biological details of interest.[8] In this review, we focus our attention on uncertainty and sensitivity analyses techniques that are suitable for use with *agent-based* models. A mathematical, agent-based model comprises several distinct *agents* that may interact with each other and their environment. In an agent-based tumour model, for example, an agent typically corresponds to one tumour cell or a group of tumour cells. This naturally allows for heterogeneity amongst tumour cells, which is useful as tumour heterogeneity is associated with many complications involved in modelling (and treating) solid tumours. Accordingly, many modellers in the field of mathematical oncology choose to work with agent-based models.[24]

There already exist multiple method papers that describe *how* to perform uncertainty and sensitivity analyses when working with agent-based models, authors Alden *et al.* even provide a free R-based software package (Spartan[1]) that enables the user to perform different such methods, including the three methods discussed in this review. However, as these methods have been developed across multiple research fields, both inside and outside of the natural sciences, it is difficult to find one comprehensive review that discusses not only how to perform these methods, but also *where* these methods come from, and *why* certain conventions are proposed and/or used. To this end, we have in this review gathered such information for three uncertainty and sensitivity analyses techniques, namely Consistency Analysis, Robustness Analysis and Latin Hypercube Analysis. Our aim is that this will allow the reader to better evaluate uncertainty and sensitivity analyses presented by other authors, and encourage the reader to consider performing these methods when suitable.

In order to understand the impact that parameter uncertainty and parameter perturbations have on results produced by a mathematical model, uncertainty and sensitivity analyses can be used. A mathematical model that comprises a set of uncertain model parameters (or *inputs*), is able to produce a range of possible responses (or *outputs*). *Uncertainty analysis* assesses the range of these outputs overall, and provides information regarding how certain (or uncertain) we should be with our model results, and the conclusions that we draw from them.[2] *Sensitivity analysis* describes the relationship between uncertainty in inputs and uncertainty in outputs. It can be used to identify which sources of input uncertainty (*i.e.* which model parameters) significantly influence the uncertainty in the output and, equally importantly, which do not.[2] Assessing how sensitive the output is to small input perturbations is a healthy way to scrutinise our mathematical models.[14] Moreover, for a well-formulated model, knowledge regarding how input uncertainty influences output uncertainty can yield insight into the biological system that has not yet been empirically observed.[1] Furthermore, if the uncertainty in some input parameter is shown to not affect output uncertainty, the modeller may consider fixing that parameter, and thus reducing model complexity in accordance with a minimal-parameter modelling approach. In *local* sensitivity analysis techniques, model parameters (inputs) are perturbed one at a time whilst other parameters remain fixed at their calibrated value. In *global* sensitivity analysis techniques, all model parameters are simultaneously perturbed.[4]

There exist several sensitivity and uncertainty analyses techniques, but here we will focus on three such techniques that are suitable to use in conjunction with agent-based mathematical models. These techniques are namely Consistency Analysis, Robustness Analysis and Latin Hypercube Analysis, which all answer important, and complementary, questions about mathematical models and their corresponding *in silico* responses.[1,23]

Question 1: How many data samples (i.e. how many *in silico* runs) are needed in order to mitigate uncertainty originating from intrinsic model stochasticity?
Answer: See Consistency Analysis.

Question 2: How robust are model reponses (outputs) to local parameter perturbations?
Answer: See Robustness Analysis.

Question 3: How robust are model reponses (outputs) to global parameter perturbations?
Answer: See Latin Hypercube Analysis.

Note that Consistency Analysis is only meaningful when analysing models with stochastic variables.

The statistical techniques described in this review have been developed and applied across multiple academic disciplines, both inside and outside of the natural sciences. Consequently, terminology and notations vary in the literature. The aim of this review is to combine pertinent literature from various academic fields whilst keeping terminology and mathematical notations consistent, unambiguous and tailored towards a mathematical and scientific audience. Therefore, when needed, certain algorithms from the literature are here reformulated into expressions that a mathematician would consider to be conventional. This review is intended to provide gentle, yet comprehensive, instructions to the modeller wanting to perform uncertainty and sensitivity analyses on agent-based models. Thorough directions on how to perform Consistency Analysis (Section 3), Robustness Analysis (Section 4) and Latin Hypercube Sampling and Analysis (Section 5) are provided. Consistency Analysis utilises *the A-measure of stochastic superiority*, which is therefore discussed in Section 2. Throughout this review, we have included some historical information that elucidates why certain statistical conventions are used. Each section also contains pictorial, step-by-step instructions on how to perform the aforementioned techniques. Worked examples of all methods discussed in this review are provided in

Section 6. These worked examples use *in silico* data produced in a previous agent-based, multiscale, mathematical oncology study.[9]

1.1. *Methods outside the scope of this review*

Note that there exist other uncertainty and sensitivity analyses techniques, suitable for agent-based models, that are outside the scope of this review. For example, Bayesian inference is a statistical technique that uses Bayes theorem, prior beliefs and observed data to infer input parameter values and their associated uncertainties.[12] When this inference is difficult to calculate, computational methods such as Approximate Bayesian Computation (ABC)[13] or, if the model is highly computationally expensive, Approximate Approximate Bayesian Computation (AABC)[3] can be used to perform inference on input model parameters. The Sobol method is a variance-based sensitivity analysis method that, using Monte-Carlo multidimensional integration, allows for the evaluation of each individual parameter's fractional contribution to the output variance.[31] Another global, variance based sensitivity analysis method that enables quantification of the input parameter's fractional contributions to the output variance is Fourier amplitude sensitivity testing (FAST) which uses an underlying algorithm that involves Fourier decomposition.[26] Although the Sobol method and FAST use different underlying techniques, they both allow us to say that "input parameter r_i contributes R_i percent to the output variance". Newer sensitivity analysis methods and tools tend to take advantage of the current abundance of computing power. The recently introduced MASSIVE (Massively parallel Agent-based Simulations and Subsequent Interactive Visualization-based Exploration) methodology, for example, combines parallel computing and interactive data visualisation to produce a graphical user interface that provides an overview of input-output relations for a broad input parameter range.[22]

2. The A-Measure of Stochastic Superiority

2.1. *The common language statistics*

In 1992, McGraw and Wong introduced the *common language statistics* (*CL*) as an intuitive way to compare two distributions of data.[18] The *CL* was initially introduced as a tool to compare data from normal distributions, but was later on approximated for use on any continuous distributions. The *CL* describes the probability that a random data sample from one of the distributions is greater than a random data sample from the other distribution. For example, if we have two continuous data distributions B and C, and we are comparing the distributions with respect to some variable X, then the *CL* is simply given by

$$CL_{BC}(X) = P(X_B > X_C), \tag{1}$$

where standard probability notations have been used so that $P(X_B > X_C)$ denotes the probability that a random data sample X_B from distribution B is greater than a random data sample X_C from distribution C.[18] Thus the subscript of X here signifies the distribution from which the data sample X was taken.

2.2. *The A-measure of stochastic superiority*

The CL was developed to compare continuous data distributions, but Vargha and Delaney[29] introduced the *A-measure of stochastic superiority* (or A-measure for short) as a generalisation of the CL that can directly be applied to compare both continuous and discrete distributions of variables that are at least ordinally scaled. When comparing two distributions B and C, with respect to the variable X, the A-measure $A_{BC}(X)$ is given by

$$A_{BC}(X) = P(X_B > X_C) + 0.5P(X_B = X_C), \tag{2}$$

where $P(X_B = X_C)$ denotes the probability that a random data sample from distribution B is equal to a random data sample from distribution C. By comparing Equations 1 and 2, it is clear that in the continuous case, where $P(X_B = X_C) = 0$, the A-measure reduces to the CL.

If two distributions that are identical with respect to the variable X are compared, then $P(X_B > X_C) = P(X_C > X_B)$ and we say that the distributions B and C are *stochastically equal* with respect to the variable X. On the other hand, if $P(X_B > X_C) > P(X_C > X_B)$, then we say that the distribution B is *stochastically greater than* distribution C, and accordingly, that distribution C is *stochastically smaller than* distribution B.[29] If distribution B is stochastically greater than distribution C with respect to the variable X, it simply occurs more often that the sample X_B is greater than the sample X_C when two random samples X_B and X_C are compared. These definitions of stochastic relationships (*stochastically equal to, stochastically greater than, stochastically smaller than*), used by Vargha and Delaney,[29] amongst others, are weaker than definitions used by some other authors, but sufficient and appropriate for our current purposes: comparing distributions of discrete data samples produced by *in silico* simulations based on stochastic, agent-based mathematical models.

When comparing two samples X_B and X_C, the possible outcomes are (i) that X_B is greater than X_C, (ii) that X_B is equal to X_C and (iii) that X_B is smaller than X_C. These three possible outcomes must sum up to one so that,

$$P(X_B > X_C) + P(X_B = X_C) + P(X_C > X_B) = 1. \tag{3}$$

In the continuous case, $P(X_B = X_C) = 0$ as previously stated, and thus it follows that

$$P(X_C > X_B) = 1 - P(X_B > X_C), \quad \text{for continuous distributions}, \tag{4}$$

and thus it suffices to know only one of the values $P(X_B > X_C)$ or $P(X_C > X_B)$ in order to determine the stochastic relationship between the distributions B and C with respect to X.

▶ *For example: if $P(X_B > X_C) = 0.4$, then it is clear that $P(X_C > X_B) = 0.6$ and thus that $P(X_B > X_C) < P(X_C > X_B)$, or equivalently, that distribution B is stochastically smaller than distribution C.*

However, in the discrete case, $P(X_B = X_C)$ is not generally equal to zero and therefore,

$$P(X_C > X_B) = 1 - P(X_B > X_C) - P(X_B = X_C) \quad \text{for discrete distributions.} \quad (5)$$

Consequently, one single value $P(X_B > X_C)$ or $P(X_C > X_B)$ alone can generally not be used to determine the stochastic relationship between the distributions B and C.

▶ *For example: if, again, $P(X_B > X_C) = 0.4$, it follows that $P(X_C > X_B) = 0.6 - P(X_B = X_C)$. This does not give us enough information to determine the stochastic relationship between the two distributions B and C.*

In order to proceed to compare the distributions B and C in this case, the *stochastic differenceδ* is introduced, where δ is given by

$$\delta = P(X_B > X_C) - P(X_C > X_B), \quad \delta \in [-1, 1]. \quad (6)$$

Via a linear transformation, the *transformed stochastic difference, $\delta' \in [0, 1]$*, can be obtained using Equation 5 so that

$$\begin{aligned}
\delta' = \frac{\delta + 1}{2} &= \frac{P(X_B > X_C) - P(X_C > X_B) + 1}{2} \\
&= \frac{P(X_B > X_C) - (1 - P(X_B > X_C) - P(X_B = X_C)) + 1}{2} \\
&= P(X_B > X_C) + 0.5 P(X_B = X_C) = A_{BC}(X),
\end{aligned} \quad (7)$$

from which we can see that the A-measure, $A_{BC}(X)$ (Equation 2), measures the stochastic difference between $P(X_B > X_C)$ and $P(X_C > X_B)$ under a linear transformation.[29]

In order to estimate the A-measure using samples from two distributions, the point estimate of the A-measure, here denoted the $\hat{A}$-measure (with a hat), is used. (In the Spartan package,[1] this is referred to as the A test score). For example, if we want to compare two discrete distributions B and C, where B comprises m data samples (of some variable X) so that $B = \{b_1, b_2, \ldots, b_m\}$ and C comprises n data samples (of some variable X) so that $C = \{c_1, c_2, \ldots, c_n\}$ then

$$\hat{A}_{BC}(X) = \frac{\#(b_i > c_j)}{mn} + 0.5 \frac{\#(b_i = c_j)}{mn}, \quad (8)$$

where $i = 1, 2, \ldots, m$ and $j = 1, 2, \ldots, n$ and $\#$(event) is the 'counting function' that simply denotes the number of times that a certain event occurs when comparing all

$$B = \{b_1, b_2, b_3, b_4, b_5\} = \{1, 2, 3, 2, 1\}, \quad m=5$$
$$C = \{c_1, c_2, c_3, c_4, c_5\} = \{6, 0, 0, 2, 1\}, \quad n=5$$

	$i=1$	$i=2$	$i=3$	$i=4$	$i=5$	sum
# $(b_i > c_j)$	2	3	4	3	2	14
# $(b_i = c_j)$	1	1	0	1	1	4
# $(b_i < c_j)$	2	1	1	1	2	7

$$\hat{A}_{BC}(X) = 14/(5 \times 5) + 0.5 \times 4/(5 \times 5) = 0.64$$

Fig. 1. Using Equation 8 to compute the point estimate of the A-measure, *i.e.* the $\hat{A}$-measure or $\hat{A}_{BC}$, of the two distributions of data samples B and C of sizes m and n respectively.

possible pairs of data samples (b_i, c_j). For clarity, Figure 1 provides an example of how the $\hat{A}$-measure can be computed by simply counting events.

Using more conventional mathematical notation, the $\hat{A}$-measure is given by

$$\hat{A}_{BC}(X) = \frac{1}{mn} \sum_{i=1}^{m} \sum_{j=1}^{n} H(b_i - c_j), \tag{9}$$

where $H(x)$ is the Heaviside step function such that

$$H(x) = \begin{cases} 1 & \text{for } x > 0, \\ \dfrac{1}{2} & \text{for } x = 0, \\ 0 & \text{for } x < 0. \end{cases} \tag{10}$$

If $\hat{A}_{BC}(X) = 0.5$, then the distributions B and C are stochastically equal with respect to the variable X. The $\hat{A}$-measure can thus be used to measure 'how equal' two discrete distributions B and C are, by assessing how much the $\hat{A}$-measure ($\in [0, 1]$) deviates from equality, *i.e.* the value 0.5. The closer the $\hat{A}$-measure is to 0.5, the 'more equal' the two compared distributions are.[29] In many applications, we are only interested in 'how equal' two distributions B and C are, and it is not important which distribution is the stochastically greater one. In such cases we are only interested in *how much* the $\hat{A}$-measure deviates from stochastic equality (*i.e.* the value

0.5) but the *direction* is not important. Or in mathematical terms: the *magnitude* of the difference between the $\hat{A}$-measure and stochastic equality is important but the *sign* is not. The magnitudal (or scaled) $\hat{A}$-measure (or $\hat{A}$-value), here denoted $\underline{\hat{A}}$ with an underscore, ignores the sign of deviation from equality and is given by

$$\underline{\hat{A}} = \begin{cases} \hat{A}_{BC}(X) & \text{if } \hat{A}_{BC}(X) \geq 0.5, \\ 1 - \hat{A}_{BC}(X) & \text{if } \hat{A}_{BC}(X) < 0.5. \end{cases} \tag{11}$$

The statistical significance is used to describe the effect of the stochastic difference between two distributions B and C. If two distributions B and C are 'fairly equal' (*i.e.* if they yield an $\underline{\hat{A}}_{BC}$-measure close to 0.5) then the statistical significance is classified as *small*. The statistical significance is classified using the magnitudal $\underline{\hat{A}}$-measure and, using guidelines from Vargha and Delaney,[29] the statistical significance is classified to be small, medium or large with respect to X according to the following threshold values for $\underline{\hat{A}}_{BC}(X)$,

$$\text{Statistical Significance} = \begin{cases} \text{small} & \text{if } \underline{\hat{A}}_{BC}(X) \in [0.5, 0.56], \\ \text{medium} & \text{if } \underline{\hat{A}}_{BC}(X) \in (0.56, 0.64], \\ \text{large} & \text{if } \underline{\hat{A}}_{BC}(X) \in (0.64, 0.71]. \end{cases} \tag{12}$$

These threshold values (that might appear somewhat arbitrary) were first introduced by psychologist and statistician Jacob Cohen[6,7] in the 1960s when comparing normal distributions, but then in terms of another statistical measurement: the effect size (Cohen's) **d** where

$$\mathbf{d} = \frac{|(\text{mean of population } B) - (\text{mean of population } C)|}{\sigma}, \tag{13}$$

and σ is the standard deviation of either B or C (as B and C here are assumed to have the same standard deviation).[7,25] Omitting details from statistics, a small **d**-value essentially corresponds to a big overlap between distributions B and C, whilst a large **d**-value corresponds to a small overlap between distributions B and C, as is illustrated in Figure 2. Cohen decided to use the threshold **d**-values for describing 'small', 'medium' and 'large' effect sizes to be 0.2, 0.5 and 0.8 respectively.[7] If we hold on to the assumption that B and C are two normal distributions with the same variability, and furthermore say that they contain the same number of data samples, we can use measures of overlap to get a further 'feel' for the previously discussed effect sizes, as illustrated in Figure 2. Cohen's **d** value can also be converted into *'the probability that a random data sample X_B from (normal) distribution B is larger than a random data sample X_C from (normal) distribution C,*[18] but that is exactly what the $\hat{A}$-measure $\hat{A}_{BC}(X)$ measures! So this is where the threshold values for the

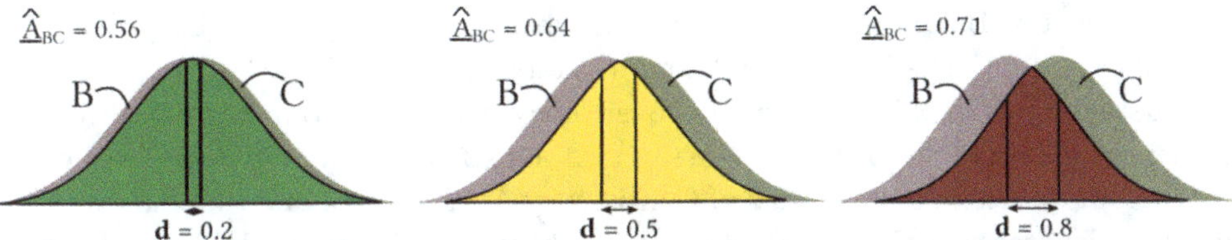

Fig. 2.　The small (left), medium (centre) and large (right) threshold values for the scaled A-measure of stochastic superiority ($\hat{\underline{A}}_{BC}$) are based on Cohen's **d**-values comparing two normal distributions B and C with the same variance. The higher the overlap between B and C, the smaller the **d**-value, and the smaller the $\hat{\underline{A}}_{BC}$-measure ($\hat{\underline{A}}_{BC} \in [0.5, 1]$).

descriptors 'small', 'medium' and 'large' statistical differences, listed in Equation 12, come from.

Now, Cohen motivated his choice of the **d**-value thresholds using a blend of intuitive 'everyday' examples and mathematical reasoning.[7] However, he did issue a warning regarding the fact that the threshold values should be determined based on the research methodology at hand. Thus the modeller should not blindly use Cohen's suggested thresholds, but instead reason what constitutes a small enough statistical significance in the study at hand. The modeller must also decide how fine the data samples in the data distributions should be before performing Consistency Analysis. In many applications, it is likely that the amount of data samples required in order to achieve a small statistical significance increases with the fineness of the data. Nonetheless, scientific conventions are useful and thus in the remainder of this review we will use the threshold values suggested by Cohen, as is done in other mathematical biology studies.[1]

3. Consistency Analysis

In silico simulations based on mathematical models with built-in stochasticity will not produce the same output data every simulation run. Consistency Analysis (also called aleatory analysis) is a stochastic technique that answers the question: how many data samples do we need to produce in order to mitigate uncertainty originating from intrinsic model stochasticity? In our case, *one data sample* is the product of *one in silico simulation*, so an equivalent question is: how many *in silico* simulations should we run before describing our results in terms of, for example, average values, standard deviations or similar?

Let us say that one *in silico* simulation produces one data sample of some output response X. This data sample can for example correspond to 'the population size at time point T', or something similar. It is up to the modeller to identify and decide what the meaningful output response(s) should be, and Consistency Analysis can be performed on multiple output responses at multiple time steps, for comprehensiveness. Before we begin, note that when performing Consistency Analysis, we always use the calibrated model parameters.

The first step involved in performing Consistency Analysis is to produce multiple distributions of data of various sizes. We say that a distribution with n data samples has a distribution size n, and the goal of Consistency Analysis is to find the smallest n value (here denoted n^*) that yields a small stochastic significance. To do this, we create various *distribution groups* that all contain 20 distributions each of some distribution size n, as is shown in **Step 1** in Section 3.1. Following the methodology described in previous work by Alden *et al.*, and the Spartan package that they developed,[1] we create one distribution group that contains 20 distributions of size $n = 1$, one distribution group that contains 20 distributions of size $n = 5$ and so on. Here, the n values 1, 5, 50, 100 and 300 are evaluated[1] and thus we must produce a total of $20 \times (1 + 5 + 50 + 100 + 300) = 9120$ *in silico* runs. (Note that if the desired accuracy is not achieved for the highest investigated n value, here $n = 300$, higher values of n can be explored).

We here let a distribution $D_{n,k}$ denote the kth distribution of distribution size n so that

$$D_{n,k} = \{d_{n,k}^1, d_{n,k}^2, \ldots, d_{n,k}^n\}, \tag{14}$$

where $d_{n,k}^h$ is the the hth data sample in distribution $D_{n,k}$ and $h = 1, 2, \ldots, n$. The $\hat{A}$-measure resulting from comparing two distributions $D_{n,k}$ and $D_{n,k'}$ with respect to the variable X is denoted by $\hat{A}_{k,k'}^n(X)$.

Now, within every distribution-group, we compare the first distribution ($k = 1$) to all other distributions ($k' = 2, 3, \ldots, 20$) using the $\hat{A}$-measure. This yields 19 $\hat{A}$-measures per distribution-group, as is shown in **Step 2** in Section 3.1. The maximum scaled $\hat{A}$-measure with respect to X, occurring in a distribution-group g that contains distributions of size n_g, is denoted $\underline{\hat{A}}_{max}^{n_g}(X)$. The smallest value n_g for which $\underline{\hat{A}}_{max}^{n_g}(X) \leq 0.56$ is denoted n^*. In other words: n^* corresponds to the smallest distribution size for which all of the 19 computed $\hat{A}$-measures yield a small stochastic significance, as is shown in **Step 3** in Section 3.1. This answers the question that we set out to answer via Consistency Analysis: n^* data samples (or *in silico* runs) are needed in order to mitigate uncertainty originating from intrinsic model stochasticity. The procedure on how to perform Consistency Analysis is outlined Section 3.1.

3.1. *Quick guide: Consistency analysis*

Here follows a quick guide for how to perform Consistency Analysis.

Consistency Analysis answers Question 1: How many data samples (or *in silico* runs) are needed in order to mitigate uncertainty originating from intrinsic model stochasticity?

Produce multiple groups of distributions. Each group (g) contains 20 distributions comprising n_g data samples each, as illustrated below.

We let $D_{n,k}$ denotes the k:th distribution of distribution size n so that

$$D_{n,k} = \{d^1_{n,k}, d^2_{n,k}, .. , d^{n-1}_{n,k}, d^n_{n,k}\}.$$

g = 1
$n_g = n_1 = 1$

$D_{1,1} = \{d^1_{1,1}\}$
$D_{1,2} = \{d^1_{1,2}\}$
$\vdots$
$D_{1,20} = \{d^1_{1,20}\}$

g = 2
$n_g = n_2 = 5$

$D_{5,1} = \{d^1_{5,1}, d^2_{5,1}, d^3_{5,1}, d^4_{5,1}, d^5_{5,1}\}$
$D_{5,2} = \{d^1_{5,2}, d^2_{5,2}, d^3_{5,2}, d^4_{5,2}, d^5_{5,2}\}$
$\vdots$
$D_{5,20} = \{d^1_{5,20}, d^2_{5,20}, d^3_{5,20}, d^4_{5,20}, d^5_{5,20}\}$

g = 3
$n_g = n_3 = 50$

$D_{50,1} = \{d^1_{50,1}, d^2_{50,1}, ..., d^{49}_{50,1}, d^{50}_{50,1}\}$
$D_{50,2} = \{d^1_{50,2}, d^2_{50,2}, ..., d^{49}_{50,2}, d^{50}_{50,2}\}$
$\vdots$
$D_{50,20} = \{d^1_{50,20}, d^2_{50,20}, ..., d^{49}_{50,20}, d^{50}_{50,20}\}$

g = 4
$n_g = n_4 = 100$

$D_{100,1} = \{d^1_{100,1}, d^2_{100,1}, ..., d^{99}_{100,1}, d^{100}_{100,1}\}$
$D_{100,2} = \{d^1_{100,2}, d^2_{100,2}, ..., d^{99}_{100,2}, d^{100}_{100,2}\}$
$\vdots$
$D_{100,20} = \{d^1_{100,20}, d^2_{100,20}, ..., d^{99}_{100,20}, d^{100}_{100,20}\}$

g = 5
$n_g = n_5 = 300$

$D_{300,1} = \{d^1_{300,1}, d^2_{300,1}, ..., d^{299}_{300,1}, d^{300}_{300,1}\}$
$D_{300,2} = \{d^1_{300,2}, d^2_{300,2}, ..., d^{299}_{300,2}, d^{300}_{300,2}\}$
$\vdots$
$D_{300,20} = \{d^1_{300,20}, d^2_{300,20}, ..., d^{299}_{300,20}, d^{300}_{300,20}\}$

We thus need to produce $20\Sigma n_g$ data samples, here $20(1+5+50+100+300)= =9120$ data samples.

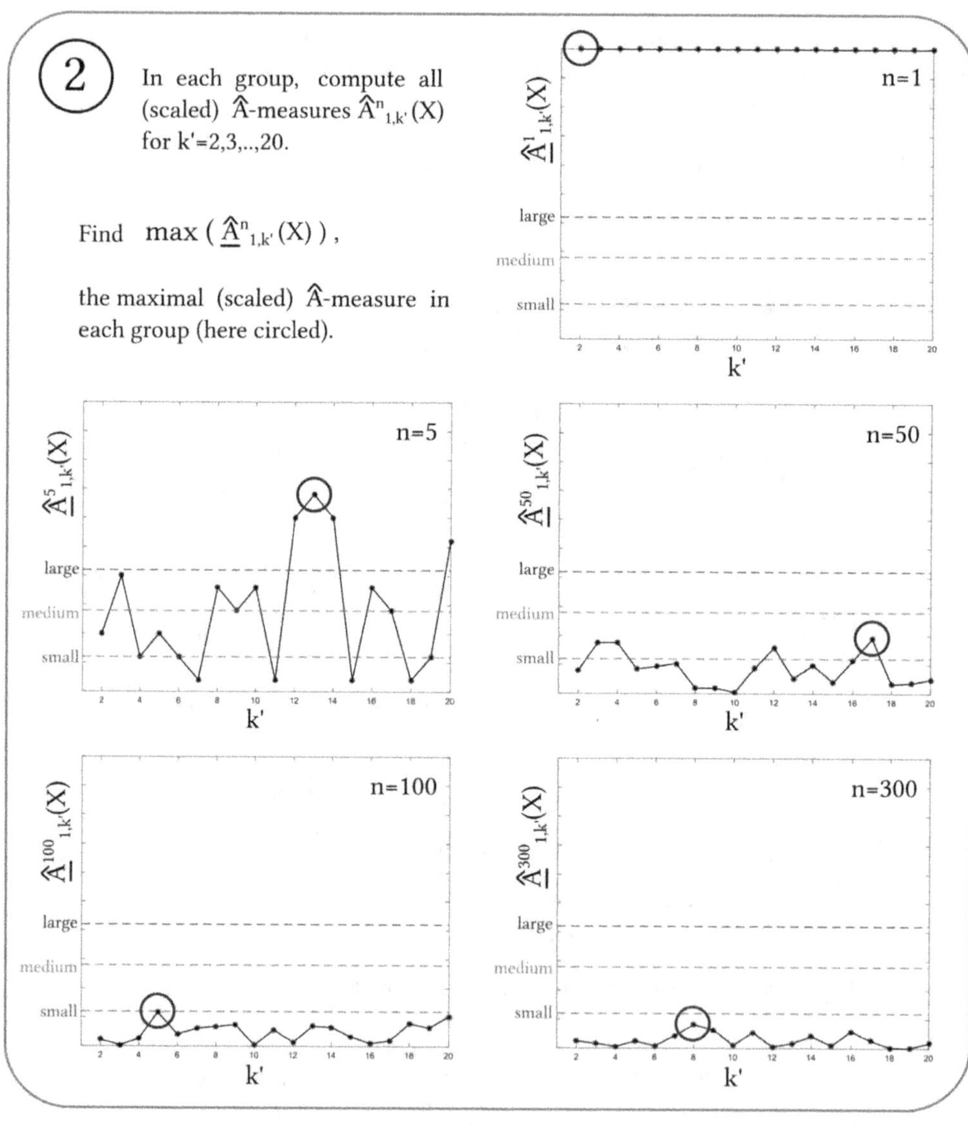
2
In each group, compute all (scaled) $\widehat{A}$-measures $\widehat{A}^n_{1,k'}(X)$ for k'=2,3,..,20.

Find $\max(\widehat{\underline{A}}^n_{1,k'}(X))$,

the maximal (scaled) $\widehat{A}$-measure in each group (here circled).

$\widehat{\underline{A}}^1_{1,k'}(X)$
n=1
large
medium
small
k'

$\widehat{\underline{A}}^5_{1,k'}(X)$
n=5
large
medium
small
k'

$\widehat{\underline{A}}^{50}_{1,k'}(X)$
n=50
large
medium
small
k'

$\widehat{\underline{A}}^{100}_{1,k'}(X)$
n=100
large
medium
small
k'

$\widehat{\underline{A}}^{300}_{1,k'}(X)$
n=300
large
medium
small
k'

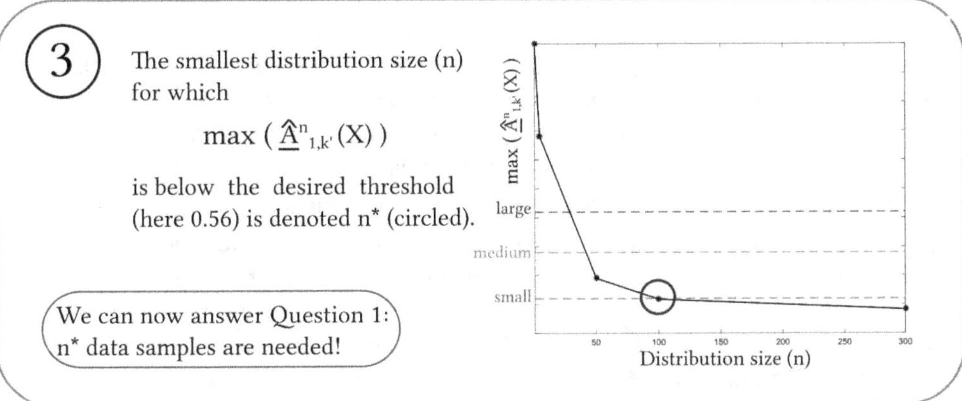

A few remarks:

- Consistency Analysis can be performed on one or multiple output variables X.

- When performing Consistency Analysis, the calibrated model parameters are used.

- The threshold values for *small, medium* and *large* statistical significance can be adjusted if appropriate.

4. Robustness Analysis

Robustness Analysis answers the question: how robust are model responses to *local* parameter perturbations? Robustness Analysis investigates if, and how, perturbing the value of *one* input parameter significantly changes an output X. Using the $\hat{A}$-measure, data distributions containing output data produced by perturbed input parameters, are compared to a data distribution containing output data produced by the calibrated input parameters. All perturbed data distributions are here of size n^*, where n^* is decided in the Consistency Analysis process, previously described in Section 3, when analysing stochastic models.

Before commencing the Robustness Analysis, we must identify the uncertain model parameters that we want to investigate the robustness of. We denote these parameters p^i, where $i = 1, 2, \ldots, q$, and thus we have a total of q parameters whose robustness we will investigate. Now, as illustrated in **Step 1** in Section 4.1, we let each such parameter p^i be investigated at $r(p^i)$ different parameter values (including the calibrated value), and thus we need to generate a total of P distributions of sample size n^* where

$$P = \sum_{i=1}^{q} r(p^i). \tag{15}$$

Note that the number of investigated parameter values, $r(p^i)$, need not be the same for every input parameter p^i. Investigated distributions of sample size n^* are here denoted D_{n^*,p^i_j}, where $i = 1, 2, \ldots, q$ denotes which parameter is being perturbed and $j = 1, 2, \ldots, r(p^i)$ denotes the specific perturbation of parameter p^i. For some perturbation $j = C$, the parameter value p^i_j equals the calibrated value for input parameter p^i. For each parameter that we are investigating, the $\hat{A}$-measure is used to compare the calibrated distribution D_{n^*,p^i_C} to all distributions D_{n^*,p^i_j}. Note that when $j = C$, the calibrated distribution is compared to itself and thus the $\hat{A}$-measure equals 0.5. These $\hat{A}$-measures provide information regarding the statistical significance, specifically if it can be described to be *small, medium* or *large* under parameter perturbations. Plotting the corresponding $\hat{A}$-measure over the parameter value p^i_j for each parameter p^i, paints an informative picture of local parameter robustness, as shown in **Step 2**, in Section 4.1. Another descriptive way to demonstrate the influence that parameter values p^i_j have on some output response X is to use boxplots. As is illustrated in **Step 3** in Section 4.1, boxplots can be used to clearly show the median, different percentiles, and outliers of some data distribution D_{n^*,p^i_j} as a function of the parameter value p^i_j. The methodology to perform Robustness Analysis is outlined in Section 4.1. Note that Robustness Analysis does not pick up on any non-linear effects between an input parameter p^i and an output X, that occur when more than one model parameter is simultaneously perturbed.[4] Such effects can however be identified using a global sensitivity analysis technique, such as Latin Hypercube Analysis, as described in Section 5.

4.1. *Quick guide: Robustness analysis*

> Robustness Analysis answers Question 2: How robust are model reponses (outputs) to **local** parameter perturbations?

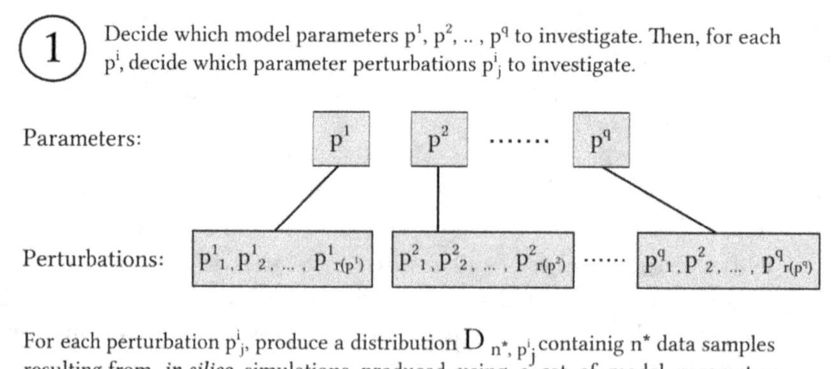

(1) Decide which model parameters $p^1, p^2, \ldots, p^q$ to investigate. Then, for each p^i, decide which parameter perturbations p^i_j to investigate.

For each perturbation p^i_j, produce a distribution $D_{n^*,\,p^i_j}$ containig n^* data samples resulting from *in silico* simulations produced using a set of model parameters in which (only) the parameter p^i is perturbed (to the value p^i_j) and all other model parameters are kept at their calibrated value.

（2） Using the $\widehat{A}$-measure, compare all distributions $D_{n^*,\,p_j^i}$ to the calibrated distribution D_{n^*,p_C^i} .

For each parameter p^i , plot these $\widehat{A}$-measures over the parameter value p_j^i . These plots determine parameter ranges in which parameter perturbations yield a statistical significance that is *small, medium* or *large* (compared to the calibrated model parameters).

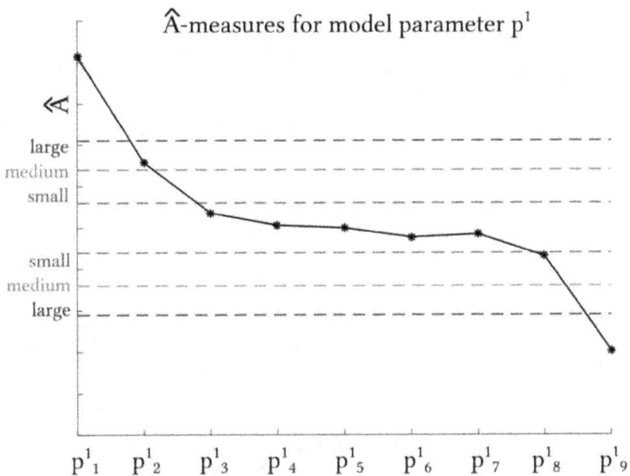

In the above plot, the statistical significance is *small* (within green lines) for the parameter value range (p^1_3, p^1_7). p^1_5 is the calibrated value for p^1.

3   For each parameter p^i, create boxplots over the perturbed value p^i_j. These plots graphically provide information regarding the influence that parameter value p^i_j has on the output X.

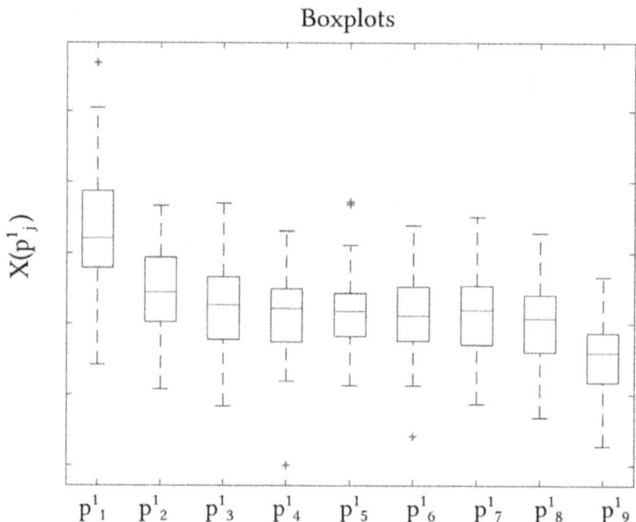

Boxplot key

The central line marks the median.

The upper box-edge (X=u) marks the 75th percentile and the bottom box-edge (X=b) marks the 25th percentile. The box has a size Q = u-b.

The top whisker stretches to the largest data sample such that $X \leq u + Q$ and the bottom whisker stretches to the smallest data sample such that $X \geq b - Q$.

Data samples outside the range [b - Q, u + Q] are outliers, these are shown with plus (+) symbols.

> **A few remarks:**
>
> - Robustness Analysis is a type of local sensitivity analysis.
>
> - Robustness Analysis can be performed on one or multiple output variables X.
>
> - When performing Robustness Analysis, model parameters are perturbed one at a time.

5. Latin Hypercube Sampling and Analysis

Latin Hypercube Analysis answers the question: how robust are model responses to *global* parameter perturbations? Latin Hypercube Analysis is a type of global sensitivity analysis that investigates the relationship between input parameters and output responses when all input parameters are simultaneously perturbed. The parameters that we want to perturb are (as in Section 4) denoted p^i, where $i = 1, 2, \ldots, q$. Thus the parameters $p^1, p^2, \ldots, p^q$ together span a parameter space of dimension q. It is impossible to test every possible combination of input parameter values if they are picked from continuous ranges. In fact, even if we select a finite number of parameter values $r(p^i)$ to test for each parameter p^i, or if we pick discrete parameter values, comparing every possible combination of parameter values may require us to produce an impractically large number of simulation runs. Thus performing *in silico* simulations for all possible combinations of input parameters will in many cases be at worst impossible, and at best impractical. In order to circumvent this issue, Latin Hypercube Sampling can be used.[1] It is a sampling technique that ensures comprehensive testing coverage over the parameter space whilst keeping the number of tested parameter combinations low enough to be applicable in practice.[19,20] After Latin Hypercube *Sampling* (Section 5.1), Latin Hypercube *Analysis* (Section 5.2) is used in order to assess global sensitivity.

5.1. *Latin hypercube sampling*

In the two-dimensional case, a *Latin Square* is an $\ell \times \ell$ square grid containing ℓ (traditionally Latin, hence the name) different symbols such that each symbol occurs exactly once in every row and exactly once in every column,[28] as illustrated in Figure 3. Analogously, in the Latin Hypercube Sampling framework, consider two parameters p^1 and p^2, spanning a parameter space of dimension $q = 2$, where both p^1 and p^2 are sectioned into ℓ intervals. We then pick ℓ combinations of input parameter values (or sampling points) (p_j^1, p_j^2), where $j = 1, 2, \ldots, \ell$, such that every p^1-interval is sampled from exactly once and every p^2-interval is sampled from exactly once. Within the parameter range of an interval, the sampled parameter value p_j^i is

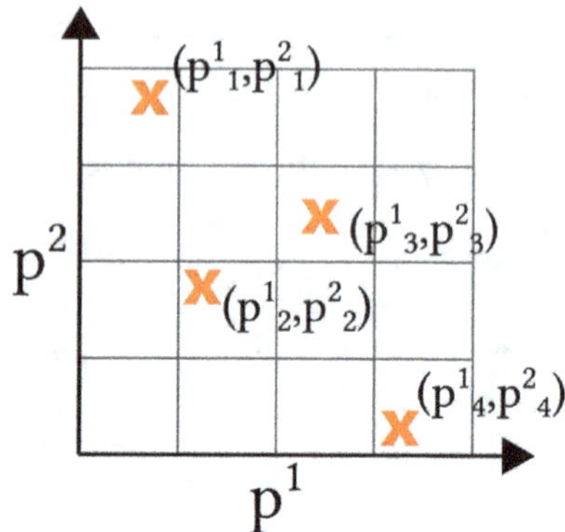

Fig. 3. Left: An $\ell \times \ell$ Latin Square in which each Latin symbol occurs ℓ times, exactly once in each row and exactly once in each column. Right (analogously): A two-dimensional parameter space spanned by the input parameters p^1 and p^2 that are both sectioned into ℓ intervals. Using Latin Hypercube sampling, ℓ parameter combinations (p_j^1, p_j^2) are sampled where $j = 1, 2, \ldots, \ell$ and each p^1-interval is sampled from exactly once and each p^2-interval is sampled from exactly once.

randomly selected (unless of course the interval contains only one possible value p_j^i). Note that the j index denotes the coordinate combination that p_j^i belongs to, not the interval from which the parameter value p_j^i was taken. Thus there is no condition demanding that the values p_j^i are ordered in a way such that $p_1^i < p_2^i < \ldots < p_\ell^i$.

The analogy between a Latin Square and Latin Hypercube Sampling from a two-dimensional parameter space is illustrated in Figure 3. The Latin Square can be extended to higher dimensions to form a Latin Cube (dimension = 3) or a Latin Hypercube (dimension > 3) and, analogously, the two-dimensional sampling space illustrated in Figure 3 can be extended to q dimensions, spanned by the input parameters $p^1, p^2, \ldots, p^q$.[28]

For each parameter p^i, the total investigated parameter range is $[min(p^i), max(p^i)]$, where $min(p^i)$ and $max(p^i)$ respectively denote the minimum and maximum values of p^i to be investigated. Now each parameter range $[min(p^i), max(p^i)]$ is sectioned into N intervals, and we denote these intervals by $u_{p^i}^1, u_{p^i}^2, \ldots, u_{p^i}^N$. Note that all input parameters p^i must be sectioned into the same number of intervals. If the intervals are of equal size, then the size of an interval, $w(p^i)$, is

$$w(p^i) = \frac{max(p^i) - min(p^i)}{N}, \tag{16}$$

and the rth interval $u_{p^i}^r$ has a parameter range such that

$$u_{p^i}^r = [min(p^i) + w \cdot (r - 1), min(p^i) + w \cdot r], \tag{17}$$

where $r = 1, 2, \ldots, N$.

Note that there are more than one way to populate Latin symbols in a Latin Square, this can be realised by regarding Figure 3 and noticing that the **A**-symbols and the **B**-symbols cover the Latin Square in different ways. Analogously, and by extension, there are multiples ways to populate sampling coordinates in a Latin Hypercube Sampling framework. Some of these ways provide better coverage of the parameter space than do others,[28] but details regarding such sampling-optimisation are outside the scope of this review. Here, we use the built-in **MATLAB** function **lhsdesign**[17] to select which parameter combinations to use according to a Latin Hypercube Sampling approach, details about the implementation are available in the Appendix. Note that, in our case, all N intervals $u_{p^i}^1, u_{p^i}^2, \ldots, u_{p^i}^N$ for a parameter p^i are uniformly spaced, but the choice of spacing can be adjusted to the specific application at hand.[17]

Now let us address the choice of intervals N, as this is not straightforward. Using the Latin Hypercube Sampling framework, every parameter p^i, where $i = 1, 2, \ldots, q$, is partitioned into N intervals and, consequently, N combinations comprising q parameter values are sampled and tested. Compared to a small N value, a large value of N will provide more data to use, and draw conclusions from, in the Latin Hypercube *Analysis* stage, however, it will also increase the computational cost in the Latin Hypercube *Sampling* stage. There is no strict rule for how to choose N, but suggested values for N in the literature are $N = 2q$ for large values of q (*i.e.* high-dimensional parameter spaces) or $N = 4q/3$ which has been described to be 'usually satisfactory'.[10,16] Authors of the Spartan package use a lot larger numbers in their provided examples.[1] In this example study, we decide to use $N = 100$ uniform intervals. At the end of the day, the choice of N is up to the modeller, who must outweigh the (computational) cost of producing a large number of data samples, with the advantage of having a vast amount of data, and thus plentiful information, in the analysis stage. Details regarding quantitative choices of N are outside the scope of this review.

5.2. *Latin hypercube analysis*

During the Latin Hypercube Sampling process, N different points in the q-dimensional parameter space spanned by the input parameters $p^1, p^2, \ldots, p^q$ are selected as *sampling points*, as shown in **Step 1** in Section 5.3. One such sampling point, C_j, can be described by its coordinates in the parameter space so that $C_j = (p_j^1, p_j^2, \ldots, p_j^q)$. Each sampling point C_j is used to generate n^* output responses $X(C_j)$, where n^* is determined using Consistency Analysis. Subsequently, the median output value, here denoted $\tilde{X}(C_j)$, is computed for every C_j. Now, our overall aim is to investigate the relationship between an input parameter p^i and an output response X. We investigate this input-output relationship in two steps, one of which is qualitative and one of which is quantitative. In the first and qualitative step, we produce two-dimensional scatterplots in which median output data,

$$\tilde{X}(C_1), \tilde{X}(C_2), \ldots, \tilde{X}(C_N)$$
$$= \tilde{X}(p_1^1, p_1^2, \ldots, p_1^q), \tilde{X}(p_2^1, p_2^2, \ldots, p_2^q), \ldots, \tilde{X}(p_N^1, p_N^2, \ldots, p_N^q),$$

are plotted over parameter values

$$p_1^i, p_2^i, \ldots, p_N^i,$$

for one of the input parameters p^i. We do this for every input parameter $i = 1, 2, \ldots, q$ and thus q scatterplots are created. By simply visually analysing the data in the scatterplots, we are able to make qualitative observations regarding the relationship between the input and the output. Examples of such observations are provided in **Step 2** in Section 5.3.

As a second step, we use a quantitative measure, such as the Pearson Product Moment Correlation Coefficient (or the correlation coefficient for short), to quantitatively describe the correlation between input parameters and output responses, as done in **Step 3** in Section 5.3. The correlation coefficient is denoted r, where $r \in [-1, +1]$. It describes the linear association between the input parameter and the output response in terms of both magnitude and direction. A positive (linear) correlation between p^i and $X(C_j)$ means that if either the input value or the output value increases, so does the other one, and thus r is positive. Conversely, a negative correlation means that if either p^i or $X(C_j)$ increases, the other one decreases, and thus r is negative. The magnitude of r describes the strength of the correlation, where a magnitude of 1 corresponds to a strong linear association, and a small magnitude corresponds to a weak correlation. An r value of approximately zero indicates that there is no linear correlation between the two investigated variables. Note that the Pearson Product Moment Correlation Coefficient picks up linear associations only, thus there may exist other, non-linear correlations that are not captured by the correlation coefficient r. Therefore it is important to, not only quantitatively compute input-output correlations, but to also qualitatively assess the relationships between inputs and outputs via data visualisation in scatterplots as previously described.

The correlation coefficient, r^i, describing the correlation between an input parameter p^i, and an output response X (in median form) is given by,[21]

$$r^i = \frac{\sum_{j=1}^{N}(p_j^i - \bar{p}^i)(X(C_j) - \bar{X})}{\sqrt{\left(\sum_{j=1}^{N}(p_j^i - \bar{p}^i)^2\right)\left(\sum_{j=1}^{N}(X(C_j) - \bar{X})^2\right)}}, \tag{18}$$

where a bar denotes the mean value.

When it comes to interpreting quantitative input-output relationships based on the correlation coefficient r, there are no all-encompassing threshold values to use for descriptors such as 'weak', 'moderate', 'strong'.[11,21,27] Relationships quantified by correlation coefficient values close to the extrema 0 or 1 may be easy to describe as 'negligible' or 'strong', respectively. However, correlation coefficient values in the middle of the [0,1] range are more difficult to label. Various 'rules of thumb' have been suggested in the literature but, at the end of the day, it is up to the modeller to

Table 1. Suggested descriptor threshold values for the magnitude of the correlation coefficient, $|r|$, reported in the literature.

Descriptor Reference	Negligible	Weak	Moderate	Strong	Very strong
Mukaka[21]	[0,0.3)	[0.3,0.5)	[0.5,0.7)	[0.7,0.9)	[0.9,1]
Schober *et al.*[27]	[0,0.1)	[0.1,0.4)	[0.4,0.7)	[0.7,0.9)	[0.9,1]
Krehbiel[11]	"A linear relationship exists if $|r| \geq 2/\sqrt{\text{number of samples}}$."				

appropriately judge what constitutes a 'weak', 'moderate' or 'strong' input-output relationship in the specific (modelling) application at hand, taking into account the research area, the number of data samples, and the range of investigated input values.[27] However, even without rigid descriptor threshold values, we can compare the correlation coefficient values for all input-output pairs and see which input values are the most influential within the ranges of regarded input values. As a guide, suggested correlation coefficient descriptor threshold values presented in the literature are listed in Table 1. The methodology to perform Latin Hypercube Sampling and Analysis is outlined in Section 5.3.

5.3. *Quick guide: Latin hypercube sampling and analysis*

Latin Hypercube Analysis answers Question 3: How sensitive are model reponses (outputs) to **global** parameter perturbations?

(1) Decide which model parameters $p^1, p^2, .., p^q$ to investigate. Then, for each p^i, decide which parameter range $[\min(p^i), \max(p^i)]$ to investigate. Split every such parameter range into N intervals $u^1_{p^i}, u^2_{p^i}, ..., u^N_{p^i}$.

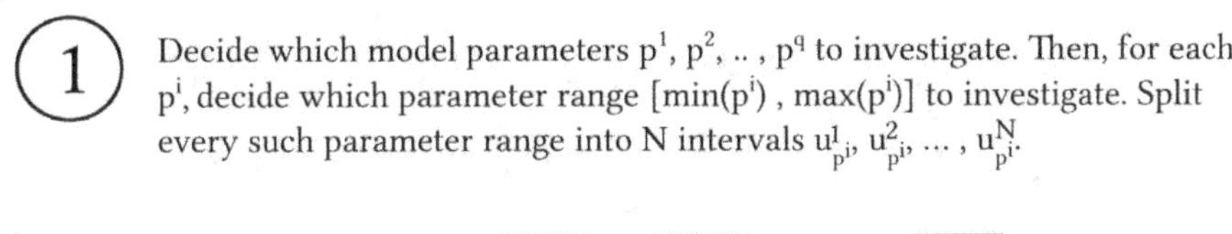

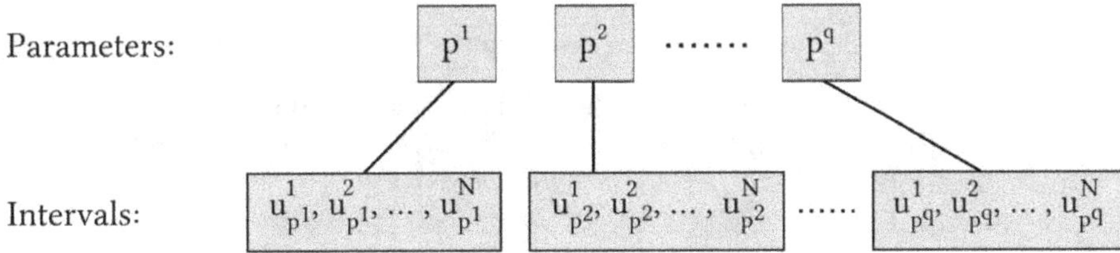

Using Latin Hypercube Sampling, select N sampling points $C_j = (p^1_j, p^2_j, ... p^q_j)$ such that every interval is sampled from exactly once.

 For every sampling point C_j, perform n^* *in silico* runs (where n^* is chosen via consistency analysis) and compute the median output response $\underset{\sim}{X}(C_j)$.

For every investigated input parameter p^i, create a scatterplot that plots the median output values

$$\underset{\sim}{X}(C_1),\ \underset{\sim}{X}(C_2),\ \ldots,\ \underset{\sim}{X}(C_N)$$

over their repective input parameter values

$$p^i_1,\ p^i_2,\ \ldots,\ p^i_N$$

Observe the data, visualised by scatter-plots, to draw qualitative conclusions about the input-output relationships.

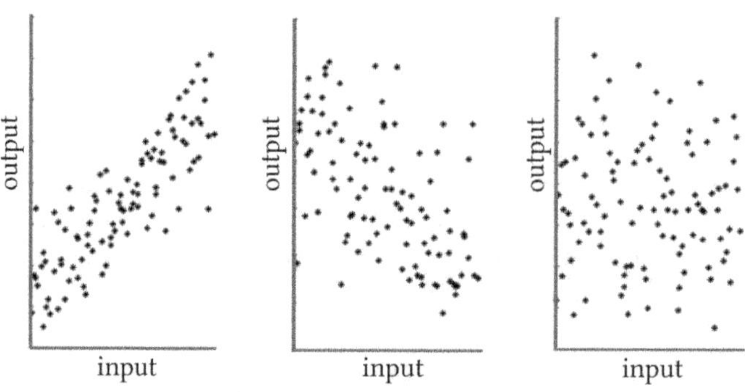

| **Qualitative remarks:** | The input and output are *strongly, positively* correlated. | The input and output are *moderately, negatively* correlated. | The input and output are *not* correlated. |

For every investigated input parameter p^i, assess the (linear) quantitative correlation between the median output values

$$\underset{\sim}{X}(\,C_1\,),\ \underset{\sim}{X}(\,C_2\,),\ \dots,\ \underset{\sim}{X}(\,C_N\,)$$

and the input parameter values

$$p^i_1,\ p^i_2,\ \dots,\ p^i_N$$

using the correlation coefficient r.

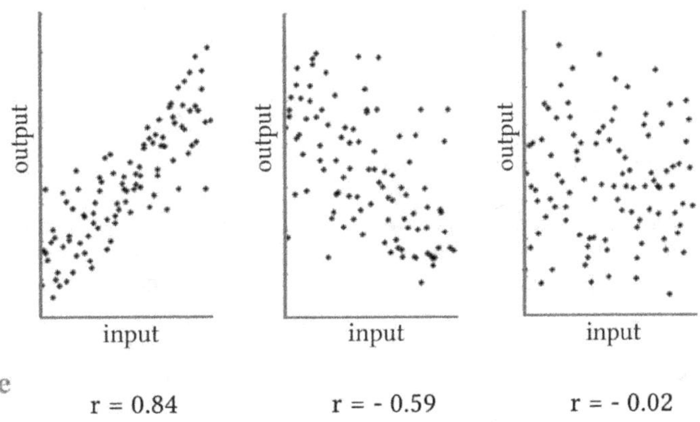

Quantitative measure: r = 0.84   r = - 0.59   r = - 0.02

A few remarks:

- Latin Hypercube Analysis is a type of global sensitivity analysis.

- Latin Hypercube Analysis can be performed on one or multiple output variables X.

- When performing Latin Hypercube Analysis, all model parameters are simultaneously perturbed.

6. A Worked Example: Analysing a Mathematical Cancer Model

In this section we will perform Consistency Analysis, Robustness Analysis and Latin Hypercube Analysis on an agent-based mathematical model that describes a population of cancer cells (*in vitro*) that are subjected to an anti-cancer drug (AZD6738) that may inhibit DNA damage repair in cells and, by extension, cause cell death. Full details of this model are available in one of our recent research papers,[9] but a pictorial model summary is provided in Figure 4. This summary contains sufficient

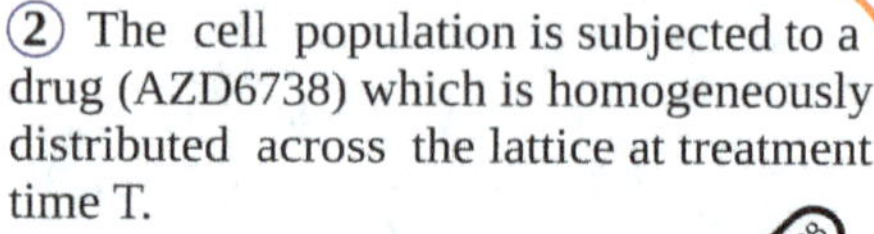
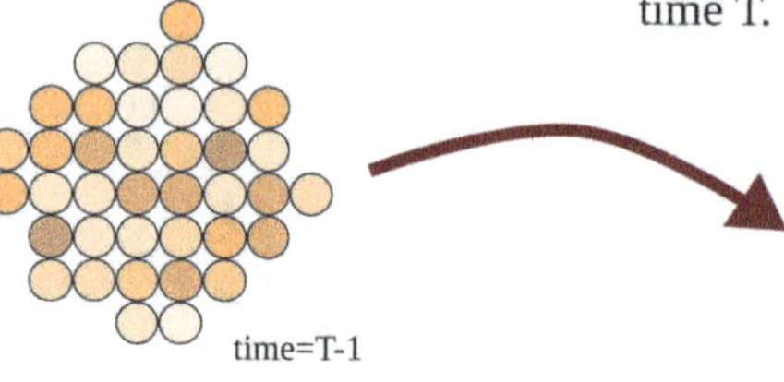
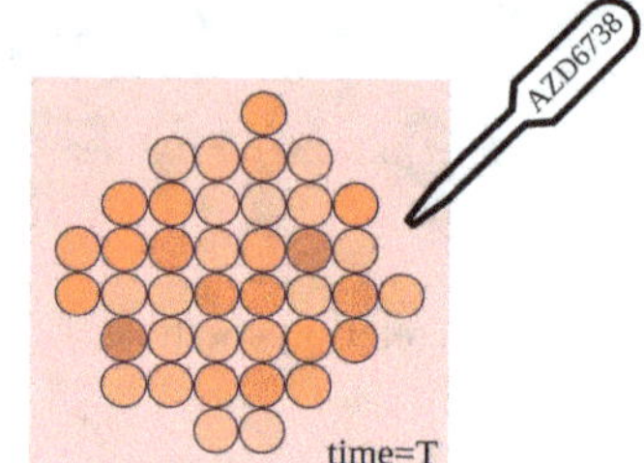

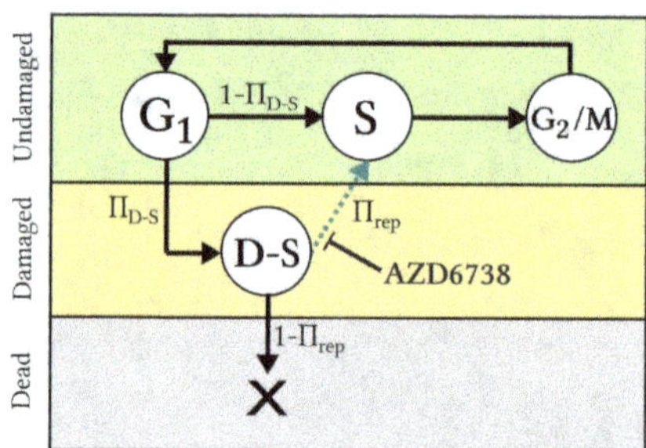

The probability that a cell repairs itself from the D-S state to the S state is given by $\prod_{\text{rep}}$, where

$$\prod\nolimits_{\text{rep}} = 1 - \frac{C^{\gamma}}{EC_{50}^{\gamma} + C^{\gamma}},$$

and C is the drug concentration.

Note that $\prod_{\text{rep}}$ decreases with increasing drug concentration and that $\prod_{\text{rep}} = 1$ in the absence of drug (when $C=0$).

The above equation is an agent-based adaptation of the sigmoidal E-max model. EC_{50} denotes the drug concentration achieving half of the maximal drug effect and γ is the Hill-exponent.

Fig. 4.   A summary of an agent-based model that is used to simulate human colon carcinoma cells *in vitro* subjected to a drug that targets cellular DNA damage responses. In this review, we perform uncertainty and sensitivity analyses on this model as a worked example. Full details are available in the research paper in which this model was first introduced.[9]

information for our current purposes: performing uncertainty and sensitivity analysis through a worked example. In order to do this, we need to specify a set of model inputs and outputs. The full model includes seven input parameters p^i, $i = 1, \ldots, 7$, three of which are mentioned in the model summary and will be investigated in this review. These input parameters are p^1: the probability Π_{D-S} that a cell enters the damaged S phase in the cell cycle, p^2: the drug's EC_{50} value and p^3: the Hill-exponent

(γ) used to compute cellular drug responses. Furthermore, we consider two *in silico* measurements as outputs, specifically X^1: the percentage of DNA-damaged (*i.e.* γ-H2AX positive) cells at the end of the simulation and X^2: the cell count at the end of the simulation.

6.1. *Worked example: Consistency analysis*

In order to perform Consistency Analysis, we follow steps 1, 2 and 3 outlined in the quick guide in Section 3.1.

Step 1: Using the calibrated model parameters, we run our *in silico* experiment $20 \times (1 + 5 + 50 + 100 + 300) = 9120$ times in order to produce 9120 data samples. Note that, in this case, we have two output responses of interest, and thus one data sample consists of an output-pair (X^1, X^2). Post the *in silico* production of data, we organise our data samples into five groups of distributions, where each group consists of 20 distributions of data samples. In the first group, each one of the 20 distributions includes only one data sample. In the second, third, forth and fifth group, each distribution respectively includes 5, 50, 100 or 300 data samples.

Step 2: In each distribution group, we compute and plot the $\hat{A}$-measure (in original and scaled form) for each of its 20 distributions, as is done in Figures 5 through to 9. These figures clearly demonstrate that the statistical significance decreases with increasing distribution size n. Note that the two output responses of interest are computed and plotted independently of each other, as we are aiming to find a distribution size that yields a small statistical significance for both X^1 and X^2.

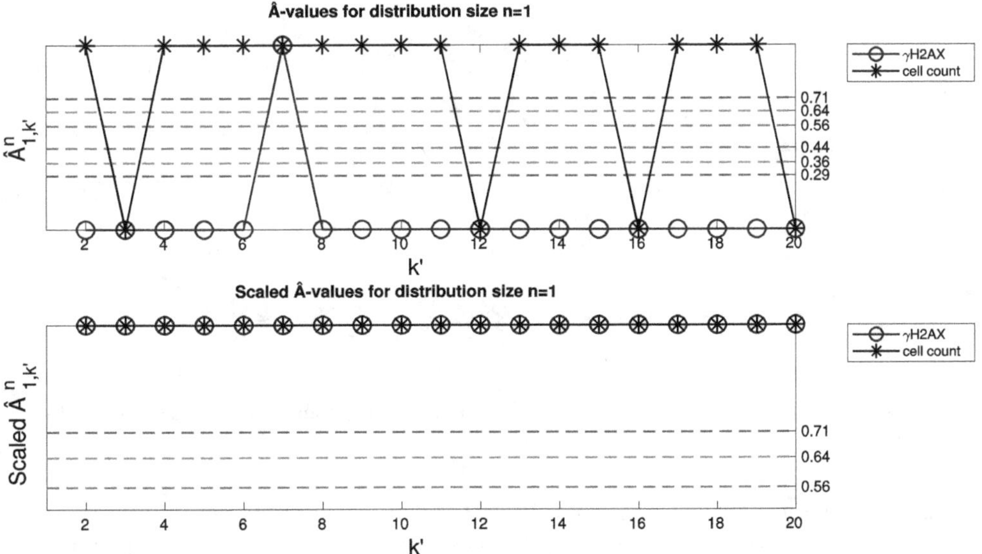

Fig. 5.   Consistency Analysis, $\hat{A}$-values in initial (top) and scaled (bottom) form for distribution size $n = 1$.

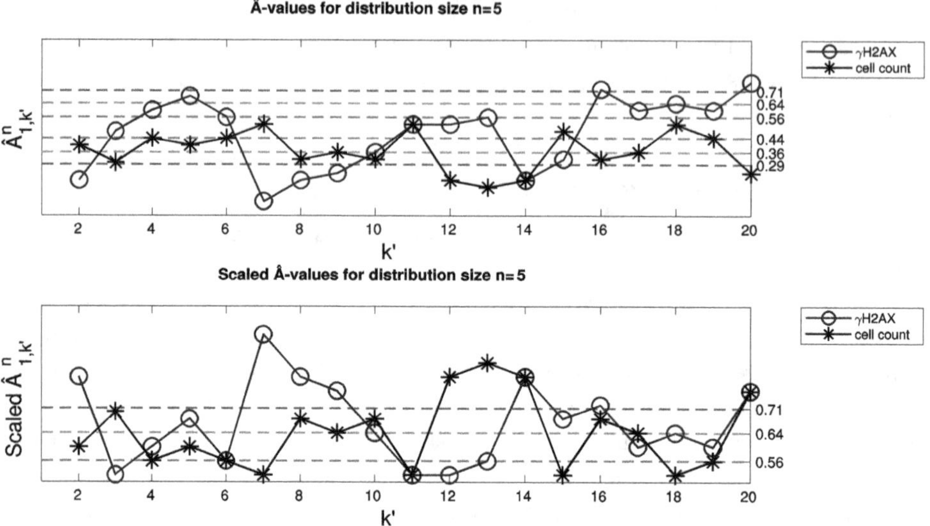

Fig. 6. Consistency Analysis, $\hat{A}$-values in initial (top) and scaled (bottom) form for distribution size $n = 5$.

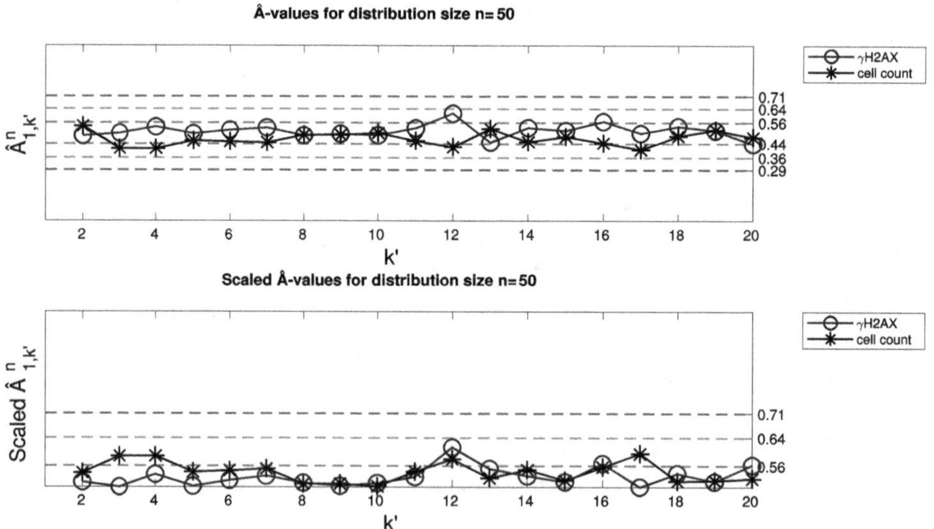

Fig. 7. Consistency Analysis, $\hat{A}$-values in initial (top) and scaled (bottom) form for distribution size $n = 50$.

Step 3: The largest scaled $\hat{A}$-measure in each distribution group is computed and plotted over the group's distribution size. The smallest distribution size for which the statistical significance is small (*i.e.* ≤ 0.56) for both X^1 and X^2 is denoted n^*. By regarding Table 2 and Figure 10, we can see that, in this case, $n^* = 100$. Accordingly,

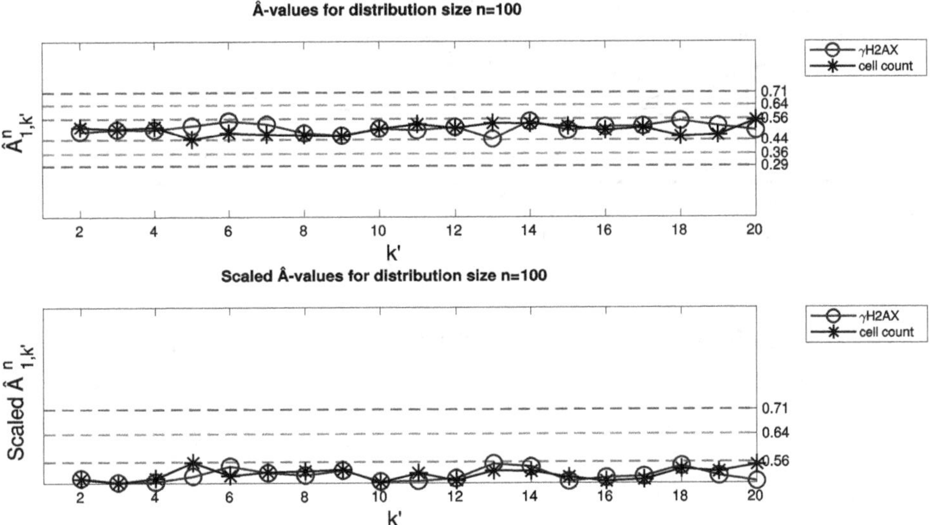

Fig. 8. Consistency Analysis, $\hat{A}$-values in initial (top) and scaled (bottom) form for distribution size $n = 100$.

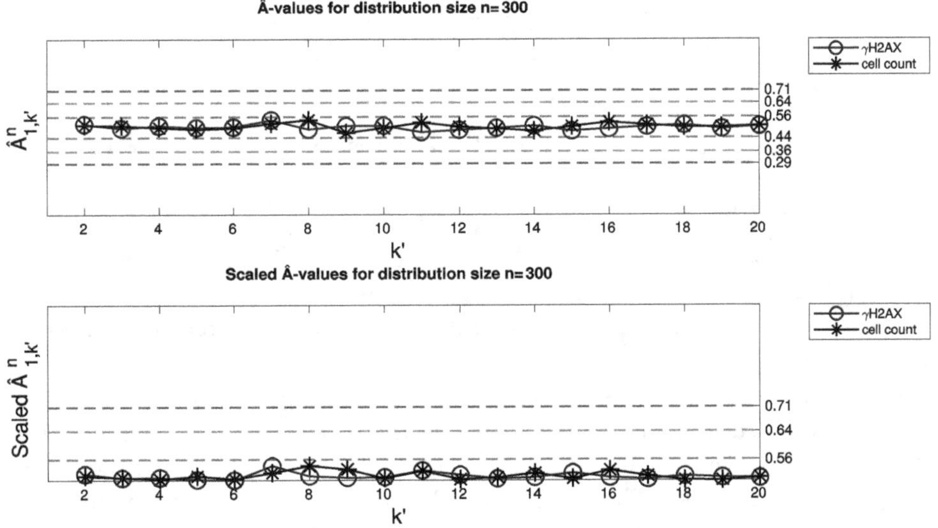

Fig. 9. Consistency Analysis, $\hat{A}$-values in initial (top) and scaled (bottom) form for distribution size $n = 300$.

we determine that 100 simulation runs are sufficient to mitigate uncertainty originating from intrinsic model stochasticity. Thus when talking about, for example, average values and standard deviations produced by this model, we should base these measures on data samples from 100 *in silico* runs.

Table 2. Maximal scaled $\hat{A}$-values for various distribution sizes n.

Distribution size Output	$n = 1$	$n = 5$	$n = 50$	$n = 100$	$n = 300$
X^1	1	0.92	0.61	0.55	0.54
X^2	1	0.84	0.59	0.56	0.54

The output responses are X^1: the percentage of γH2AX-positive (*i.e.* DNA-damaged) cells, and X^2: the cell count at the end of the simulation.

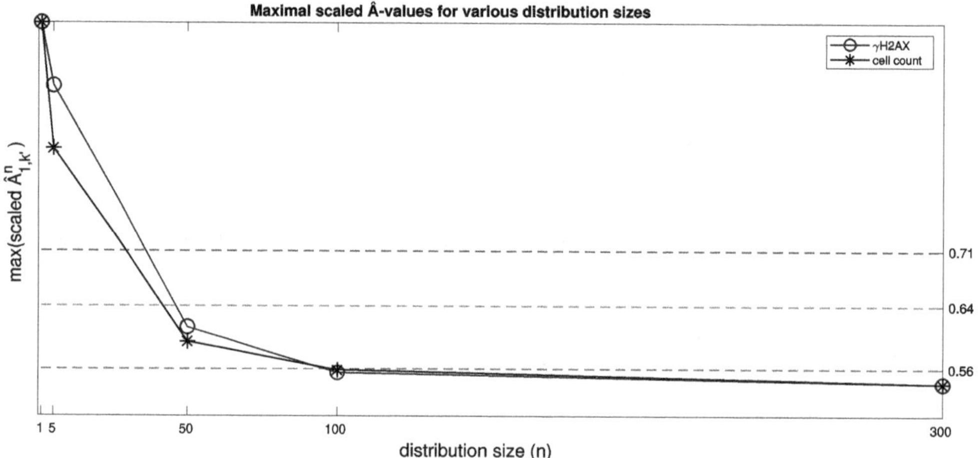

Fig. 10. Consistency Analysis, maximal scaled $\hat{A}$-values for various distribution sizes n.

6.2. *Worked example: Robustness analysis*

We now set out to perform Robustness Analysis using steps 1, 2 and 3 described in the quick guide in Section 4.1. Here, our goal is to investigate how sensitive model output responses are to local (one at a time) parameter perturbations of the inputs Π_{D-S}, EC_{50} and γ.

Step 1: The first thing we must do here is to decide appropriate parameter ranges to investigate for each input parameter. This decision should ideally be guided by information from experimental data or biological/mathematical knowledge about the modelling scenario at hand. The mathematical model summarised in Figure 4 is driven by *in vitro* data, and from this data suitable parameter ranges can be deduced and established (full details are available in the original model paper[9]).

For example, recall that the parameter EC_{50} corresponds to the drug concentration C that achieves half of E_{max}, the maximal drug effect. Now the *in vitro* data shows that $C = 1\,\mu$M yields $\approx 0.5E_{max}$ and that $C = 3\,\mu$M (or higher) yields $\approx E_{max}$, whilst a drug concentration of $C = 0.3\,\mu$M evokes roughly the same drug response as the control case (when no drug at all is applied and thus $C = 0$).[9,5] From this, we can reason that it is appropriate to investigate EC_{50} parameter values in the range

$(1 \pm 0.75)\,\mu$M. Similarly, we may decide to investigate Π_{D-S} values in the parameter range $(75 \pm 10)\%$, and γ values in (2 ± 1).

Now that we have established our parameter ranges of interest, it remains to decide how many (here evenly spaced) parameter values in these ranges to test. The more parameter values we test, the more detailed information we get, but recall that for every investigated parameter we need to create n^* data samples (where n^* is determined in the Consistency Analysis). The decision of how many parameter values to include in the Robustness Analysis should be informed by the fineness of available *in vitro/in vivo* data, computational costs, scientific reasoning and model application. If a model is being used in a pharmaceutical setting, for example, a detailed analysis may be of extra importance. Here, we respectively choose to investigate 9,7 and 9 evenly spaced parameter values (including the calibrated parameter values) in the parameter ranges for Π_{D-S}, EC_{50} and γ. Thus we need to produce $n^* \times (9 + 7 + 9) = 100 \times (25)$ *in silico* data samples.

Steps 2 and 3: Post *in silico* simulations, the $\hat{A}$-measures are computed and plotted over all parameter values p^i_j for each input parameter p^i. Such plots are here shown for input parameters Π_{D-S} (Figure 11), EC_{50} (Figure 12) and γ (Figure 13).

The data samples produced in the *in silico* experiments are also represented using boxplots in these figures. The $\hat{A}$-measures and the boxplots demonstrate the effect that local perturbations of the input parameters have on both output responses X^1 and X^2.

Figure 11 illustrates that increasing the probability Π_{D-S} that a cell enters the damaged S state increases the percentage of γH2AX-positive cells and decreases the cell count, as is to be expected. Further, Figure 12 demonstrates that the model output is highly sensitive to perturbations of EC_{50}. Increasing EC_{50} results in a higher percentage of γH2AX-positive cells and a lower cell count. Finally, Figure 13

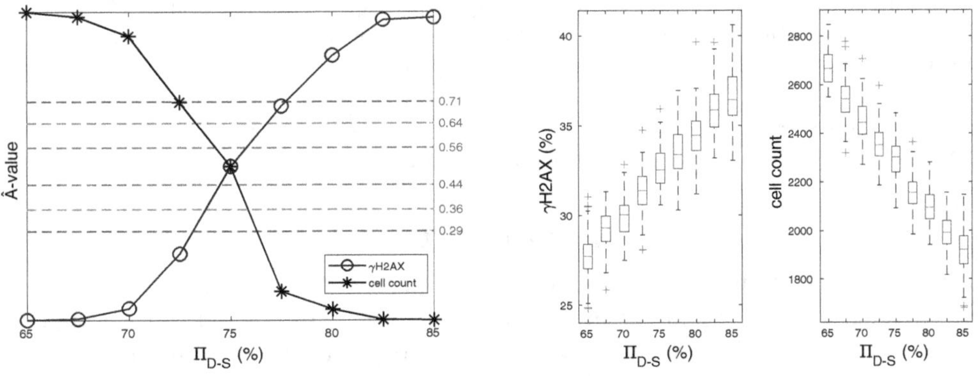

Fig. 11.   Robustness Analysis, Left: $\hat{A}$-values resulting from comparisons between distributions of data samples produced with perturbed Π_{D-S} values, and the distribution produced with the calibrated (unperturbed) Π_{D-S} value. Right: Output responses, in terms of percentage of γH2AX-positive (*i.e.* damaged) cells, and cell count as a result of perturbations to the input variable Π_{D-S}.

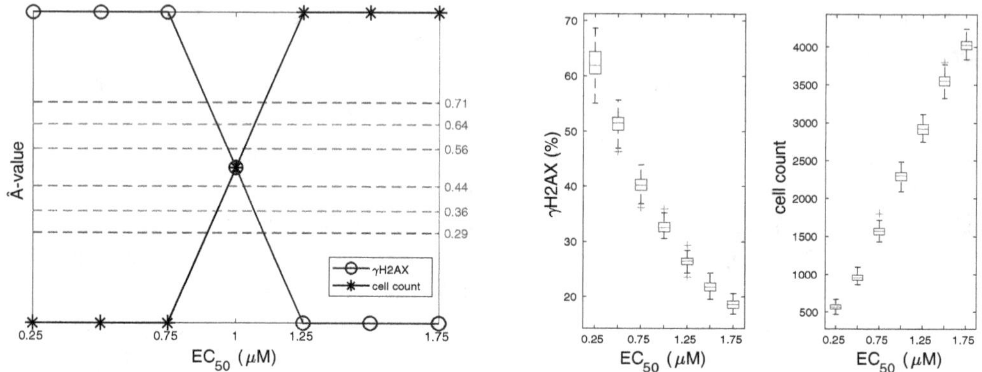

Fig. 12.   Robustness Analysis, Left: $\hat{A}$-values resulting from comparisons between distributions of data samples produced with perturbed EC_{50} values, and the distribution produced with the calibrated (unperturbed) EC_{50} value. Right: Output responses, in terms of percentage of γH2AX-positive (*i.e.* damaged) cells, and cell count as a result of perturbations to the input variable EC_{50}.

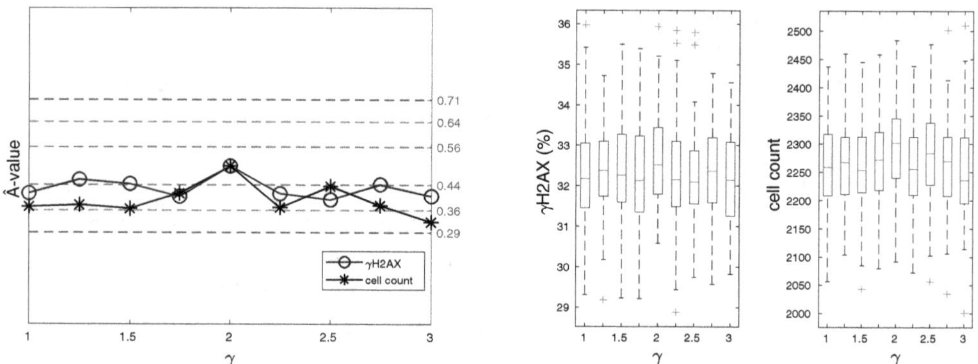

Fig. 13.   Robustness Analysis, Left: $\hat{A}$-values resulting from comparisons between distributions of data samples produced with perturbed γ values, and the distribution produced with the calibrated (unperturbed) γ value. Right: Output responses, in terms of percentage of γH2AX-positive (*i.e.* damaged) cells, and cell count as a result of perturbations to the input variable γ.

shows that the regarded output responses are less sensitive to small perturbations of the Hill-exponent γ than to small perturbations of Π_{D-S} and EC_{50}.

6.3. *Worked example: Latin hypercube analysis*

Following steps 1, 2 and 3, as described in the quick guide in Section 5.3, Latin Hypercube Analysis is here performed in order to investigate how sensitive output responses are to *global* parameter perturbations. We here investigate parameter values within parameter ranges that we consider to be 'plausible' post Robustness Analysis.

Step 1: For each input parameter, we decide to split the investigated parameter range into $N = 100$ intervals. Why 100 you might ask? Well, in the original paper

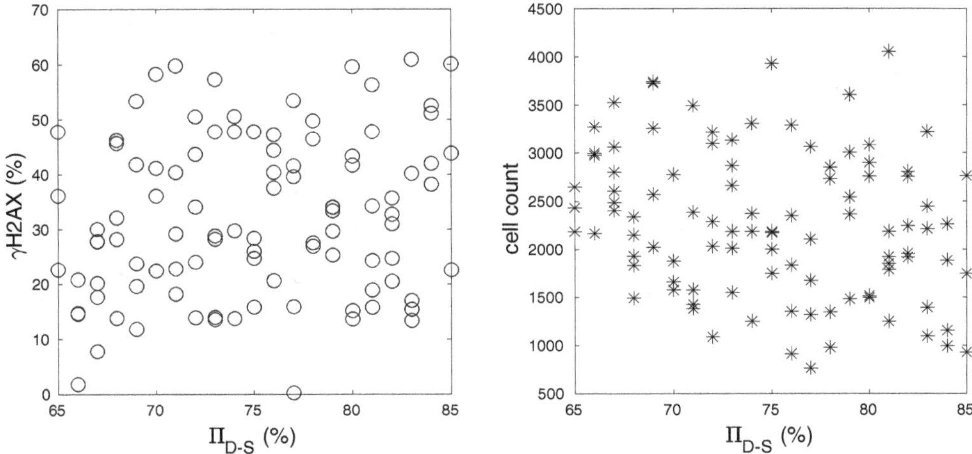

Fig. 14.   Latin Hypercube Analysis. Output responses in terms of γH2AX-positive cells (left) and cell count (right) when global parameter perturbations are performed. The scatterplots show the correlation between outputs and the input value of Π_{D-S}.

where we first introduced this model,[9] the model took seven input parameters and thus $q = 7$. Accordingly, we tried using $N = 4q/3 \approx 10$ and $N = 2q = 14$ intervals at first (following the suggestions discussed in Section 5) but neither of these options produced enough data samples to yield meaningful information in steps 2 and 3 below. Therefore, we decided to use $N = 100$ instead, as this choice covered a larger range of the input parameter space whilst coming at a feasible computational cost. Now, we can use the built-in **MATLAB** function lhsdesign[17] to create combinations of input parameter values (represented by a point C_p in input parameter space) that shall be used to produce the *in silico* data samples needed for Latin Hypercube Analysis.

Step 2: For each C_p, median output responses (X^1 and X^2) of $n^* = 100$ *in silico* runs are computed. These are plotted in '*output-over-input*' scatterplots for each investigated input parameter in Figures 14, 15 and 16. From these figures we can make some qualitative remarks: Figure 14 indicates that the relationships between the input variable Π_{D-S} and the output responses X^1 and X^2 are, respectively, positively and negatively correlated. This agrees with the intuitive notion that if the probability that a cell enters the D-S state increases, so does the percentage of damaged cells (X^1) whilst the cell count (X^2) decreases as more cells will be susceptible to the drug and potentially die. The scatterplots in Figure 15 demonstrate that the input variable EC_{50} impacts the output responses more than do other investigated input parameters (within the regarded ranges). EC_{50} is negatively, linearly correlated with X^1 and positively, linearly correlated with X^2. In Figure 16, however, there is no visually apparent correlation between the input parameter γ and the output.

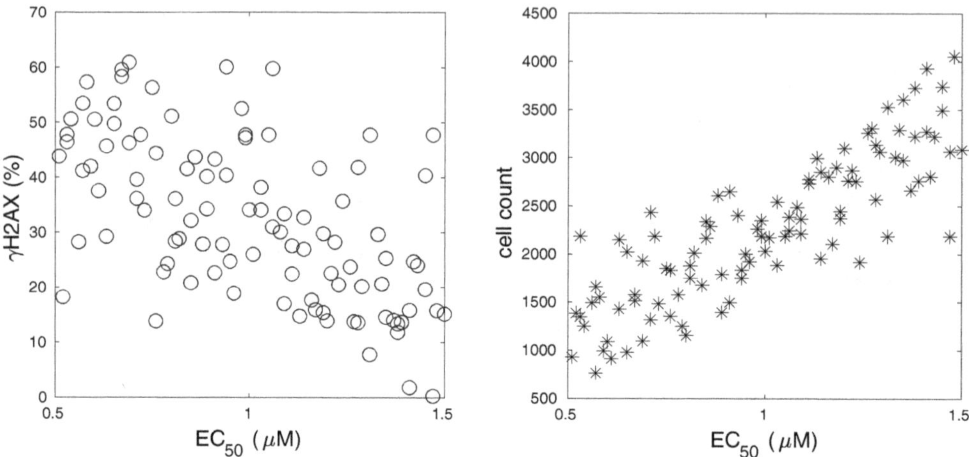

Fig. 15.   Latin Hypercube Analysis. Output responses in terms of γH2AX-positive cells (left) and cell count (right) when global parameter perturbations are performed. The scatterplots show the correlation between outputs and the input value of EC_{50}.

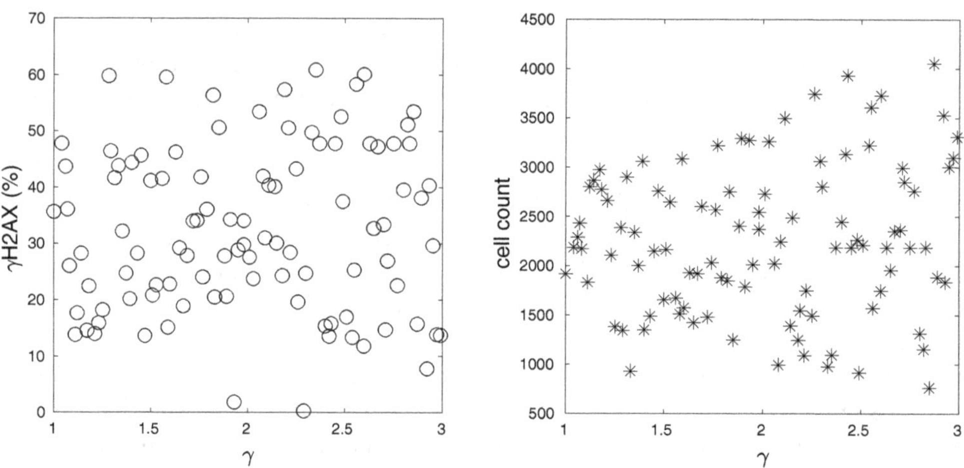

Fig. 16.   Latin Hypercube Analysis. Output responses in terms of γH2AX-positive cells (left) and cell count (right) when global parameter perturbations are performed. The scatterplots show the correlation between outputs and the input value of γ.

Step 3: After making some qualitative remarks in Step 2, we now compute the Pearson Product Moment Correlation Coefficients between the various input-output pairs for a quantitative analyses. These correlation coefficients are listed in Table 3. To decide threshold values for correlation coefficient descriptors, we here compromise between threshold values suggested by other authors (listed in Table 1) whilst taking into account that we are only regarding parameter values within 'plausible' ranges. With this as a guide, we here decide to refer to the linear input-output relationship as

Table 3. The Pearson Product Moment Correlation Coefficients, computed in the Latin Hypercube Analysis, quantitatively describe input–output relationships.

Input / Output	Π_{D-S}	EC_{50}	γ
X^1	0.19	-0.59	0.05
X^2	-0.24	0.84	0.12

being 'negligible' for γ, where the obtained correlation coefficients are 0.05 and 0.12 for X^1 and X^2 respectively. We further say that for Π_{D-S} the linear input-output relationship is positively/negatively weak for X^1 and X^2 respectively. For EC_{50}, however, the linear input-output relationship is moderately negative for X^1 and strongly positive for X^2. Clearly, we must be careful when choosing our EC_{50} value in the model, as this highly influences the model output!

7. Conclusion

This review is intended as a gentle, introductory review to three uncertainty and sensitivity analyses methods, namely, Consistency Analysis, Robustness Analysis and Latin Hypercube Analysis. Information on how to implement these methods in **MATLAB** are available in the Appendix. Alternatively, all methods discussed in this review can be implemented using the R-based software package Spartan, developed by Alden *et al.*[1] In fact, many of the proceedings and conventions used in this review follow those suggested by Alden *et al.* in order to allow the reader to, as easily as possible, use Spartan if desired. Scrutinising mathematical models using uncertainty and sensitivity analyses methods is an important part in model development. In many applications, knowledge about a model's robustness is crucial.[30] In the context of quantitative pharmacology, for example, a mathematical model may be used to guide preclinical or, ultimately, clinical proceedings. In such cases, understanding how confident we can be with model results, and how sensitive a model is to parameter perturbations, is of the utmost importance.

Acknowledgments

SH was supported by the Medical Research Council [grant code MR/R017506/1] and Swansea University PhD Research Studentship. SS was supported by an STFC studentship under the DTP grant ST/N504464/1.

Appendix – MATLAB Code Snippets

Computing measure of stochastic superiority

We here list two different **MATLAB** functions that can be used in order to compute the point estimate of the A-measure of stochastic superiority in the original form,

$\hat{A} \in [0,1]$, and in the scaled form, $\underline{\hat{A}} \in [0.5,1]$. The function getA_measure_naive, listed below, uses direct implementations of Equations 9 and 11 to compute and return values for $\hat{A}_{x_0,x_1}$ and $\underline{\hat{A}}_{x_0,x_1}$, given two input vectors x_0 and x_1. The function getA_measure uses the built-in MATLAB function ranksum to do the same.

```matlab
function [A_measure, scaled_A_measure] = getA_measure(x0, x1)
    [p,h,stats] = ranksum(x0,x1);
    % Compute the A measure
    A_measure=(stats.ranksum/length(x0) - (length(x0)+1)/2)/length(x1);
    % Compute the scaled A measure
    scaled_A_measure=0.5+abs(0.5 -A_measure);
end
```

```matlab
function [A_measure, scaled_A_measure] = getA_measure_naive(x0, x1)
    % Compute the A measure
    A_measure = 0;
    for i = 1:length(x0)
        for j = 1:length(x1)
            if(x0(i)>x1(j))
                A_measure = A_measure + 1;
            elseif(x0(i)==x1(j))
                A_measure = A_measure + 0.5;
            elseif(x0(i)<x1(j))
                A_measure = A_measure + 0;
            end
        end
    end
    A_measure = A_measure/(length(x0)*length(x1));
    % Compute the scaled A measure
    if(A_measure>=0.5)
        scaled_A_measure = A_measure;
    else
        scaled_A_measure = 1-A_measure;
    end
end
```

Creating boxplots

The MATLAB function boxplot can be used to create boxplots. The input data in one column is represented by one box in the boxplot. For details regarding labeling and style alternatives, please see the MATLAB documentation.[17]

```matlab
boxplot(M);
```

Choosing Latin Hypercube Sampling Points

A Latin Hypercube Sampling matrix can be created using the **MATLAB** function **lhsdesign**, which returns a matrix of size $n \times q$, where n denotes the number of samples to be tested, and q denotes the number of input parameters to investigate (and thus perturb).

```
LHC_Matrix=lhsdesign(n,q)
```

Each row i, in the created matrix (here denoted **LHC_Matrix**), corresponds to the ith sampling point. Each element (i, j) corresponds to the parameter value of the jth input parameter in sampling point i, where each parameter ranges between 0 and 1. For different criteria on how to chose the specific parameter values within each sampled interval, please refer to the **MATLAB** documentation.[17] Sampling points can, for example, be chosen in a way that maximises the distance between sampling points in the q-dimensional sampling space.

Qualitative and Quantitative Latin Hypercube Sampling Analysis

In order to qualitatively asses the correlation between an input parameter p, and an output response X, one can use the **MATLAB** function **scatter**. In the below listings, **p** and **X** are two data vectors.

```
scatter(p,X)
```

Further, to quantify the linear correlation between p and X, the **MATLAB** function **corrcoef** can be used to compute correlation coefficients.

```
R=corrcoef(p, X);
```

References

1. K. Alden, M. Read, J. Timmis, P. S. Andrews, H. Veiga-Fernandes and M. Coles, Spartan: A comprehensive tool for understanding uncertainty in simulations of biological systems, *PLoS Comput. Biol.* **9**(2), e1002916 (2013).
2. S. M. Blower and Hadi Dowlatabadi, Sensitivity and uncertainty analysis of complex models of disease transmission: An hiv model, as an example. *International Statistical Review*, **62**, 08 (1994).
3. E. O. Buzbas and N. A. Rosenberg, AABC: Approximate approximate Bayesian computation for inference in population-genetic models, *Theor. Popul. Biol.* **99**, 31–42, Feb2015.
4. A. Charzyńska, A. Nałęcz, M. Rybiński and A. Gambin, Sensitivity analysis of mathematical models of signaling pathways, *BioTechnologia.* **93**(3), 291–308 (2012).
5. S. Checkley, L. MacCallum, J. Yates, P. Jasper, H. Luo, J. Tolsma and C. Bendtsen, Bridging the gap between in vitro and in vivo: Dose and schedule predictions for the ATR inhibitor AZD6738, *Sci. Rep.* **5**, 13545 (2015).
6. J. Cohen, The statistical power of abnormal-social psychological research: A review, *J. Abnorm. Soc. Psychol.* **65**, 145–153, Sep 1962.

7. O. Cohen, *Statistical Power Analysis for the Behavioral Sciences* (Second Edition), Lawrence Erlbaum Associates, 1988.

8. J. L. Gevertz and J. R. Wares, Developing a Minimally Structured Mathematical Model of Cancer Treatment with Oncolytic Viruses and Dendritic Cell Injections, *Comput. Math. Methods Med.* **2018**, 8760371 (2018).

9. S. Hamis, J. Yates, M. A. J. Chaplain and G. G. Powathil, Bridging in vitro and in vivo research via an agent-based modelling approach: predicting tumour responses to an atr-inhibiting drug, *Preprint: bioRxiv*, doi:10.1101/841270.

10. R. L. Iman and J. C. Helton, Comparison of uncertainty and sensitivity analysis techniques for computer models, *Report NUREGICR-3904, SAND 84-1461, Sandia National Laboratories, Albuquerque, New Mexico*, 3 1985.

11. Timothy Krehbiel, Correlation coefficient rule of thumb, *Decision Sciences Journal of Innovative Education* **2**, 97–100, 01 (2004).

12. B. Lambert, A. L. MacLean, A. G. Fletcher, A. N. Combes, M. H. Little and H. M. Byrne, Bayesian inference of agent-based models: A tool for studying kidney branching morphogenesis, *J Math. Biol.* **76**(7), 1673–1697, 06 (2018).

13. J. Liepe, P. Kirk, S. Filippi, T. Toni, C. P. Barnes and M. P. Stumpf, A framework for parameter estimation and model selection from experimental data in systems biology using approximate Bayesian computation, *Nat. Protoc.* **9**(2), 439–456, Feb 2014.

14. A. Ligmann-Zielinska, D. B. Kramer, K. Spence Cheruvelil and P. A. Soranno, Using uncertainty and sensitivity analyses in socioecological agent-based models to improve their analytical performance and policy relevance, *PLoS ONE* **9**(10), e109779 (2014).

15. Shenglin Lin, Wei Li, Xiaochao Qian, Ping Ma and Ming Yang, A Simulation Model Validation and Calibration Platform, pages 687–693, 12 2018.

16. G. Manache and C. Melching, Sensitivity of Latin Hypercube Sampling to sample size and distributional assumptions, 07 2007.

17. MATLAB. *Version 1.8.0_202 (R2019n)*. The MathWorks Inc., Natick, Massachusetts, 2019.

18. K. O. McGraw and S. P. Wong, A common language effect size statistic, *Psychological Bulletin* **111**(2), 361–365 (1992).

19. M. D. McKay, R. J. Beckman and W. J. Conover, Comparison of three methods for selecting values of input variables in the analysis of output from a computer code, *Technometrics* **21**(2), 239–245 (1979).

20. Michael D. McKay, Latin hypercube sampling as a tool in uncertainty analysis of computer models, In *Proceedings of the 24th Conference on Winter Simulation, WSC '92*, pages 557–564, New York, NY, USA, 1992. ACM.

21. M. M. Mukaka, Statistics corner: A guide to appropriate use of correlation coefficient in medical research, *Malawi Med. J.* **24**(3), 69–71, Sep 2012.

22. A. Niida, T. Hasegawa and S. Miyano, Sensitivity analysis of agent-based simulation utilizing massively parallel computation and interactive data visualization, *PLoS ONE*, **14**(3), e0210678 (2019).

23. M. Read, P. S. Andrews, J. Timmis and V. Kumar, Techniques for grounding agent-based simulations in the real domain: a case study in experimental autoimmune encephalomy-elitis, *Mathematical and Computer Modelling of Dynamical Systems* **18**(1), 67–86 (2012).

24. K. A. Rejniak and A. R. Anderson, Hybrid models of tumor growth, *Wiley Interdiscip Rev. Syst. Biol. Med.* **3**, 115–125 (125).

25. J. Ruscio and T. Mullen, Confidence Intervals for the Probability of Superiority Effect Size Measure and the Area Under a Receiver Operating Characteristic Curve, *Multivariate Behavioral Research* **47**(2), 201–223 (2012).

26. Andrea Saltelli and Ricardo Bolado, An alternative way to compute fourier amplitude sensitivity test (fast), *Comput. Stat. Data Anal.* **26**(4), 445–460, February 1998.

27. P. Schober, C. Boer and L. A. Schwarte, Correlation Coefficients: Appropriate Use and Interpretation, *Anesth. Analg.* **126**(5), 1763–1768, 05 (2018).

28. Razi Sheikholeslami and Saman Razavi, Progressive latin hypercube sampling: An efficient approach for robust sampling-based analysis of environmental models, *Environmental Modelling and Software* **93**, 109–126, 07 (2017).

29. A. Vargha and H. D. Delaney, A Critique and Improvement of the CL Common Language Effect Size Statistics of McGraw and Wong, *Journal of Educational and Behavioral Statistics*, **25**(2), 101–132 (2000).

30. S. A. Visser, D. P. de Alwis, T. Kerbusch, J. A. Stone and S. R. Allerheiligen, Implementation of quantitative and systems pharmacology in large pharma, *CPT Pharmacometrics Syst. Pharmacol.* **3**, e142, Oct 2014.

31. X. Y. Zhang, M. N. Trame, L. J. Lesko and S. Schmidt, Sobol Sensitivity Analysis: A Tool to Guide the Development and Evaluation of Systems Pharmacology Models, *CPT Pharmacometrics Syst. Pharmacol.* **4**(2), 69–79, 02 (2015).

Chapter 2

Game Theory Cancer Models of Cancer Cell–Stromal Cell Dynamics Using Interacting Particle Systems

Yinan Zheng[1], Yusha Sun[1], Gonzalo Torga[2],
Kenneth Pienta[2] and Robert Austin[1,*]

[1]*Department of Physics, Princeton University
Princeton, NJ, USA*

[2]*Johns Hopkins Medical Institute
Baltimore, MD, USA*
austin@princeton.edu

We describe an evolutionary game theory model that has been used to predict the population dynamics of interacting cancer and stromal cells. We first consider the mean field case assuming homogeneous and nondiscrete populations. Interacting Particle Systems (IPS) are then presented as a discrete and spatial alternative to the mean field approach. Finally, we discuss cases where IPS gives results different from the mean field approach.

Keywords: Cancer; evolutionary game theory; population dynamics; interacting particle systems.

1. Introduction

Cancer cells and stromal cells interact within a tumor to give both cooperative and competitive behaviors that have been attributed to various molecular signaling pathways. Competitive behavior manifests directly as cancer cell invasion and proliferation in a microenvironment already occupied by stromal cells, while there is also evidence that interactions between cancer cell and stromal cells can be cooperative.[1,2] In fact, a wealth of evidence even indicates that cancer development and progression are *supported* by the host stroma, which includes cancer-associated fibroblasts, extracellular matrix, immune cells, and the vasculature.[3] In particular, there is known to be cross-talk between host stroma and tumor cells: cancer cells in a growing tumor recruit stromal cells, changing their phenotype to become reactive stroma (the typical phenotype during wound healing and other inflammatory processes) and potentially further transforming them into tumor-associated stromal cells.[4] These stromal cells may secrete growth or other tumor-promoting factors that modulate the immune response or support the tumor microenvironment and the pro-metastatic niche.[4,5]

Evolutionary game theory (EGT), which studies the strategic interactions of biological agents based on frequency-dependent fitness functions, has been purported to provide a lens in which to understand cancer-stroma interactions as well as to interpret counter-intuitive cooperative behaviors amongst cells in the tumor microenvironment.[6] Further, EGT has been employed clinically in the form of adaptive therapy to take advantage of competition between various cancer clones.[7,8]

EGT allows for a top–down approach to infer large-scale population dynamics of varying cell types, without necessarily requiring the dissection of specific signaling pathways.[9] Thus, EGT has been adopted as a model to describe the interactions between cancer and stromal cells and as a means to predict future behavior.[9–12] The power of this methodology lies in its ability to understand current population-level behavior through experimental fitting and simultaneously predicting future dynamics. In this chapter, we build on the mean-field EGT models described in Wu *et al.* (2014)[9] and Wu *et al.* (2018)[10] by considering a spatial and discrete interacting particle system (IPS) model of EGT, based on the work of Durrett and Levin (1994).[13] The broader applicability of the IPS perspective in the range of behaviors it captures may be important for the continued role of EGT in cancer research.

2. Mean Field Approaches

Typically, mean-field EGT describes the interactions between two well-mixed populations of game "players" using a 2×2 payoff matrix: $\begin{bmatrix} a & b \\ c & d \end{bmatrix}$. The coefficients (a, b, c, d) are called payoff coefficients. The game players can be cells or organisms from any two interacting populations; for example, EGT has been used to describe interactions between different strains of bacteria,[14] but for the purposes of this chapter, the players are cancer and stromal cells. Specifically, when a cancer cell interacts with another cancer cell, it receives payoff a. When a cancer cell encounters a stromal cell, it receives payoff b. A stromal cell receives payoff c when encountering a cancer cell and payoff d when interacting with another stromal cell. We define cancer cell population $u(t)$, stromal cell population $v(t)$, and a density-dependent death rate κ. Then, the population dynamics of both populations can be described using the following coupled differential equations.[10]

$$\begin{aligned}
\frac{du}{dt} &= u\left(a \cdot \frac{u}{u+v} + b \cdot \frac{v}{u+v} - \kappa \cdot (u+v) \right), \\
\frac{dv}{dt} &= v\left(c \cdot \frac{u}{u+v} + d \cdot \frac{v}{u+v} - \kappa \cdot (u+v) \right).
\end{aligned} \tag{1}$$

These nonlinear differential equations comprise a Mean Field Theory (MFT) picture of population dynamics; that is, we look only at relative population ratios in the large-N limit,[13] and we assume both populations to be homogeneously distributed ("well-mixed"), so cells do not exhibit any spatial structure. While this approach makes the nonlinear population differential equations tractable, it also imposes limits.

We will first lay out the implications of the MFT approach. Then, we describe the limitations of this approach and introduce three alternatives to MFT as described in work by Durrett and Levin.[13] Finally, we use one of these alternatives, IPS, to describe phenomena that arise when cells are discrete and not homogeneous.

2.1. *Fundamentals*

We lay out here the basics of the mean field approach. We have cancer and stromal populations $u(t)$ and $v(t)$. We define cancer population fraction $p(t) = u/(u+v)$ and total cell population $s(t) = u + v$. κ is the death rate, the rate at which each individual cell dies. Equations (1) can be rewritten as

$$\frac{du}{dt} = u(ap + b(1-p) - \kappa s),$$
$$\frac{dv}{dt} = v(cp + d(1-p) - \kappa s). \tag{2}$$

We can now make the change of variable to write a single-variable ODE for p.[13]

$$\begin{aligned}
\dot{p} &= \frac{(u+v)\dot{u} - u(\dot{u}+\dot{v})}{(u+v)^2} \\
&= \frac{\dot{u} - p(\dot{u}+\dot{v})}{u+v} \\
&= \frac{\begin{aligned}uap + ub(1-p) - \kappa u(u+v) - uap^2 - ub(1-p)p + up\kappa(u+v) \\ - vcp^2 - vdp(1-p) + vp\kappa(u+v)\end{aligned}}{u+v} \\
&= \frac{u[ap + b(1-p) - ap^2 - bp(1-p)] - v[cp^2 - dp(1-p)]}{u+v} \\
&= p(ap + b(1-p) - ap^2 - bp(1-p)) - (1-p)(cp^2 - dp(1-p)) \\
&= ap^2(1-p) + bp(1-p)^2 - cp^2(1-p) - dp(1-p)^2) \\
&= p(1-p)(ap + b(1-p) - cp - d(1-p)) \\
&= p(1-p)[(a-c)p + (b-d)(1-p)] \\
&= p(1-p)(p - p_0)(a - b - c + d), \tag{3}
\end{aligned}$$

where we have defined $p_0 = \frac{b-d}{b-d+c-a} = \frac{1}{1-\frac{c-a}{d-b}}$. This differential equation has fixed points at $p = 0, 1, p_0$. Similarly, we derive an ODE for s in terms of s, p, the payoff coefficients and κ.

$$\begin{aligned}
\dot{s} &= \dot{u} + \dot{v} \\
&= u(ap + b(1-p) - ks) + v(cp + d(1-p) - \kappa s) \\
&= sp(ap + b(1-p)) + s(1-p)(cp + d(1-p)) - \kappa s^2 \\
&= s(ap^2 + bp - bp^2 + cp - cp^2 + d - 2dp + dp^2) - \kappa s^2 \\
&= s(p^2(a - b - c + d) + p(b + c - 2d) + (d - \kappa s)). \tag{4}
\end{aligned}$$

Now, in order to determine the stability of the fixed points, we find $\frac{d\dot{p}}{dp}$ at $p = 0, 1, p_0$

$$\dot{p} = p(1-p)(a-b-c+d)(p-p_0)$$

$$= p(1-p)\left(\frac{d-b}{p_0}\right)(p-p_0)$$

$$= (d-b)(p-p^2)\left(\frac{p}{p_0}-1\right)$$

$$= (d-b)\left(\frac{p^2-p^3}{p_0}-p+p^2\right)$$

$$\Rightarrow \frac{d\dot{p}}{dp} = (d-b)\left(\frac{2p-3p^2}{p_0}-1+2p\right). \tag{5}$$

Thus, we have

$$\left.\frac{d\dot{p}}{dp}\right|_{p=p_0} = (d-b)(2-3p_0-1+2p_0)$$

$$= (d-b)(1-p_0),$$

$$\left.\frac{d\dot{p}}{dp}\right|_{p=0} = b-d, \tag{6}$$

$$\left.\frac{d\dot{p}}{dp}\right|_{p=1} = (d-b)\left(1-\frac{1}{p_0}\right).$$

We now do casework to describe the time evolution of p given certain criteria for the payoff coefficients (a, b, c, d). We seek for each case the steady state cancer fraction $p_s = p(t = \infty)$.

2.2. *Case 1: $a < c$; $b < d$*

This is the so-called "Prisoner's Dilemma" case.[10] Here, regardless of initial conditions, stromal cells will proliferate while cancer cells will die out. That is, $p(\infty) = 0$ for any initial condition with $p(t = 0) \in [0, 1)$. We show this by analyzing stability of the fixed points.

First, we have $\frac{c-a}{d-b} > 0$, so $p_0 = \frac{1}{1-\frac{c-a}{d-b}} \notin [0, 1]$. Since $b < d$, we have from Eq. (6) that $\frac{d\dot{p}}{dp}\big|_{p=0} = b - d < 0$, so $p = 0$ is a stable fixed point. Since $p_0 \notin [0, 1]$, we have $(1-\frac{1}{p_0}) > 0$, so $\frac{d\dot{p}}{dp}\big|_{p=1} = (d-b)(1-\frac{1}{p_0}) > 0$, so $p = 1$ is an unstable fixed point. Thus, the fixed points in $[0, 1]$ are the stable fixed point $p = 0$ and the unstable fixed point $p = 1$. Then, for any initial value $p(t = 0) \in [0, 1)$, $p(t)$ is attracted to the stable fixed point 0, so we have steady state value cancer fraction $p_s = 0$.

We now find total population s_0 at equilibrium by taking $\dot{s} = 0$ and using our steady state value $p_s = 0$

$$
\begin{aligned}
0 = \dot{s}|_{s=s_0} &= s_0(p^2(a - b - c + d) + p(b + c - 2d) + (d - \kappa s_0)) \\
&= s_0(d - \kappa s_0) \\
\Rightarrow s_0 &= 0, d/\kappa.
\end{aligned}
\tag{7}
$$

Thus our equilibrium total population (assuming positive s_0) is $s_0 = d/\kappa$. Since steady state cancer population fraction $p_s = 0$, we have equilibrium cancer population $u_0 = 0$ and equilibrium stromal cell population $v_0 = d/\kappa$.

2.3. *Case 2: $a > c$; $b > d$*

This is the so-called "Harmony" case.[10] Here, regardless of initial conditions, cancer cells will proliferate while stromal cells will die out. That is, $p(\infty) = 1$ for any initial condition with $p(t = 0) \in (0, 1]$.

First note that $\frac{c-a}{d-b} > 0$, so $p_0 = \frac{1}{1 - \frac{c-a}{d-b}} \notin [0, 1]$. Since $b - d > 0$, $\frac{d\dot{p}}{dp}|_{p=0} = b - d > 0$, so $p = 0$ is an unstable fixed point. Furthermore, since $p_0 \notin [0, 1]$, we have $(1 - \frac{1}{p_0}) > 0$. Thus, $\frac{d\dot{p}}{dp}|_{p=1} = (d - b)(1 - \frac{1}{p_0}) < 0$, and so $p = 1$ is a stable fixed point. Thus, the fixed points in $[0, 1]$ are the stable fixed point $p = 1$ and the unstable fixed point $p = 0$. So, for any initial value $p(t = 0) \in (0, 1]$, $p(t)$ is attracted to the stable fixed point 1, so we have steady state value cancer fraction $p_s = 1$.

We now find total population s_0 at equilibrium by taking $\dot{s} = 0$ and using our steady state value $p_s = 1$

$$
\begin{aligned}
0 = \dot{s}|_{s=s_0} &= s_0(p^2(a - b - c + d) + p(b + c - 2d) + (d - \kappa s_0)) \\
&= s_0(a - \kappa s_0) \\
\Rightarrow s_0 &= 0, a/\kappa.
\end{aligned}
\tag{8}
$$

Thus our equilibrium total population (assuming positive s_0) is $s_0 = a/\kappa$. Since steady state cancer population fraction $p_s = 1$, we have equilibrium cancer population $u_0 = a/\kappa$ and equilibrium stromal cell population $v_0 = 0$.

2.4. *Case 3: $a > c$; $b < d$*

This is the so-called "Hawk Dove" case.[10] This case is characterized by stable fixed points $p = 0, 1$ and an unstable fixed point $p = p_0 \in (0, 1)$. First, note that $p_0 = \frac{b-d}{b-d+c-a} = \frac{d-b}{d-b+a-c}$, where $d - b > 0$ and $a - c > 0$, so $p_0 \in (0, 1)$. To show stability

$$
\begin{aligned}
\left.\frac{d\dot{p}}{dp}\right|_{p=p_0} &= (d - b)(1 - p_0) > 0, \\
\left.\frac{d\dot{p}}{dp}\right|_{p=0} &= b - d < 0, \\
\left.\frac{d\dot{p}}{dp}\right|_{p=1} &= (d - b)\left(1 - \frac{1}{p_0}\right) < 0.
\end{aligned}
\tag{9}
$$

Thus, for initial populations $p(0) > p_0$, the steady state cancer fraction p_s is given by $p_s = 1$ and for initial populations $p < p_0$, the steady state cancer fraction is $p_s = 0$. Thus, $p(t = \infty)$ is now determined by the initial conditions.

We now find total population s_0 at equilibrium by taking $\dot{s} = 0$.

Case 3a: $p(0) > p_0$. Then, $p_s = 1$, so

$$
\begin{aligned}
0 = \dot{s}|_{s=s_0} &= s_0(p^2(a - b - c + d) + p(b + c - 2d) + (d - \kappa s_0)) \\
&= s_0(a - \kappa s_0) \\
\Rightarrow s_0 &= 0, a/\kappa.
\end{aligned}
\tag{10}
$$

Thus, equilibrium total population is $s_0 = a/\kappa$. Since steady state cancer population fraction $p_s = 1$, we have equilibrium cancer population $u_0 = a/\kappa$ and equilibrium stromal cell population $v_0 = 0$.

Case 3b: $p(0) < p_0$. Then, $p_s = 0$, so

$$
\begin{aligned}
0 = \dot{s}|_{s=s_0} &= s_0(p^2(a - b - c + d) + p(b + c - 2d) + (d - \kappa s_0)) \\
&= s_0(d - \kappa s_0) \\
\Rightarrow s_0 &= 0, d/\kappa.
\end{aligned}
\tag{11}
$$

Thus equilibrium total population is $s_0 = d/\kappa$. Since steady state cancer population fraction $p_s = 0$, equilibrium cancer population is $u_0 = 0$ and equilibrium stromal cell population is $v_0 = d/\kappa$.

2.5. *Case 4: $a < c$; $b > d$*

This is the "Stag Hunt" case, characterized by unstable fixed points $p = 0, 1$ and a stable fixed point $p = p_0 \in (0, 1)$. We have that $p_0 = \frac{b-d}{b-d+c-a}$ where $b - d > 0$ and $c - a > 0$, so $p_0 \in (0, 1)$. We check the stability of all three fixed points

$$
\begin{aligned}
\frac{d\dot{p}}{dp}\Big|_{p=p_0} &= (d - b)(1 - p_0) < 0, \\
\frac{d\dot{p}}{dp}\Big|_{p=0} &= b - d > 0, \\
\frac{d\dot{p}}{dp}\Big|_{p=1} &= (d - b)\left(1 - \frac{1}{p_0}\right) > 0.
\end{aligned}
\tag{12}
$$

Thus, for any initial $p \in (0, 1)$, we have steady state cancer fraction $p_s = p_0$, so cancer and stromal cells will coexist with equilibrium cancer fraction p_0.

We now find total population s_0 at equilibrium by taking $\dot{s} = 0$ and using our steady state value $p_s = p_0$

$$
\begin{aligned}
0 = \dot{s}|_{s=s_0} &= s_0(p_0^2(a - b - c + d) + p_0(b + c - 2d) + (d - \kappa s_0)) \\
\Rightarrow s_0 &= (p_0^2(a - b - c + d) + p_0(b + c - 2d) + d)/\kappa \\
&= (p_0(d - b) + p_0(b + c - 2d) + d)/\kappa \\
&= (p_0(c - d) + d)/\kappa.
\end{aligned}
\tag{13}
$$

2.6. *Limitations*

The MFT approach is not discrete, since our cancer and stromal populations $u(t)$ and $v(t)$ are real-valued continuous functions. The MFT approach is also not spatial, since we assume that cancer and stromal cells are mixed homogeneously in space. In reality, however, cell populations are both discrete and spatial.

In fact, the MFT predictions that we have just discussed are sometimes inaccurate. Durrett and Levin[13] showed that models that are discrete and/or spatial often exhibit dynamics that are quite different from the dynamics predicted by MFT. They give three alternatives to MFT that we now describe.

3. Three Alternatives to Mean Field Theory

Durrett and Levin[13] discuss three alternatives to the mean field approach: patch models, reaction-diffusion equations, and IPSs. It is important to note that in the following discussion, the word rate is often used. When we say an event occurs at some rate λ, we mean that the times t_i between successive occurrences of the event have the exponential probability distribution $P(t_i < t) = 1 - e^{-\lambda t}$.[13]

3.1. *Reaction-diffusion equations*

Reaction-Diffusion Systems are a spatial, nondiscrete alternative to MFT, in which we consider infinitesimal cells distributed spatially in $\mathbb{R}^n$.[13] In this approach, we add diffusion terms $k_1 \nabla^2 u$ and $k_2 \nabla^2 v$ to our MFT differential equations to yield new partial differential equations for cancer populations $u(x,t)$ and stromal cell populations $v(x,t)$ for $x \in \mathbb{R}^n$

$$\frac{\partial u(x,t)}{\partial t} = k_1 \nabla^2 u + u(ap(x,t) + b(1 - p(x,t)) - \kappa s),$$
$$\frac{\partial v(x,t)}{\partial t} = k_2 \nabla^2 v + v(cp(x,t) + d(1 - p(x,t)) - \kappa s). \tag{14}$$

These diffusion terms account for migration of cancer and stromal cells through space, allowing for spatial behavior that is obscured when the population is assumed to be homogeneous as in MFT.

3.2. *Patch models*

Patch models are discrete and nonspatial.[13] As described in Durrett and Levin, they describe the dynamics of discrete cells that are divided into a collection of N subpopulations, or "patches" $S = \{0, 1, \ldots, N - 1\}$. Note that we label patches starting from 0. We then have local integer-valued functions $u(x,t)$ and $v(x,t)$ that represent the number of cancer and stromal cells, respectively, in patch x at time t. Time evolution of the system is described by the following algorithm, as described by Durrett and Levin. Essentially, the algorithm replicates the mean field approach on a local scale; instead of considering global population fractions and cell counts, cells only interact with other cells within their patch, so only the local cancer fraction in the patch is considered when applying the payoff coefficients. This algorithm comprises three steps: migration, death, and a "game step" in which payoff coefficients are taken into account.

Migration: Every discrete cell has a probability μ of migration (leaving its current patch); we call μ the migration rate. The cell is allowed to migrate to any other patch. Explicitly, if there are N patches and a cell is in patch $j \in S$, it undergoes one of the following disjoint events during each migration step: (i) It stays in its current patch with probability $1 - \mu$; (ii) It migrates to a patch $k \in \{0, \ldots, j - 1, j + 1, \ldots, N - 1\}$ with probability $\mu/(N - 1)$. Note that the cell migrates to each destination patch with equal probability, so there is no spatial structure introduced by migration.

Death: Analogous to death in the mean field approach, each cell in patch $x \in S$ dies at rate $\kappa(u(x, t) + v(x, t))$; we call κ the death rate.

Game Step: Analogous to the global cancer population fraction $p(t)$ used in the mean field approach, we define a local cancer population fraction $p(x, t) = u(x, t)/(u(x, t) + v(x, t))$. Then, each cancer cell experiences a birth/death rate $ap(x, t) + b(1 - p(x, t))$ and each stromal cell experiences a birth/death rate $cp(x, t) + d(1 - p(x, t))$. By birth/death rate, we mean the following: if $ap + b(1 - p) > 0$, each cancer cell in patch x produces another cancer cell at rate $ap + b(1 - p)$; if $ap + b(1 - p) < 0$, each cancer cell in patch x dies at rate $|ap + b(1 - p)|$. The same applies for stromal cells and the stromal birth/death rate $cp + d(1 - p)$. Again, this mirrors the dynamics of the mean field approach, except now, we treat each cell as discrete, and have a cell interact only with the other cells in its patch. However, note that the set of patches still has no spatial structure; from the perspective of a cell in a given patch, all other patches appear identical.

3.3. *Interacting particle systems: A simple example*

The IPS approach is similar to the patch model approach; the population of cancer and stromal cells is again partitioned into subpopulations called patches, except the collection of patches is now given a spatial structure. We give a simple example of an IPS to build intuition; more exotic examples are given in the next section. We define the set of patches spatially oriented in a square lattice $S = \mathbb{Z}^2$. Then, we can represent local cancer and stromal cell populations at patch $x \in \mathbb{Z}^2$ at time t as integer-valued functions $u(x, t)$ and $v(x, t)$, respectively. Just like the patch model, the time evolution of the system is described by a three-part algorithm comprising migration, death, and a "game step" where payoff coefficients contribute to the dynamics.

Migration: Every discrete cell migrates (leaves its current patch) with probability μ; we call μ the migration rate. Each cell in a patch $x \in \mathbb{Z}^2$ is only allowed to migrate to any of its four neighboring patches $\{x + (1, 0), \ x + (-1, 0), \ x + (0, 1),$

$x + (0, -1)\}$. The cell migrates to each of these patches with equal probability. Note that this definition of migration introduces a spatial aspect to the set of patches, since we can now speak of some patches being closer together than other patches. For example, patches $(2, 3)$ and $(2, 4)$ are "close" since cells are allowed to migrate between the two patches.

Death: Identical to death in the patch model approach, each cell in patch x dies at rate $\kappa(u(x, t) + v(x, t))$, where we call κ the death rate.

Game Step: The IPS approach is differentiated from patch model approaches by the notion of an *interaction neighborhood* $\mathcal{N}(x)$ of a patch x. Note that our notation here is slightly different from that in the Durrett and Levin description. Essentially, the interaction neighborhood $\mathcal{N}(x)$ determines which patches are included when calculating the local cancer population fraction of a patch x.

There are many choices for $\mathcal{N}(x)$. The simplest possible interaction neighborhood is defined by $\mathcal{N}_1(x) = \{x\}$ for any patch x, where the interaction neighborhood of a patch comprises only the patch itself. A more complex interaction neighborhood is $\mathcal{N}_2(x) = \{(z_1, z_2) \in \mathbb{Z}^2 : |z_1 - x_1| + |z_2 - x_2| \leq 1\}$ for any patch $x = (x_1, x_2) \in \mathbb{Z}^2$. Then, for any patch x, $\mathcal{N}_2(x)$ is the 3×3 square of patches centered at x.

Given our choice of $\mathcal{N}(x)$, we can now describe the game step. We define the local cancer population $u^*(x, t)$ of a patch x as the count of all cancer cells within the interaction neighborhood $\mathcal{N}(x)$. Then, $u^*(x, t) = \sum_{z \in \mathcal{N}(x)} u(z, t)$. Similarly, the local stromal population $v^*(x, t)$ is given by $v^*(x, t) = \sum_{z \in \mathcal{N}(x)} v(z, t)$. We define local cancer fraction $p(x, t)$, given by $p(x, t) = u^*(x, t)/(u^*(x, t) + v^*(x, t))$. Then, each cancer cell experiences a birth/death rate $ap(x, t) + b(1 - p(x, t))$ and each stromal cell experiences a birth/death rate $cp(x, t) + d(1 - p(x, t))$. By birth/death rate, we mean the following: if $ap + b(1 - p) > 0$, each cancer cell in patch x produces another cancer cell at rate $ap + b(1 - p)$; if $ap + b(1 - p) < 0$, each cancer cell in patch x dies at rate $|ap + b(1 - p)|$. The same applies for stromal cells and the stromal birth/death rate $cp + d(1 - p)$.

The interaction neighborhood further provides a spatial nature to the set of patches. If patch $y \in \mathcal{N}(x)$, we can think of x and y as "close", since the cancer and stromal cell populations of patch y are included when calculating the local cancer fraction $p(x, t)$ for patch x; this local cancer fraction determines the time evolution of the patch x populations in the game step.

3.4. *General IPS using graphs*

To describe more general spatial arrangements of patches, we can represent the migration and interaction neighborhoods of patches as graphs. We introduce some basic graph theory terminology. A graph G is an ordered triple (V, E, ψ). V is a set of vertices, or "nodes", usually drawn as dots. E is a set of edges; each edge connects two vertices (not necessarily distinct) and are drawn as lines. ψ is an incidence function

$\psi : E \to V \times V$ that maps each edge to the two vertices it connects. Two vertices x, y are called adjacent if there is an edge connecting x and y.

Graphs are natural representations of connectivity, and thus are well suited to represent connections between patches. We can identify the set of N patches of an IPS with a vertex set V, so that each patch is represented by a node. Then, we label the nodes $0, 1, \ldots, N - 1$, just as we labeled the patches in the Patch Model. The spatial characteristics of a general IPS are then represented by two graphs that share the vertex set V: (i) a *migration graph M*, which describes the patches to which a cell can migrate, and (ii) an *interaction graph I*, which describes the interaction neighborhood of each patch.

To be completely general, each patch $x \in V$ can also be assigned its own set of payoffs $(a(x), b(x), c(x), d(x))$ and its own death rate $\kappa(x)$. This generality is especially important in modeling experimental setups where a gradient of payoffs or death rates is introduced, as the Austin lab has recently accomplished.[9] The three-step IPS algorithm given in (3.3) is now generalized as follows.

Migration: Every discrete cell migrates (leaves its current patch) with probability μ; we call μ the migration rate. Each cell in a patch $x \in V$ is only allowed to migrate to patches that are adjacent to x in the migration graph, i.e. to any patch in the set $\{y \in V : y, x \text{ are adjacent in } M\}$). The cell migrates to each of these patches with equal probability.

Death: Each cell in patch x dies at rate $\kappa(x)(u(x, t) + v(x, t))$, where $\kappa(x)$ is the patch's death rate.

Game Step: The interaction graph I defines the interaction neighborhood of any patch $x \in V$; we define $\mathcal{N}(x) = \{x\} \cup \{y \in V : y, x \text{ are adjacent in } I\}$. As in (3.3), local cancer and stromal populations are given by $u^*(x, t) = \sum_{z \in \mathcal{N}(x)} u(z, t)$ and

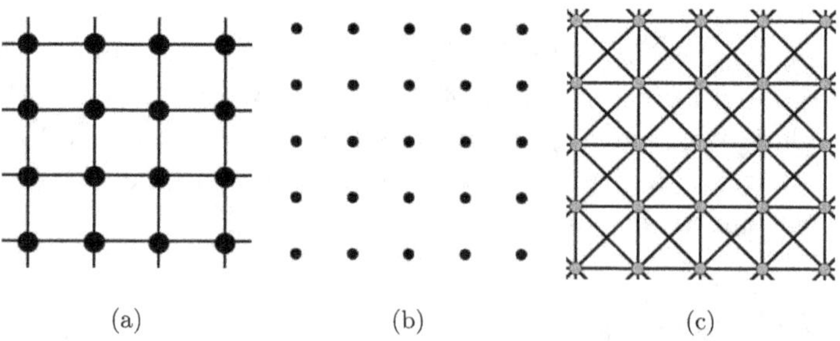

(a)   (b)   (c)

Fig. 1. (a) This infinite square lattice graph is the migration graph M of the example IPS as described in (3.3), where migration is allowed to any of four neighboring patches $\{x + (1, 0), x + (-1, 0), x + (0, 1), x + (0, -1)\}$. (b) This square lattice with no edges is the interaction graph of the interaction neighborhood $\mathcal{N}_1(x) = \{x\}$ described in the game step in (3.3). Here, a patch's interaction neighborhood comprises only itself. (c) This square lattice with diagonals is the interaction graph of the interaction neighborhood $\mathcal{N}_2(x)$ as described in (3.3); for any patch x, $\mathcal{N}_2(x)$ is the 3×3 square of patches centered at x.

$v^*(x,t) = \sum_{z \in \mathcal{N}(x)} v(z,t)$, respectively. Local cancer fraction is $p(x,t) = u^*(x,t)/(u^*(x,t) + v^*(x,t))$. Then, each cancer cell in patch x experiences a birth/death rate $a(x)p(x,t) + b(x)(1 - p(x,t))$ and each stromal cell experiences a birth/death rate $c(x)p(x,t) + d(x)(1 - p(x,t))$. For clarification of birth/death rates, see the discussion in the Game Step for the Patch Model.

Using this graph representation of migration and interaction neighborhoods, we can describe very general spatial configurations of patches. For example, Fig. 1 gives migration and interaction graphs for the example IPS described in (3.3).

4. Single-Node IPS: IPS and MFT Agree

For the rest of the chapter, we will focus on the IPS as a spatial, discrete model to describe population dynamics of interacting cancer and stromal cells. We will first describe a few simple cases where IPS and MFT should agree. Let's consider the simplest case of an IPS comprising only one node. Then, there is no migration of cells and the interaction neighborhood $\mathcal{N}$ of the node is trivially the node itself. Thus, each cell interacts with every other cell in the system, so the system is a homogeneous mixture of cancer and stromal cells. We can then expect the time evolution of the system determined by the IPS algorithm to match the theoretical results from the mean field differential equations; the IPS acts as a discrete approximation of the MFT results. We consider a few of the regimes considered for the mean field approach.

4.1. *Case* 1: *a < c; b < d*

We run an IPS simulation taking $(a,b,c,d) = (0,0,0.4,0.4)$, $\kappa = 0.0005$, with initial cancer population $u(t = 0) = 100$ and initial stromal population $v(t = 0) = 100$. The simulated time evolution of the populations over 200 time steps is shown in Fig. 2(i). The IPS shows cancer cells tending to zero and stromal cell population displaying noisy behavior around some mean steady state population $\bar{v}$, thus agreeing qualitatively with the mean field results.

We then ran the simulation for 10^4 time steps to determine the mean and variance of the stromal cell population in the steady state (we approximate that the steady state to be reached around $t = 50$, and so disregard population data for times before $t = 50$). The mean stromal cell population was found to be $\bar{v} = 800$ and the standard deviation was $\sigma = 33$. MFT gives that the stromal cell population in the steady state should be $v_0 = d/\kappa = 0.4/0.0005 = 800$ and steady state cancer population is $u_0 = 0$. We then have $v_0 \in (\bar{v} - \sigma, \bar{v} + \sigma)$, so IPS and MFT agree within an error bar of $\pm\sigma$.

4.2. *Case* 2: *a > c; b > d*

We run an IPS simulation taking $(a,b,c,d) = (0.5,0.4,0.1,0.1)$, $\kappa = 0.0005$, with initial cancer population $u(t = 0) = 100$ and initial stromal population $v(t = 0) = 100$. The simulated time evolution of the populations over 200 time steps

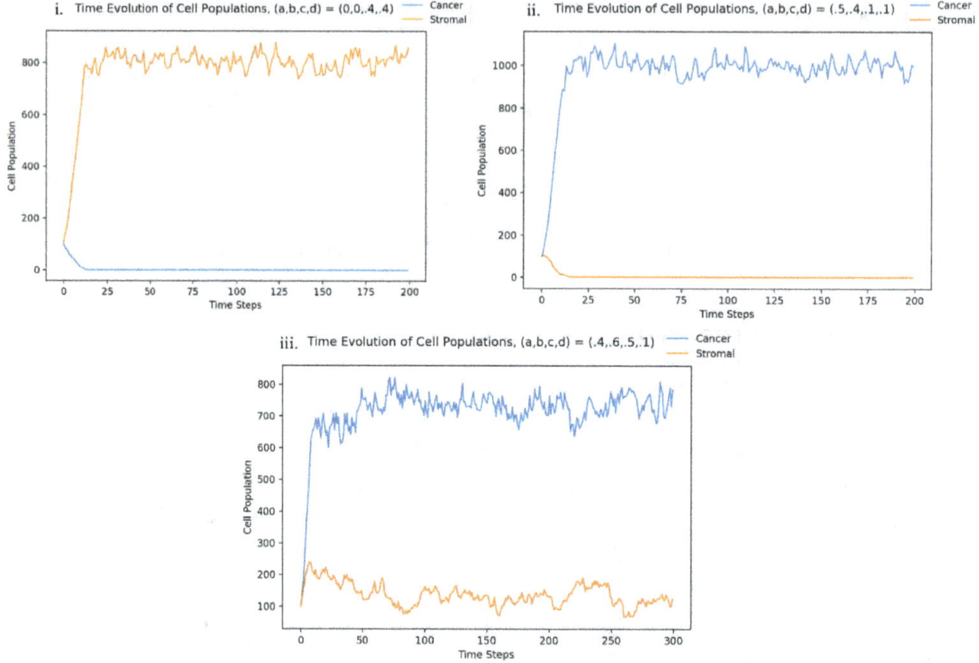

Fig. 2. Time evolution of cell populations from single-node IPS simulations in different payoff regimes. $\kappa = 0.0005$ throughout. (i) Simulation results in the $a < c; b < d$ regime, taking $(a, b, c, d) = (0, 0, 0.4, 0.4)$. (ii) Simulation results in the $a > c; b > d$ regime, taking $(a, b, c, d) = (0.5, 0.4, 0.1, 0.1)$. (iii) Results in the $a < c; b > d$ regime, taking $(a, b, c, d) = (0.4, 0.6, 0.5, 0.1)$.

is shown in Fig. 2(ii). The IPS shows stromal cells tending to zero and cancer cell population displaying noisy behavior around some mean steady state population $\bar{u}$, thus agreeing with the mean field results.

We then ran the simulation for 10^4 time steps to determine the mean and variance of the cancer cell population in the steady state. The mean cancer cell population was found to be $\bar{u} = 1000$ and the standard deviation was $\sigma = 37$. MFT gives that the cancer cell population in the steady state should be $u_0 = a/\kappa = 0.5/0.0005 = 1000$ and steady state stromal cell population is $v_0 = 0$. We have $u_0 \in (\bar{u} - \sigma, \bar{u} + \sigma)$, so again, IPS and MFT agree within an error bar of $\pm\sigma$.

4.3. *Case 3: $a < c; b > d$*

We run an IPS simulation taking $(a, b, c, d) = (0.4, 0.6, 0.5, 0.1)$, $\kappa = 0.0005$, with initial cancer population $u(t = 0) = 100$ and initial stromal population $v(t = 0) = 100$. The simulated time evolution of the populations over 300 time steps is shown in Fig. 2(iii). The IPS shows both cancer and stromal cells displaying noisy behavior around steady state mean populations $\bar{u} > 0$ and $\bar{v} > 0$, thus agreeing with the MFT prediction of cancer and stromal populations approaching stable equilibrium values $u_0 > 0$ and $v_0 > 0$, with equilibrium cancer fraction $p_0 = \frac{b-d}{b-d+c-a} = 5/6$.

Running the simulation for 10^4 time steps to determine the mean and variance of cancer and stromal cell population in the steady state, the mean cancer and stromal populations were found to be $\bar{u} = 720$ and $\bar{v} = 140$, respectively. This gives a steady state cancer fraction of $\bar{p} = \bar{u}/(\bar{u} + \bar{v}) = 0.838 \approx 5/6 = p_0$.

5. IPS on a Cycle: IPS and MFT Differ

Now, we present cases where we found IPS and MFT to differ. The parameter space for a general IPS is truly tremendous; one must choose interaction and migration graphs, payoff coefficients for each node, death rates $\kappa(x)$ for each node, and a migration rate μ. We first explore some effects that migration rate has on the qualitative results generated by the IPS. Then, we describe a case where Durrett and Levin[13] found the IPS to generate novel behavior, and attempt to describe this behavior more in depth than in the original treatment.

For these simulations, we consider a slightly more complex IPS that we call Q_n, comprising n patches and defined as follows: the interaction graph of Q_n is the edgeless graph with n vertices. The migration graph of Q_n is the cycle C_n. A cycle is a graph where the nodes and edges form a closed ring. To make these explicit, we draw the interaction and migration graphs of Q_6 in Fig. 3.

Thus, in Q_n, the n patches are arranged in a circle, and cells are only allowed to migrate to adjacent patches. The interaction neighborhood of a cell comprises only the patch that it resides in. Simply put, Q_n is a 1-D lattice with n linked patches and a periodic boundary condition (patch 0 is connected to patch $n - 1$). Thus, Q_n may be used to model an experimental setup comprising n wells arranged in a circle, where between each pair of adjacent wells there is a channel through which cancer and stromal cells can migrate at some migration rate μ.

5.1. *Effects of migration*

The spatial nature of an IPS distinguishes it from mean field approaches, which treats populations as homogeneous and mixed. The migration of cells as defined by μ and the migration graph can yield qualitative changes to the population dynamics modeled in an IPS.

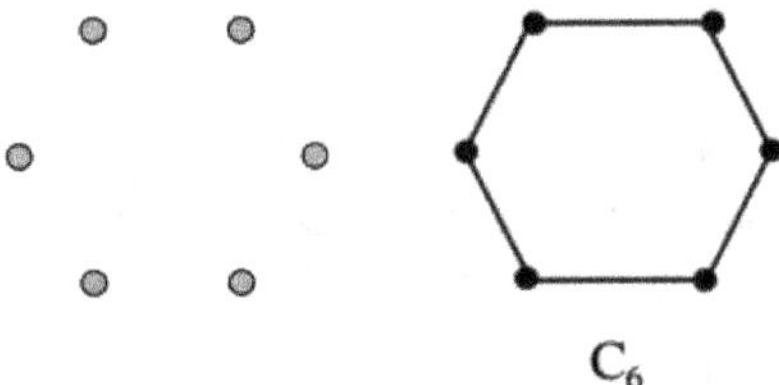

Fig. 3.   Interaction and migration graphs for Q_6. On the left is the interaction graph, the empty graph with six vertices and no edges. On the right is the migration graph, the cycle C_6 with six vertices.

5.1.1. *Kappa landscape*

We simulated the IPS Q_{20}, defining payoff coefficients of $(a, b, c, d) = (0, 0, 0.3, 0.3)$ for all patches. From a MFT perspective, these coefficients are in the Prisoner's Dilemma regime, where stromal cells dominate and cancer cell fraction $p \to 0$. We then defined the death rate $\kappa(j)$ as follows:

$$\kappa(j) = \begin{cases} 0.001, & j \in [0, 4] \cup [15, 19], \\ 0.002, & j \in [5, 9], \\ 0.01, & j \in [10, 14]. \end{cases} \tag{15}$$

We first took $\mu = 0$, running simulations for 10^4 time steps to determine the mean stromal cell population *for each patch* in the steady state; as before, we estimate that the steady state is reached around $t = 50$, and thus ignore population data from times $t < 50$. We then repeated these simulations for $\mu = 0.3$ and $\mu = 1$. The results are shown in Fig. 4.

When $\mu = 0$, the patches within the IPS have no spatial structure; that is, the dynamics of each patch are not affected by the states of the other patches. Thus, the system functions as a collection of 20 isolated single-node IPSs. As discussed in (4.1), the behavior of a single-node IPS with $a < c$; $b < d$ agree with the dynamics predicted by the mean field approach. MFT predicts that stromal cells achieve an equilibrium population $v_0 = d/\kappa$, while cancer cells die out. Thus, MFT gives the following

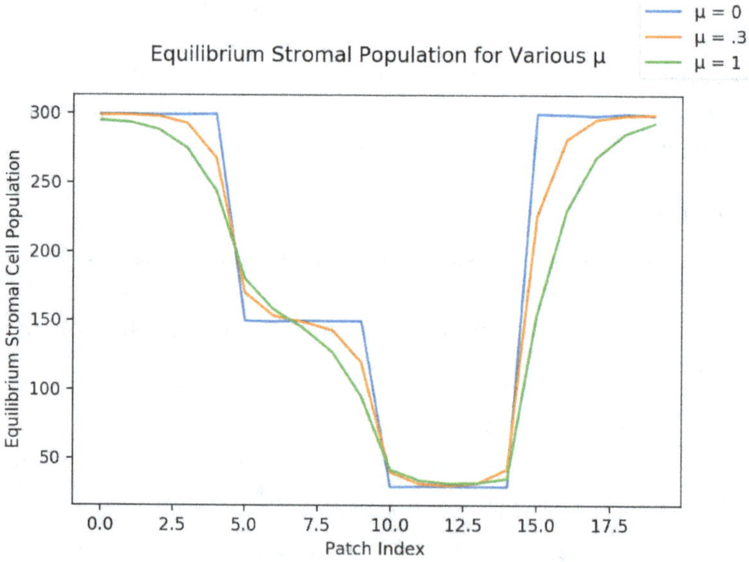

Fig. 4. Effects of migration on the equilibrium stromal cell populations for a given κ landscape. We simulate a Q_{20} IPS with payoffs constant throughout with $(a, b, c, d) = (0, 0, 0.3, 0.3)$, so we are in the $a < c, b < d$ case, where cancer cell population goes to zero for all choices of μ. $\kappa(x)$ varies as described in Eq. (15).

predictions for $v_0(j)$:

$$v_0(j) = d/\kappa(j) = \begin{cases} 300, & j \in [0,4] \cup [15,19], \\ 150, & j \in [5,9], \\ 30, & j \in [10,14]. \end{cases} \tag{16}$$

Indeed, our IPS simulation with $\mu = 0$ give similar equilibrium stromal cell populations, resulting in a step-function appearance. When migration is turned on, the individual patches are no longer isolated from adjacent patches. The net qualitative effect is a smoothing of the steps, an intuitive result since migration allows patches with higher stromal populations to donate to adjacent patches with lower populations. The smoothing effect is greater for higher migration rates, and is most evident at the interface between patches 14 and 15, where the greatest population gap in the $\mu = 0$ simulation is observed.

5.1.2. *Payoff landscape*

We then simulated the IPS Q_{20}, defining $\kappa = 0.001$ for all patches. We define payoffs as follows: $(b,c,d) = (0.4, 0.4, 0.1)$ for all patches and

$$a(j) = \begin{cases} 0.1, & j \in [0,4] \cup [15,19], \\ -0.1, & j \in [5,14]. \end{cases} \tag{17}$$

We thus have $a < c; b > d$ for all patches, so all patches are in the stag hunt regime. As in (5.1.1), we run IPS simulations for $\mu = 0, 0.3, 1$. For each simulation, we simulate for $2 * 10^4$ time steps to determine mean cancer cell population fraction for each patch in the steady state (as usual, we take the steady state to be $t \geq 50$). The results are displayed in Fig. 5.

When $\mu = 0$, the dynamics of each patch are again independent of the state of the other patches, so each patch represents an isolated single-node IPS. As discussed in (4.3), the single-node IPS agrees with MFT results in the stag-hunt regime. That is, the cancer cell and stromal cell populations tend toward positive equilibrium populations u_0 and v_0, respectively, while cancer population fraction tends toward the stable fixed point $p_0 = \frac{b-d}{b-d+c-a}$. MFT predicts

$$p_0(j) = \frac{b-d}{b-d+c-a(j)} = \begin{cases} 0.5, & j \in [0,4] \cup [15,19], \\ 0.375, & j \in [5,14]. \end{cases} \tag{18}$$

Indeed, the $\kappa = 0$ simulations agree with the MFT predictions, giving a step function result. When migration is turned on, we again see a smoothing of the step function; the smoothing effect is greater for larger choices of μ. This result is intuitive; consider two adjacent patches j and $j+1$ where $p(j,t) > p(j+1,t)$ at some time. The set of cells that migrate from j to $j+1$ in a time step will then have cancer fraction roughly equal to $p(j,t) > p(j+1,t)$, and thus will increase the cancer fraction in patch $j+1$.

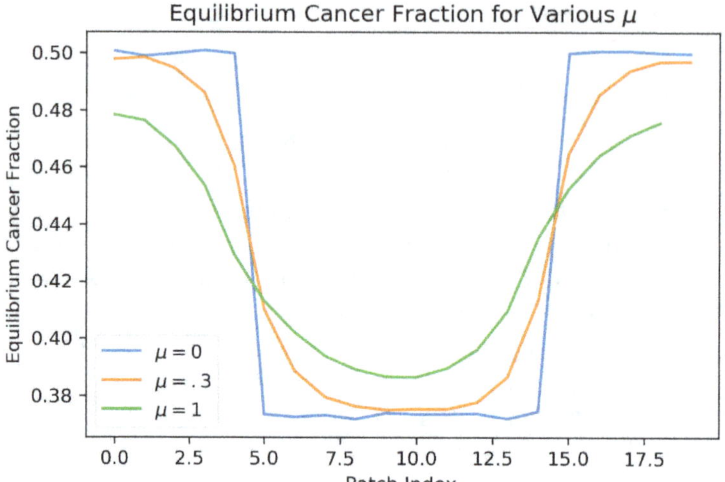

Fig. 5. Effects of migration on the equilibrium stromal cell populations for a given payoffs landscape. We simulate a Q_{20} IPS with $\kappa = 0.001$ for all patches. Payoffs $(b, c, d) = (0.4, 0.4, 0.1)$ are kept constant for all patches. Payoff coefficient $a(j)$ varies as described in Eq. (17). We are in the $a < c, b > d$ case in every patch.

5.2. *Migration in the $b = c$; $a > c$; $d > b$ regime*

The spatial nature of IPSs gives rise to unexpected behavior that is qualitatively different from MFT predictions in the hawk-dove regime $(a > c, b < d)$. We are particularly interested in the special case where $b = c$.[13] In this quadrant, MFT gives that cancer cell fraction $p(t)$ tends away from some unstable fixed point $p_0 \in (0, 1)$. Specifically, if $p(t = 0) < p_0$, $p(t)$ tends toward zero as stromal cells dominate, and stromal cell population $v(t)$ tends toward steady state population $v_0 = d/\kappa$. If $p(t = 0) > p_0$, $p(t)$ tends toward 1 as cancer cells dominate, and cancer population $u(t)$ tends toward steady state population $u_0 = a/\kappa$.

5.2.1. *Reaction-diffusion predictions*

Durrett and Levin[13] maintain that spatial models (reaction-diffusion equations and IPS) give different qualitative results from nonspatial models (MFT and patch models) in the $b = c; a > c; d > b$ regime. They first consider reaction-diffusion equations in 1-D describing dynamics of cancer and stromal populations $u(x, t), v(x, t)$

$$\begin{aligned}
\frac{\partial u}{\partial t} &= k_1 \frac{\partial^2 u}{\partial x^2} + u(ap + b(1 - p) - \kappa s), \\
\frac{\partial v}{\partial t} &= k_2 \frac{\partial^2 v}{\partial x^2} + v(cp + d(1 - p) - \kappa s).
\end{aligned} \tag{19}$$

We cite the following result from Durrett and Levin[13]: Consider the initial condition $u(x, t = 0) = f(x)$, $v(x, t = 0) = g(x)$ where $f(x)$ is a monotonically decreasing function with $f(-\infty) = \bar{u}$ and $f(\infty) = 0$, and $g(x)$ is a monotonically increasing function with $g(-\infty) = 0$ and $g(\infty) = \bar{v}$. Then, u and v will have traveling wave solutions $u(x,t) = f(x - \rho t)$ and $v(x,t) = g(x - \rho t)$, where ρ has the same sign as the quantity $(a - d)$. These wave solutions imply that if ρ is positive, the cancer population at any point x will monotonically increase with time, while the stromal population will decrease. If ρ is negative, the opposite occurs. Qualitatively, this result suggests that given a region of high cancer fraction adjacent to a region of mostly stromal cells, the cancer cell region will displace the stromal cell region at a linear rate if $a > d$; the opposite will happen if $d > a$.

It makes sense that ρ and the quantity $(a - d)$ have the same sign. Consider a point with an equal number of cancer and stromal cells, so $u = v$ and the cancer fraction is $p = 0.5$. Since $b = c$, if $a - d > 0$, we have at this point $u(ap + b(1 - p)) = 0.5u(a + b) > 0.5u(c + d) = v(c(p) + d(1 - p))$, so the payoffs favor the cancer population over the stromal population, a conclusion consistent with positive ρ. Likewise, if $d - a > 0$, the payoffs will favor the stromal population over the cancer population at this point, consistent with a negative value of ρ.

5.2.2. *IPS predictions*

Durrett and Levin[13] suggest that an IPS will display similar qualitative behavior to the reaction-diffusion equations given an interface between a region with high cancer fraction and a region with low cancer fraction. That is, the interface will move to favor the cancer population if $a - d > 0$ and will favor the stromal population if $d - a < 0$. In this section, we will give a much more explicit demonstration of this qualitative behavior using an IPS initialized with a region with primarily cancer cells and a region comprising primarily stromal cells.

We ran the IPS Q_{40} with payoff coefficients $(a, b, c, d) = (0.8, 0.4, 0.4, 0.6)$ and $\kappa = 0.0008$ for all patches. We initialize cancer and stromal cell populations $u(j, t)$ and $v(j, t)$ as follows:

$$u(j, t = 0) = \begin{cases} 0, & j \in [0, 19] \cup [25, 39], \\ 1000, & j \in [20, 24], \end{cases} \tag{20}$$

$$v(j, t = 0) = \begin{cases} 750, & j \in [0, 19] \cup [25, 39], \\ 0, & j \in [20, 24]. \end{cases} \tag{21}$$

Thus, we initialize patches $j \in [0, 19] \cup [25, 39]$ with only stromal cells and initialize a "bubble" of cancer cells in the patches $j \in [20, 24]$. First suppose migration were turned off. Then, each patch would function as an isolated single-cell IPS containing only one cell type, and thus would display dynamics that match MFT predictions. Then, in the patches containing only cancer cells, MFT predicts equilibrium cancer population $u_0 = a/\kappa = 1000$; in the patches containing stromal cells, MFT predicts equilibrium stromal population $v_0 = d/\kappa = 750$. In our IPS simulation, we have thus specifically initialized patches $j \in [0, 19] \cup [25, 39]$ with v_0 stromal cells and patches $j \in [20, 24]$ with u_0 cancer cells. Thus, this IPS with our chosen initial conditions is the steady state of *any* Q_{40} IPS where $\mu = 0$ and patches $j \in [0, 19] \cup [25, 39]$ were initialized with only stromal cells and patches $j \in [20, 24]$ were initialized with only cancer cells. Thus, we can view our simulation as first allowing *any* such no-migration IPS to reach equilibrium, and then turning on migration between patches at time $t = 0$.

Note that our IPS as a whole has initial cancer cell population 5000 and initial stromal population 26,250, giving global initial cancer fraction $p(t = 0) = 0.16 < 1/3 = p_0$, where $p_0 = \frac{b-d}{b-d+c-a}$ is the unstable fixed point given by MFT predictions. Thus, since global cancer fraction is below p_0, MFT would predict stromal cells to dominate and cancer fraction to tend to zero. We will see that this is certainly not the case for the spatial IPS simulation. We run this IPS for various values of μ and graph the cancer cell fraction for each patch in Fig. 6.

Instead of observing the cancer cells die out, the IPS simulations show that the "bubble" of cancer cells grows linearly in time, ultimately taking over the entire system. For $\mu = 0$, we see that the initial configuration of cancer cell fraction for all

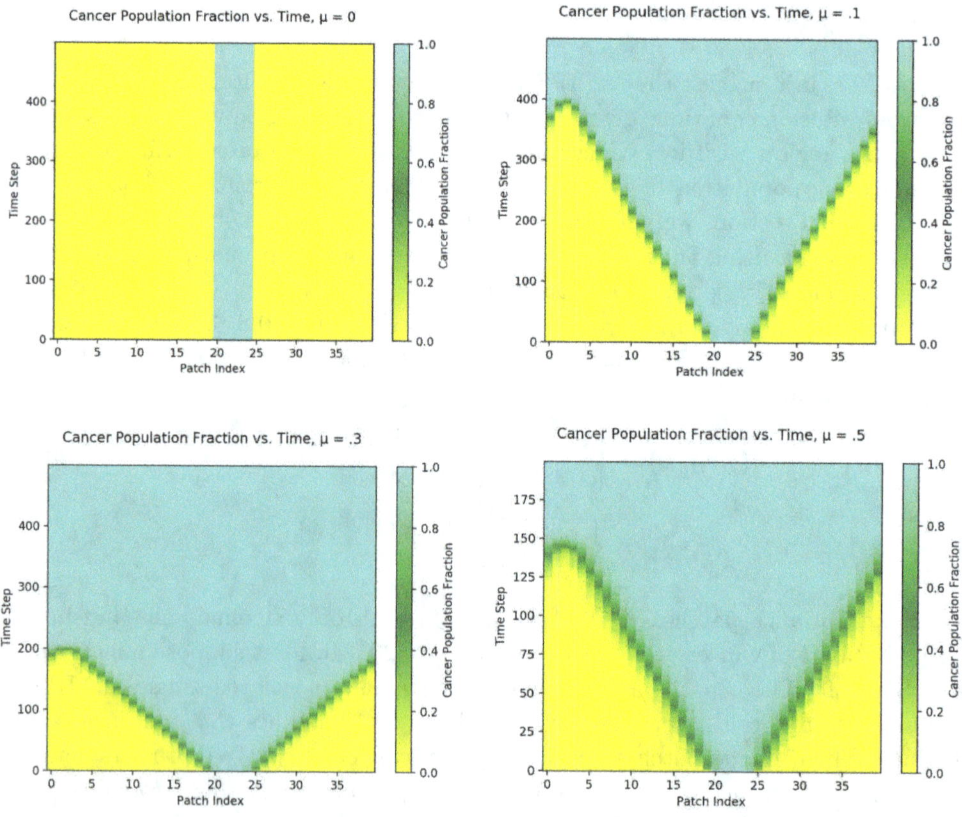

Fig. 6. Q_{40} IPS simulations of cancer fraction time evolution for various migration rates μ and a special initial condition in which cancer and stromal cells are segregated as described in Eqs. (20) and (21). We have payoffs $(a, b, c, d) = (0.8, 0.4, 0.4, 0.6)$ and death rate $\kappa = 0.0008$ for all patches.

40 patches is maintained through time. For positive μ, the rate of growth of the cancer bubble increases as migration rate increases, an intuitive result.

The linear expansion of the cancer bubble is analogous to the linear velocity ρ of the traveling wave solution to the related reaction-diffusion equations discussed earlier in (5.2.1). The spatial nature of migration in this IPS explains this linear expansion. As seen in (5.1.2), migration terms have the qualitative effect of smoothing a discontinuity in cancer fraction between adjacent cells. In other words, migration replaces the cancer fraction of a patch with a stochastic weighted average of the cancer fraction of the patch and its adjacent patches. To see this, Fig. 7 shows a magnified view of cancer fractions around the cancer-stromal cell interface between patches 19 and 20.

Patch 19 is initialized to all stromal cells, but receives migrating cancer cells from patch 20, increasing the patch 19 cancer fraction. Eventually, migration pushes the patch 19 cancer fraction to exceed the MFT unstable fixed point $p_0 = 1/3$. In the Q_{40} IPS, the interaction neighborhood of patch 19 is the patch itself, so once the patch cancer fraction exceeds p_0, the game step (where only payoffs are considered) functions as one would expect from MFT, directing p away from the fixed point p_0 and so $p(19, t) \to 1$. Indeed, we see that cancer fractions $p(x, t)$ in all patches that were initialized to all stromal cells are increasing functions of time with some noise due to stochasticity of migration, death, and the payoffs.

In general, the dynamics of a patch in the IPS consider only the *local* cancer population near the patch; if the local cancer fraction around a patch is high enough

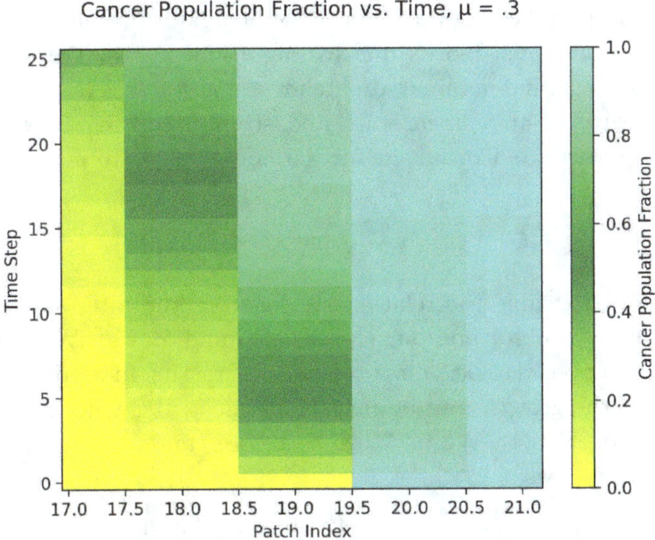

Fig. 7. Q_{40} IPS simulations of cancer fraction time evolution for $\mu = 0.3$ and a special initial condition in which cancer and stromal cells are segregated as described in Eqs. (20) and (21). We have zoomed in to observe behavior around patch 19. We have payoffs $(a, b, c, d) = (0.8, 0.4, 0.4, 0.6)$ and death rate $\kappa = 0.0008$ for all patches.

to exceed p_0, the patch can tend toward $p = 1$, even if the *global* cancer fraction is much lower than p_0.

The intuition why the cancer bubble *grows* in this case is the same as the reasoning behind why the velocity ρ of the traveling wave solutions for $u(x,t)$ and $v(x,t)$ has the same sign as the quantity $a - d$ in the related reaction diffusion equation (see (5.2.1)); this IPS can be seen as a discrete approximation of the 1D reaction diffusion equations with periodic boundary conditions (since Patch 39 and Patch 0 are connected). We have $b = c$ and $a > d$, so cancer is favored at the interface regions between cancer-dominated and stromal-dominated patches.

We now consider the effects of changing the sign and magnitude of $a - d$ on the growth/shrinkage of the cancer bubble by running more simulations on the IPS Q_{40} with $\kappa = 0.008$ and $\mu = 0.1$. Patches are initialized with either all cancer cells or all stromal cells, as before, but this time, we initialize an equal number of cancer and stromal patches

$$u(j, t = 0) = \begin{cases} 0, & j \in [0, 9] \cup [30, 39], \\ 1000, & j \in [10, 29], \end{cases} \tag{22}$$

$$v(j, t = 0) = \begin{cases} 1000, & j \in [0, 9] \cup [30, 39], \\ 0, & j \in [10, 29]. \end{cases} \tag{23}$$

We then run these simulations for varying payoffs. We hold $(a, b, c) = (0.8, 0.4, 0.4)$ constant while dialing d from 0.6 to 1.2. The results are shown in Fig. 8.

When $a > d$, we obtain the familiar result where the cancer bubble grows linearly until it overtakes the entire system. When $a = d$, the cancer and stromal cell patches coexist in the original spatial configuration, with some mixing around initial interface of the cancer and stromal cell blocks due to migration; the spatial separation of cancer-dominated and stromal-dominated patches seems stable. For the $a < d$ cases, the cancer bubble shrinks at a linear rate and stromal cells overtake the entire system. The rate of growth and shrinkage furthermore appears to increase as $|a - d|$ increases.

5.2.3. *Rate of growth*

We now attempt to determine the relationship between the rate of growth r of a cancer or stromal cell "bubble" and our choices of μ and $|a - d|$. We have already seen that growth rate appears to be a monotonically increasing function of both μ and $|a - d|$. We simulate the growth rate by initializing a Q_{40} IPS with $\kappa = 0.0008$ and payoffs $(a, b, c) = (0.8, 0.4, 0.4)$. We initialize as before cancer and stromal populations $u(j, t), v(j, t)$ as follows:

$$u(j, t = 0) = \begin{cases} 0, & j \in [0, 9] \cup [30, 39], \\ 1000, & j \in [10, 29], \end{cases} \tag{24}$$

$$v(j, t = 0) = \begin{cases} 1000, & j \in [0, 9] \cup [30, 39], \\ 0, & j \in [10, 29]. \end{cases} \tag{25}$$

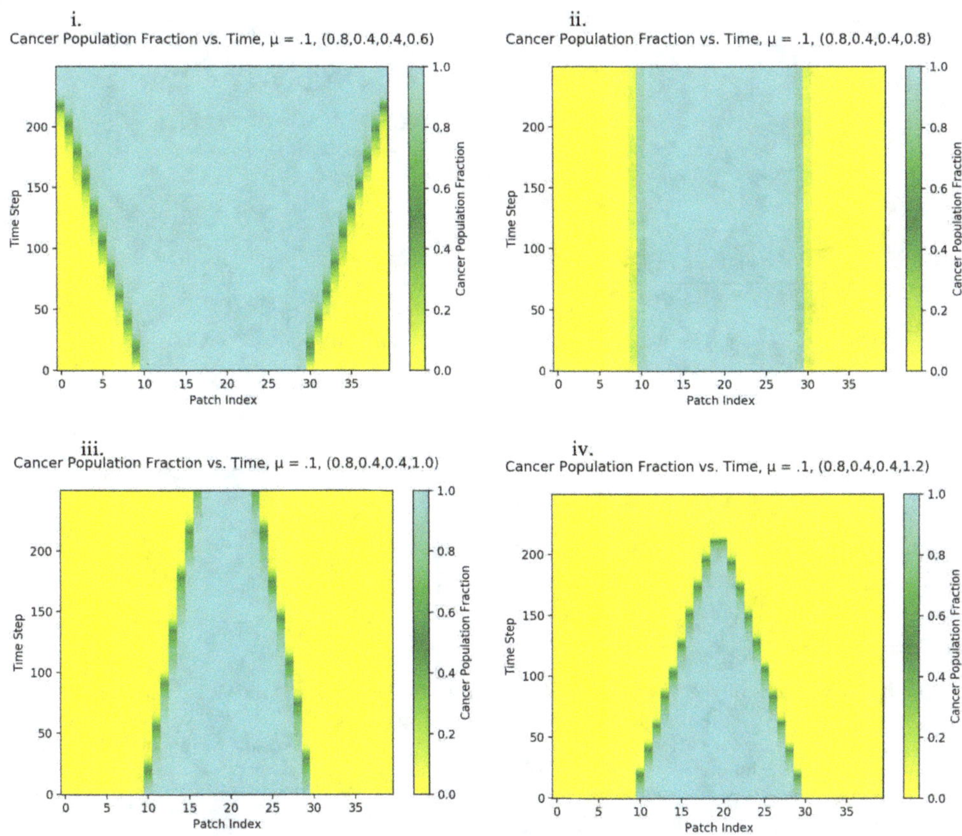

Fig. 8. Q_{40} IPS simulations of cancer fraction time evolution for various payoff coefficients d and a special initial condition in which cancer and stromal cells are segregated as described in Eqs. (22) and (23). We hold constant the payoffs $(a, b, c) = (0.8, 0.4, 0.4)$, migration rate $\mu = 0.1$, and death rate $\kappa = 0.0008$ for all patches. (i) Payoff coefficient $d = 0.6$. (ii) $d = 0.8$. (iii) $d = 1$. (iv) $d = 1.2$.

We then simulate the number of time steps $\tau(\mu, d)$ it takes for cancer to overtake the entire system ($p = 1$) for various values of $\mu \in [0.1, 1]$ and $d \in [0.42, 0.7]$. Note that we have chosen d values such that $a - d > 0$, so we expect the cancer bubble to grow and eventually overtake the entire system. Since growth of the bubble is linear, we expect $r \propto 1/\tau(\mu, d)$. Thus, the qualitative behavior of $r(\mu, d)$ will be similar to that of $1/\tau(\mu, d)$. Figure 9 shows results of these simulations.

Figure 9(i) shows that for fixed d (and thus fixed $a - d$), $\tau(\mu)$ is mostly decreasing in μ, with some kinks in the graph due to noise. τ appears to be asymptotic as $\mu \to 0$; we expect this since when $\mu = 0$, the cancer and stromal patches remain isolated (see Fig. 6), so growth rate $r \approx 0$. Thus, the system is never overtaken by the cancer cells and $\lim_{\mu \to 0} \tau(\mu) = \infty$. Meanwhile, as seen in Fig. 9(iii), $1/\tau$ is monotonically increasing in μ, so $r \propto 1/\tau(\mu, d)$ should be a monotonic increasing function of μ as well.

In Fig. 9(ii), we see that for fixed μ, $\tau(d)$ is increasing for $d \in [0.42, 0.7]$, and thus $\tau(|a - d|)$ is a decreasing function of $|a - d|$. Note that as $d \to a = 0.8$, τ appears to

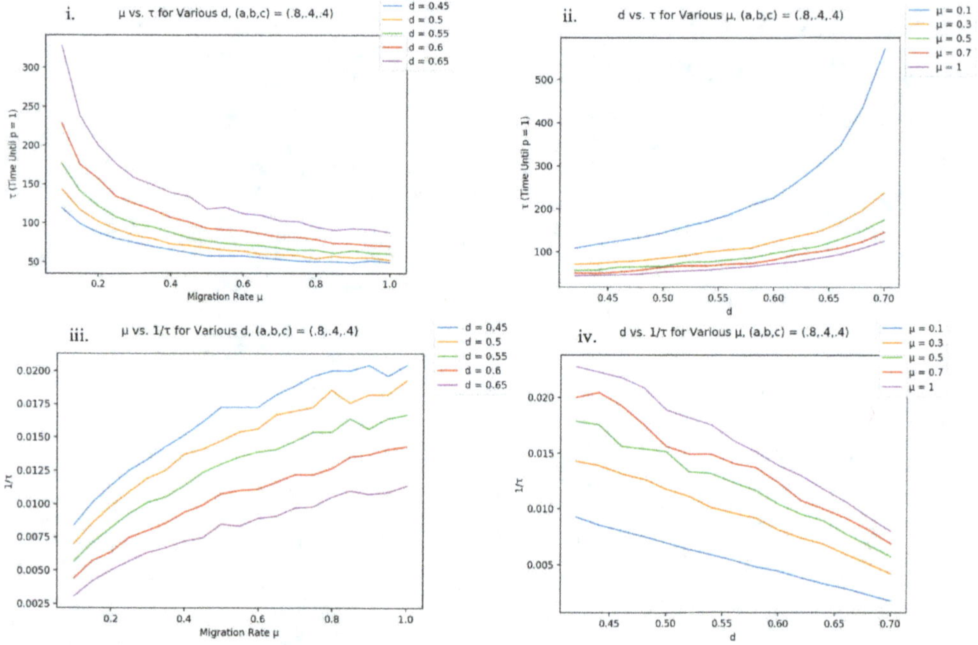

Fig. 9. Q_{40} IPS simulations of $\tau(\mu, d)$ and $1/\tau(\mu, d)$ for various values of μ and d. We hold constant the payoffs $(a, b, c) = (0.8, 0.4, 0.4)$ and death rate $\kappa = 0.0008$ for all patches. (i) We plot μ vs. τ for various values of d. (ii) We plot d vs. τ for various values of μ. (iii) We plot μ vs. $1/\tau$ for various values of d. (iv) We plot d vs. $1/\tau$ for various values of μ.

be asymptotic, corresponding to a growth rate $r \approx 0$. We expect this behavior, since when $d = a$, neither cancer cells nor stromal cells dominate (see Fig. 8), so we expect $\lim_{d \to a} \tau(d) = \infty$. As seen in Fig. 9(iv), $1/\tau$ is close to linearly decreasing in d, so we expect $r \propto 1/\tau(\mu, d)$ to almost be a linearly decreasing function of d as well. Thus, r should be a mostly linearly increasing function of $|a - d|$.

This particular spatial behavior may hold interesting implications. A recent paper considered an IPS in the $a > c; d > b$ regime, concluding that for populations with cancer fractions just under the mean field unstable fixed point p_0, migration can create small perturbations to a patch's cancer fraction that could be greatly amplified, since a patch with local cancer fraction that is perturbed to exceed p_0 would see p tending away from p_0, toward the fixed point $p = 1$.[10] The spatial behavior we observe for the special case $b = c$ is even more drastic: given an initialization of patches where most patches contain entirely stromal cells but a small region of patches contains mostly cancer cells, this region of cancer cells can invade the surrounding stromal cell patches at a linear rate, even if the global cancer fraction is far less than the mean field unstable fixed point p_0. We have shown that this behavior is particularly aggressive when migration rate is high and $|a - d|$ is large.

6. Conclusion

EGT is a promising model that can describe the population dynamics of interacting cancer and stromal cell populations. In this chapter, we first described the MFT approach to modeling game interactions, using analysis of fixed points to describe the behavior of the associated nonlinear differential equations. We then described IPSs as a spatial and discrete alternative. Finally, taking the Durrett and Levin paper[13] as a guide, we described cases where IPS and MFT disagree, exploring in detail a regime where linear growth of a region of cancer cells is possible.

The IPS parameter space is truly gigantic; much work is needed to explore the behavior of systems with more exotic spatial arrangements of patches and various spatial gradients of payoffs and death rates. Experimental work is underway involving construction of minuscule spatially separated subpopulations of interacting cancer and stromal cells.[9,15] We expect much of the observed dynamics will likely not fit the MFT picture and that the IPS model will be necessary. Thus, a deeper understanding of IPS-specific behaviors will be essential in analyzing the observed population dynamics, and determine to what extent EGT can predict the outcome of cancerous tumor progression and the extent to which it can be used in clinical treatments.

Acknowledgment

This work was supported by NSF PHY-1659940.

References

1. R. Axelrod, D. E. Axelrod and K. J. Pienta, *Proc. Natl. Acad. Sci.* **103**(36), 13474 (2006).
2. S. Hummert, K. Bohl, D. Basanta, A. Deutsch, S. Werner, G. Theiben ..., S. Schuster, *Mol. BioSyst.* **10**(12), 3044 (2014).
3. R. M. Bremnes, T. Dønnem, S. Al-Saad, K. Al-Shibli, S. Andersen, R. Sirera, ..., L. T. Busund, *J. Thoracic Oncol.* **6**(1), 209 (2011).
4. K. M. Bussard, L. Mutkus, K. Stumpf, C. Gomez-Manzano and F. C. Marini, *Breast Cancer Res.* **18**(1), 84 (2016).
5. D. F. Quail and J. A. Joyce, *Nat. Med.* **19**(11), 1423 (2013).
6. M. Archetti and K. J. Pienta, *Nat. Rev. Cancer* **19**(2), 110 (2019).
7. J. Zhang, J. J. Cunningham, J. S. Brown and R. A. Gatenby, *Nat. Commun.* **8**(1), 1 (2017).
8. K. Staňková, J. S. Brown, W. S. Dalton and R. A. Gatenby, *JAMA Oncol.* **5**(1), 96 (2019).
9. A. Wu, D. Liao, T. D. Tlsty, J. C. Sturm and R. H. Austin, *Interface Focus* **4**(4), 20140028 (2014).
10. A. Wu, D. Liao, V. Kirilin, K. C. Lin, G. Torga, J. Qu, ..., R. Austin, *Cancer Converg.* **2**(1), 1 (2018).
11. D. Liao and T. D. Tlsty, *Interface Focus* **4**(4), 20140037 (2014).
12. D. Liao and T. D. Tlsty, *Interface Focus* **4**(4), 20140038 (2014).
13. R. Durrett and S. Levin, *Theor. Popul. Biol.* **46**(3), 363 (1994).
14. G. Lambert, S. Vyawahare and R. H. Austin, *Interface Focus* **4**(4), 20140029 (2014).
15. K. C. Lin, Y. Sun, G. Torga, P. Sherpa, Y. Zhao, J. Qu and R. H. Austin, *Lab on a Chip* **20**(14), 2453 (2020).

Chapter 3

Occupancy and Fractal Dimension Analyses of the Spatial Distribution of Cytotoxic (CD8[+]) T Cells Infiltrating the Tumor Microenvironment in Triple Negative Breast Cancer

Juliana C. Wortman[1], Ting-Fang He[2], Anthony Rosario[2], Roger Wang[2], Daniel Schmolze[3], Yuan Yuan[4], Susan E. Yost[4], Xuefei Li[5], Herbert Levine[5], Gurinder Atwal[6], Peter Lee[2] and Clare C. Yu[1,*]

[1]*Department of Physics and Astronomy*
University of California, Irvine, CA 92697, USA

[2]*Cancer Immunotherapeutics & Tumor Immunology*
City of Hope Comprehensive Cancer Center and Beckman Research Institute
1500 East Duarte Road, Duarte, CA 91010, USA

[3]*Department of Pathology*
City of Hope Comprehensive Cancer Center and Beckman Research Institute
1500 East Duarte Road, Duarte, CA 91010, USA

[4]*Department of Medical Oncology and Molecular Therapeutics*
City of Hope Comprehensive Cancer Center and Beckman Research Institute
1500 East Duarte Road, Duarte, CA 91010, USA

[5]*Department of Bioengineering and the Center for Theoretical Biological*
Physics Rice University, Houston, TX 77030, USA

[6]*Cold Spring Harbor Laboratory*
Cold Spring Harbor, NY 11724, USA
**cyu@uci.edu*

Favorable outcomes have been associated with high densities of tumor infiltrating lymphocytes (TILs) such as cytotoxic (CD8[+]) T cells. However, the clinical significance of the spatial distribution of TILs is less well understood. We have developed novel statistical techniques to characterize the spatial distribution of TILs at various length scales. These include a box counting method that we call "occupancy" and novel applications of fractal dimensions. We apply these techniques to the spatial distribution of CD8[+] T cells in the tumor microenvironment of tissue resected from 35 triple negative breast cancer patients. We find that there is a distinct difference in the spatial distribution of CD8[+] T cells between good clinical outcome (no recurrence within at least 5 years of diagnosis) and poor clinical outcome (recurrence within 3 years of diagnosis). The statistical significance of the difference between good and poor outcome in the occupancy, fractal dimension (FD), and FD difference of CD8[+] T cells is comparable to that of the CD8[+] T cell density. Even when we randomly exclude some of the cells so that the images have the same cell density, we still find that the fractal dimension at short length scales is correlated with cancer recurrence, implying that the actual spatial distribution of $CD8^+$ cells, and not just the $CD8^+$ cell density, is associated with clinical outcome. The occupancy and FD difference indicate that the $CD8^+$ T cells are more spatially dispersed in good outcome and more aggregated in poor outcome. We discuss possible interpretations.

Keywords: Triple negative breast cancer; tumor infiltrating lymphocytes; T cells; tumor microenvironment; spatial distribution; fractal dimension.

1. Introduction[a]

Tumors are heterogeneous, consisting of numerous components including cancer cells, collagen fibers, blood vessels, lymph vessels, fibroblasts, and various types of immune cells. The cancer cells cluster to form cancer cell islands that are typically hundreds of microns in diameter. These cancer cell islands are surrounded by stroma where most of the immune cells, such as lymphocytes, reside. The two main types of lymphocytes are B cells, which perform a variety of functions such as making antibodies, and T cells. $CD8^+$ T cells are cytotoxic or "killer" T cells.

Numerous studies have found that high densities of tumor infiltrating lymphocytes (TILs) correlate with favorable clinical outcomes.[3–5] For example, higher densities of $CD3^+$ and $CD8^+$ T cells were associated with a lower rate of recurrence in hepatocellular carcinoma.[6] In addition, immunotherapy is more effective when T cells infiltrate the tumor, e.g., a higher density of $CD8^+$ T cells in melanoma tumors is directly correlated with a better prognosis and clinical response to checkpoint inhibitors.[7]

However, analyzing only the density of immune cells neglects the heterogeneous distribution of immune cells in the tumor tissue. This heterogeneity could even skew density measurements by over/underestimating the density, depending on the location of the samples in the tissue. There are indications of the clinical relevance of the spatial distribution of T cells[8] in and around[9] tumors. In this chapter, we present novel ways to analyze the heterogeneity and spatial distribution of TILs to determine whether they provide additional information, beyond what is captured by the density alone.

There have been a number of efforts to quantify spatial heterogeneity of the tumor microenvironment based on comparing populations of cells.[10] For example, Saltz *et al.* used deep learning to identify 50×50 micron patches with a high density of TILs in digitized H&E stained images of 13 different tumor types from The Cancer Genome Atlas (TCGA).[8] They used a technique known as affinity propagation to group the patches into clusters, and found that the number and spatial extent of the clusters is correlated with clinical outcome. However, there is no unique way to define which patch belongs to which cluster. In addition, the typical size of the clusters depends on an input parameter. In another example, the Morisita–Horn index was used to quantify the spatial colocalization of tumor and immune cells, and it was

---

[a] *Note*: Some of the material in this chapter is in our archived papers, Refs. 1 and 2, where more details and results can be found.

found that significant colocalization was associated with a higher disease-specific survival in Her2-positive breast cancers.[10,11] The Getis–Ord analysis[12] was used to locate immune hotspots where the clustering of immune cells was significantly above background. A combined immune-cancer hotspot score was found to be associated with good prognosis in ER-negative breast cancer.[13] A quantitative measure of the infiltration of immune cells into a tumor is the intratumor lymphocyte ratio (ITLR) which is defined as the ratio of the number of intratumor lymphocytes to the total number of cancer cells in a histological sample.[14] A high ITLR was found to be associated with good disease specific survival in ER-negative/Her2-negative breast cancer.[14,15]

Natrajan *et al.*[16] quantified the spatial heterogeneity in breast tumors with regard to cancer cells, lymphocytes and stromal cells by calculating the Shannon entropy in different regions of the tumor and using Gaussian mixture models to fit the distribution of Shannon entropies. Their ecosystem diversity index (EDI) was the number of Gaussians needed to fit the distribution. They found that high EDI values were associated with high micro-environmental diversity and poor prognosis. Somewhat ironically, with this measure, if most of the regions have high Shannon entropies such that a single Gaussian can be used to fit the distribution, then the EDI is low.

Fractal dimensions[17] have been used to characterize the irregular morphology of tumors[18–20] and vasculature[21,22] as well as subcellular structures such as mitochondria[23] and nuclei.[24] There are numerous ways to calculate fractal dimensions. In the box counting method, the number $N(L)$ of squares (each with area L^2) needed to cover the 2D image of, e.g., a tumor, is proportional to L^{-d}, where d is the fractal dimension in the limit that L goes to zero (or a very small value). More irregular shapes correspond to higher fractal dimensions and poorer prognoses.[18,19]

Assuming that the structure of tumor tissue is reflected in the arrangement of cancer cell nuclei, Waliszewski *et al.* calculated several different fractal dimensions as well as the Shannon entropy and lacunarity to characterize the spatial distribution of cancer cell nuclei in prostate tumor tissue and compared the results to the corresponding Gleason scores in an attempt to find a more objective way to classify prostate tumor tissue.[25,26]

Unlike these previous approaches that quantified the spatial distribution at a single length scale, our techniques use coarse graining to characterize the spatial distribution at different length scales. As an example, we analyze the spatial distribution of T cells in images of tumor tissue from a small group of patients with triple negative breast cancer (TNBC). Our results indicate that T cells tend to be more spatially dispersed in cases where TNBC does not recur and to be more clustered when TNBC does recur.

2. Methods

Occupancy[1,2]: The goal of quantifying the spatial distribution of some entity is to map it to a scalar (number). In our case, we start with 2D images of tissue samples with various types of cells stained with chromophores. We then overlay the image

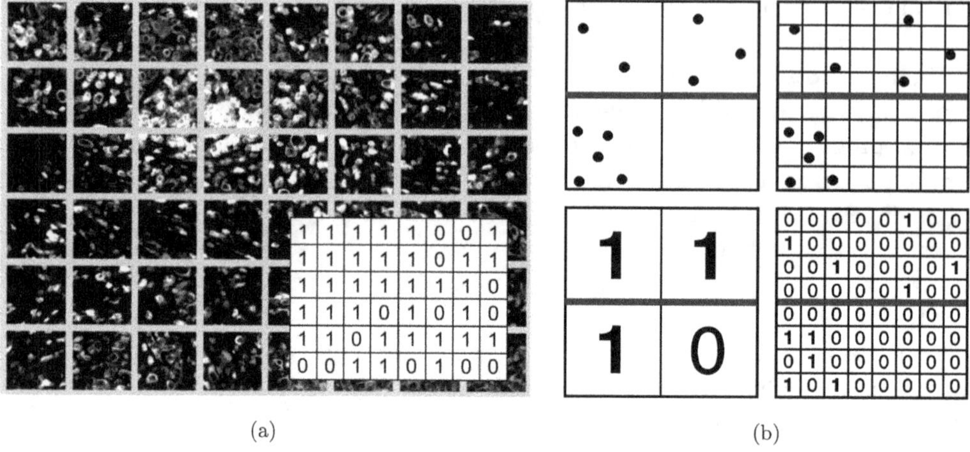

(a)   (b)

Fig. 1.   (a) Diagrams illustrating occupancy analysis. (a) Grid of squares superimposed on tissue. (a, inset) Grid of squares with 0's and 1's indicating the answer to a yes–no question. (b) Cartoon illustrating how FD difference can determine the spatial dispersion of cells. The upper halves (above the thick black lines) of the two images show points that are spread out while the lower halves show clustered points. At long length scales (big boxes) the FD is 2 in the upper half but not in the lower half. At small length scales (small boxes), both halves have the same number of boxes with points and hence the same fractal dimension. Thus, the difference in FD between large and small length scales is greater for points that are more spread out.[1,2]

with a grid of squares as in Fig. 1(a). Each square has an area of L^2. For each square, we ask a binary (yes–no) question, e.g., "Is there at least one CD8$^+$ T cell in the square?" If the answer is yes, we assign a 1 to that square. If the answer is no, we assign a 0 to the square (see Fig. 1(b)). The occupancy p is the fraction of squares with 1's, i.e., it is an estimate of the probability that a square will have a 1. To characterize the spatial distribution at different length scales, we varied the size of the squares in the grid and computed the occupancy as a function of L, the length of one side of a square. We can calculate the area under the curve (AUC) of occupancy versus L as a way to characterize the curve.

Fractal Dimension[1,2]: While there are a number of different ways to define the fractal dimension, we use a variation of the box counting method.[27] We again imagine overlaying the image with a grid of squares as described above, assigning a '1' (or '0') to the square if the answer to a binary question is yes (or 'no'). Let each square have size $L \times L$. The number $n(L)$ of squares with '1' will be proportional to $(1/L^d)$, where d is an exponent less than or equal to the dimension of the image, i.e., 2. In fact, d is one type of fractal dimension. If the system is self-similar and fractal over a range of length scales L, $n(L)$ should follow a power law: $n(L) \sim (1/L^d)$ and the exponent d should be independent of L. So we plot $\log [n(L)]$ versus $\log[A/L]$, where A is a constant, and use linear regression to fit a least squares line through the points. The fractal dimension x is defined to be the slope of the line: $x = -d[\log n(L)]/d[\log L]$. We use a different variable, x, rather than d in the event that $n(L)$ does not follow a simple power law as L is varied. The fractal dimension is found for each patient and then averaged over patients with a given clinical outcome, i.e., good clinical outcome, poor clinical outcome and normal tissue.

Relation of Occupancy and Fractal Dimension[2]: There is a simple relation between the dependence of occupancy on L and fractal dimension d. Suppose the total number $N(L)$ of squares covering the image of the tissue goes as $(1/L^D)$. Then if $n(L) \sim (1/L^d)$, the occupancy $p = n(L)/N(L) \sim L^{D-d}$. Note that D need not be equal to 2 since the image of the tissue may be irregular or there may be regions that were not imaged.

Fractal Dimension Difference[1,2]: A way to determine whether cells are clustered or spread out is to calculate the difference Δs in fractal dimension between large and small length scales: $\Delta s = s_{\text{Large}} - s_{\text{small}}$, where s_{Large} is the fractal dimension at large length scales and s_{small} is the fractal dimension at small length scales. The small and large length scales should roughly bracket the typical, or median, nearest neighbor distance between cells of the same type, e.g., CD8$^+$ T cells. In all the cases we examined, $\Delta s > 0$. If Δs is large, it means that the cells are more dispersed, i.e., more spatially spread out because they appear more two-dimensional at large length scales and more zero-dimensional (point-like) at small length scales (see Fig. 1(b)). If $\Delta s = 0$, the fractal dimension does not change with length scale and the system is self-similar, i.e., fractal. If Δs is small, then the system is closer to being fractal and the cells are more aggregated. In this chapter, the large length scale range is 200–600 microns and the small length scale range is 10–40 microns.

3. Breast Cancer Tissue Samples

We obtained primary tumor tissue resected from 37 triple negative breast cancer patients[1,2]: 24 patients had a good clinical outcome (no recurrence within 5 years) and 13 patients had poor clinical outcome (recurrence within 3 years). As a control, we also analyzed the normal breast tissue from 9 patients who underwent reduction mammoplasty (breast reduction surgery). None of the patients had received chemotherapy or radiation at the time of resection. The baseline patient information is presented in Table 1.

Tissue preparation[1]: Specimens were identified through an IRB-approved protocol via the City of Hope (COH) Biospecimen Repository which is funded in part by the National Cancer Institute. Other investigators may have received specimens from the same patients. Samples from patients diagnosed with triple negative breast cancer and treated at COH from January 1, 1994 to March 4, 2015 were retrieved. Eligible patients had the following features: stages I–III breast cancer; at least one tumor biospecimen was available from the initial surgical resection or biopsy; clinical outcome data was available for identification of relapse free survival; no prior treatment at the time of surgical biopsy. Archived formalin-fixed paraffin-embedded (FFPE) tumor tissues were sectioned (i.e., 3–5 microns per slide) and baked onto glass microscope slides. Immunostaining was performed with anti-pan cytokeratin (clone AE1/AE3, Dako), anti-CD8 (clone SP16, Biocare), and anti-CD20

Table 1. Baseline patient information. In the TNM classification, T denotes tumor burden, N denotes lymph node classification, and M denotes the metastasis status. No patients were known to have any metastasis at the time of their initial surgery.

	Good outcome	Poor outcome	Normal tissue
Number of patients	24	13	9
Mean Age	55	58	24
Age Range	27–76	46–79	18–45
Stage I	5	7	
Stage II	17	6	
Stage III	2	0	
Stage IV	0	0	
T1	7	7	
T2	16	6	
T3	1	0	
N0	19	11	
N1	3	2	
N2	2	0	
M0	3	0	
M1	0	0	
MX	21	13	
Grade 1	0	0	
Grade 2	3	1	
Grade 3	21	12	
Mastectomy	15	2	
Breast conserving surgery	9	11	

(L26, Dako) antibodies using the Opal TSA (PerkinElmer). Samples were further counterstained with DAPI to visualize the nuclei of all cells. Prior to imaging, the tissue sections were coverslipped with ProLong® Gold Antifade mounting media (Cat. # P36930, Life Technologies). All the images were acquired using the Vectra 3.0 Automated Quantitative Pathology Imaging System (PerkinElmer). We used commercial (inForm Image Analysis Software, PerkinElmer and TIBCO Spotfire Software) and in-house custom developed software (R) and algorithms to, at a minimum, identify each cell, define its type (cancer or specific immune), and assign it Cartesian coordinates, allowing the image processing described below. Using an automated tissue segmenter algorithm built in inForm®, we further divided the images into areas of cancer islands and stroma based on anti-pan cytokeratin antibody staining.

Multispectral staining: The following cell phenotypes were identified on the same slide: cytotoxic (killer) T cells (CD8$^+$), B cells (CD20$^+$), epithelial/cancer cells (PanCK), or "other". The cells were also counterstained with DAPI to show the location of nuclei.

Regions of Interest[1]: A pathologist delineated tumor regions that were deemed representative of the entire tumor in terms of cellularity and the TIL distribution, and that were free of artifacts such as tissue folding. Obvious large swaths of necrotic tumor were avoided, as were peri-tumoral lymphoid aggregates not in close proximity to the tumor cells. Given these constraints, regions were chosen to maximize tumor area.

Image Analysis[1]: We tiled each image with a grid of identical squares. We omitted squares that had no cells, since they lay outside the tissue. However, we included squares that were within the tissue but contained no cells of interest. The length L of one side of a square varies from 10 μm to 600 μm. To determine if a square received a '1' or a '0', we asked a yes-no question, e.g., "Is there at least one $CD8^+$ T cell in this square?" We processed the data using the R packages "spatstat" for point patterns and "EBImage" for raster images.[28,29]

For comparison, we calculated the occupancy and fractal dimension of a random (Poisson) distribution of points with uniform density.[1]

To ascertain whether a quantity, such as $CD8^+$ T cell density, is clinically significant, we calculated the p-value under the null hypothesis and receiver operating characteristic (ROC) area under the curve (AUC), both standard statistical measures of a binary classifier.[1,30] In our case, for the binary classifier, we assumed the clinical outcome was either good or poor. Typically, a p-value less than 0.05 indicates that the result is statistically significant. The ROC curve is a parametric curve where the parameter is a cutoff value for the quantity of interest, e.g., $CD8^+$ T cell density, the x-axis is the false positive rate, and the y-axis is the true positive rate. The ROC AUC varies from 0 to 1. High ROC AUC values indicate that the quantity is a good predictor of clinical outcome. A ROC AUC value of 0.5 means the quantity does no better than random chance.

4. Results

Occupancy, fractal dimension and FD difference of $CD8^+$ T cells are associated with clinical outcome.[1,2] Figure 2 is a plot of the average occupancy for $CD8^+$ T cells, i.e., the fraction of boxes with 1's, as a function of the square size L. The occupancy for points randomly placed according to a Poisson distribution is shown as a solid black line in Fig. 2; clearly, T cells are not randomly distributed. The difference between good and poor clinical outcome reflects, to some extent, the difference in T cell density. The average $CD8^+$ T cell density in the tumor is higher in patients with good outcome ($4.5 \times 10^4/cm^2$) than poor outcome ($2.1 \times 10^4/cm^2$). The flattening of the curves at large square sizes is reminiscent of scale invariance, suggesting the use of fractal dimensions.

The slope of the plot of $\ln(n(L))$ versus $\ln(1/L)$ in Fig. 3(a) gives the fractal dimension (FD).[1,2] Larger fractal dimensions, especially those close to two, correspond to more spatially uniform distributions of cells. Randomly placed points (Poisson distribution) are two-dimensional at large length scales as expected. For T cells at long length scales (where the data can be fit to a straight line) good outcome has a larger fractal dimension and therefore is more area-filling than poor outcome

                                   J. C. Wortman et al.

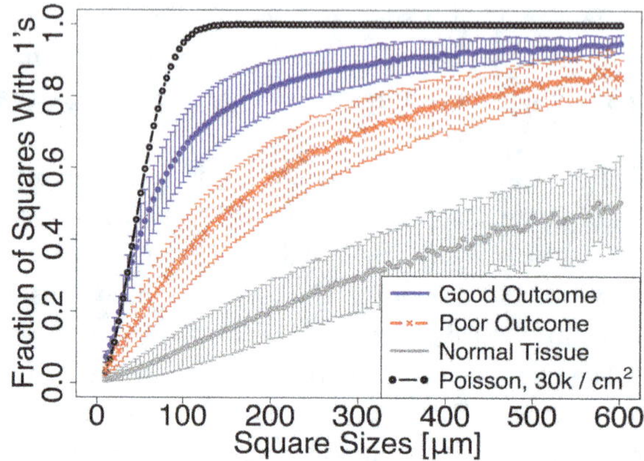

Fig. 2.   Plot of CD8$^+$ T cell occupancy versus square size for good clinical outcome (blue, solid), poor clinical outcome (red, dashed) and points randomly distributed according to a uniform Poisson process with a density of 3×10^4 points/cm^2 (solid black line). The error bars indicate 95% confidence intervals.[1,2]

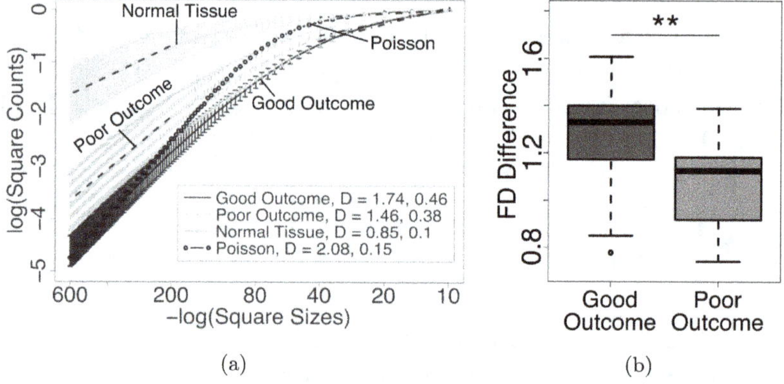

(a)   (b)

Fig. 3.   (a) Log–log plot of the number of squares with at least one CD8$^+$ T cell versus the inverse box size in microns. (Logarithms are base e.) At long length scales (200–600 microns on the left side of plot), the mean fractal dimension x (slope) is 1.74 for good outcome (dark, solid), 1.46 for poor outcome (light, dashed), 0.85 for normal tissue (light, solid) and 2.08 for Poisson (black). The p-value for good versus poor outcome is 3×10^{-4} at long length scales. At short length scales (10–40 microns on the right side of the plot), the mean fractal dimension is 0.46 for good outcome, 0.38 for poor outcome, 0.1 for normal tissue and 0.15 for Poisson. The p-value for good versus poor outcome is 0.09 at short length scales. Black dashed lines show the least squares linear regression fit at long and short length scales. Because different images had different numbers of squares containing tissue (cells of any type), for each image, we normalized the number $n(L)$ of boxes with 1's by the total number $N(L)$ of boxes with cells in computing the fractal dimension. Thus, the y-axis values are negative. The error bars correspond to 95% confidence intervals. The slope used to find the fractal dimension is from a least squares fit made using linear regression. This is true of all subsequent plots. (b) Box and Whisker plot showing that the median FD difference between large (200–600 microns) and small (10–40 microns) length scales for CD8$^+$ T cells is clinically significant ($p = 1.5 \times 10^{-5}$, ROC AUC = 0.88). The line in the middle of each box is plotted at the median, and the inferior and superior limits of the box correspond to the 25th and 75th percentiles, respectively. The whiskers correspond to the minimum and maximum values. Since the difference in median values is significant, the bar at the top is marked with two asterisks corresponding to $p < 0.01$.

which may be due to the higher CD8$^+$ T cell density in patients with a good outcome. The median FD difference between large (200–600 microns) and small length scales (10–40 microns) is larger for good outcome and clinically significant as shown in Fig. 3(b). This indicates that the CD8$^+$ T cells are more spatially dispersed for good outcome and more aggregated for poor outcome.

To determine whether the difference in fractal dimension and FD difference between good and poor clinical outcome is solely the result of the difference in cell density, we thinned the density by randomly eliminating CD8$^+$ T cells in the various images until the densities were the same value (2549 cells/cm^2) in all the tissue images.[1,2] Figure 4(a) is a log–log plot of the number of boxes with at least one CD8$^+$ T cell versus the logarithm of the inverse box size. With the same density, there is still a statistically significant difference between the fractal dimension of good and poor outcome patients at all length scales, indicating that the spatial distribution of the CD8$^+$ T cells is correlated with clinical outcome. At long length scales (200–600 microns), the thinned fractal dimension is slightly higher for poor

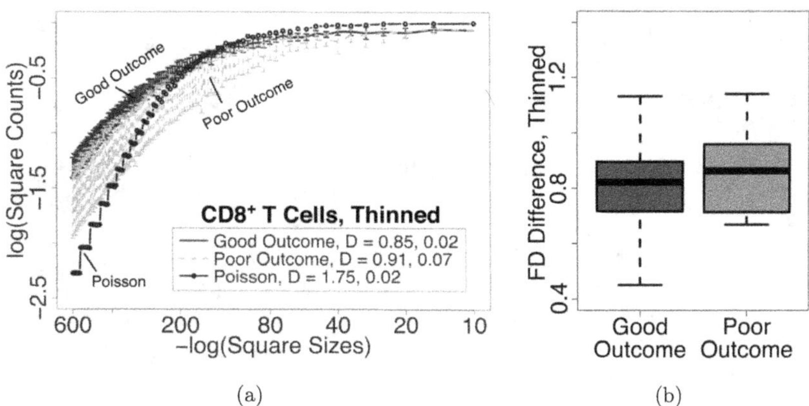

(a)   (b)

Fig. 4. (a) Log–log plot of the number of squares with at least one CD8$^+$ T cell versus the inverse box size (in microns) for images where the T cell density has been reduced to 2,549 cells/cm^2 in each image. The fractal dimension x at large length scales (200–600 microns on the left side of the plot) is 1.75 for the Poisson distribution, 0.85 for good outcome, and 0.91 for poor outcome. The p-value for good versus poor outcome is 0.41 at long length scales. The fractal dimension at short length scales (10–40 microns) is 0.02 for Poisson, 0.02 for good outcome, and 0.07 for poor outcome. The p-value for good versus poor outcome is 0.035 for short length scales. At intermediate length scales (50 to 200 microns), the fractal dimension for Poisson is 0.22, and the fractal dimension for poor outcome is higher ($x = 0.33$) than for good outcome ($x = 0.18$), with a p-value of 1.6×10^{-3}. Black dashed lines show the least squares linear regression fit at long and short length scales. Because different images had different numbers of squares with tissue (cells of any type), for each image, we normalized the number $n(L)$ of boxes with 1's by the total number $N(L)$ of boxes with cells in computing the fractal dimension. Thus, the y-axis values are negative. The error bars correspond to 95% confidence intervals. (b) Box and whisker plot showing that the median FD difference between large (200–600 microns) and small (10–40 microns) length scales for thinned CD8$^+$ T cells is not clinically significant ($p = 0.83$, ROC AUC $= 0.47$). The line in the middle of each box is plotted at the median, and the inferior and superior limits of the box correspond to the 25th and 75th percentiles, respectively. The whiskers correspond to the minimum and maximum values.

outcome (0.91) than for good outcome (0.85). The fractal dimension of the thinned point patterns at short length scales between 10 and 40 microns is significantly higher for poor outcome ($x = 0.07$) than for good outcome ($x = 0.02$), with a p-value of 0.035 (see Fig. 4(a)). However, the median FD difference for thinned CD8$^+$ T cells is not clinically significant, indicating that the spatial dispersion of thinned CD8$^+$ T cells is not substantially different between good and poor outcome as shown in Fig. 4 (b).

Fractal dimension and FD difference of CD8$^+$ T cells in cancer cell islands is clinically relevant.

While most CD8$^+$ T cells are found in the stroma, a few are located in cancer cell islands. The fractal dimension, as well as the FD difference, of CD8$^+$ T cells in cancer cell islands is relevant to clinical outcome (Fig. 5). In contrast, the fractal dimension

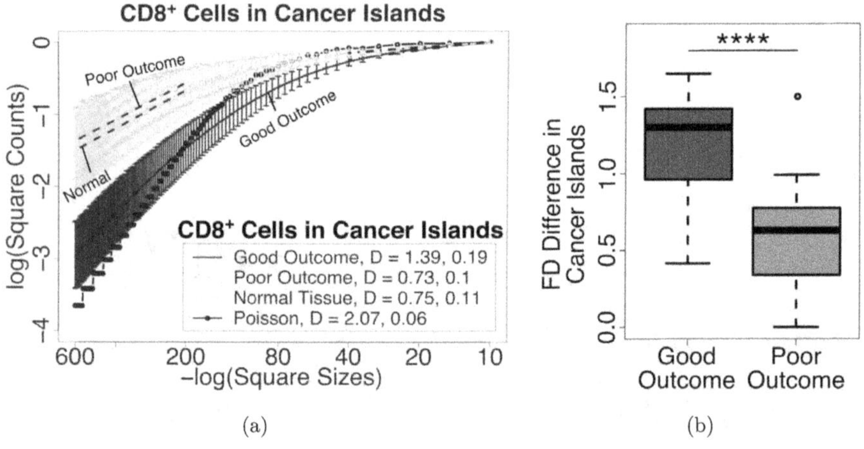

Fig. 5. (a) Log–log plot of the number of squares with at least one CD8$^+$ T cell that is in a cancer cell island/epithelial tissue versus the inverse box size in microns. At long length scales (200–600 microns on the left side of plot), the mean fractal dimension x (slope) is 1.39 for good outcome (dark, solid), 0.73 for poor outcome (light, dashed), 0.75 for normal tissue (light) and 2.07 for Poisson (black). The p-value for good versus poor outcome is 1.1×10^{-5} at long length scales. At short length scales (10–40 microns), the mean fractal dimension x (slope) is 0.19 for good outcome (dark, solid), 0.10 for poor outcome (light, dashed), 0.11 for normal tissue (light, solid) and 0.06 for Poisson (black). The p-value for good versus poor outcome is 0.014 at short length scales. Black dashed lines show the least squares linear regression fit at long and short length scales. Because different images had different numbers of squares with tissue (cells of any type), for each image, we normalized the number $n(L)$ of boxes with 1's by the total number $N(L)$ of boxes with cells in computing the fractal dimension. Thus, the y-axis values are negative. The error bars correspond to 95% confidence intervals. (b) Box and whisker plot showing that the median FD difference between large (200–600 microns) and small (10–40 microns) length scales for CD8$^+$ T cells in cancer cell islands is clinically significant ($p = 1.5 \times 10^{-5}$, ROC AUC $= 0.88$). The line in the middle of each box is plotted at the median, and the inferior and superior limits of the box correspond to the 25th and 75th percentiles, respectively. The whiskers correspond to the minimum and maximum values. Since the difference in median values is significant, the bar at the top is marked with four asterisks corresponding to $p < 0.0001$.

and the FD difference of $CD8^+$ T cells in the stroma are not as strongly associated with outcome but are still statistically significant (see Table 2).

Fractal dimensions of cells (of all types) in cancer cell islands and the stroma do not explain the difference in fractal dimension of $CD8^+$ T cells in good and poor outcome.

The fractal dimension of the $CD8^+$ T cell distribution differs significantly between good and poor clinical outcome but this is not necessarily true for other types of cells. For example, the fractal dimension of the distribution of cells (of any type) occupying cancer cell islands is not significantly different between good and poor outcome as shown in Fig. 6(a). In the case of cells in stroma (Fig. 6(b)), there is only a modest difference between good and poor outcome at shorter (10–40 microns) length scales. Thus, the differences in $CD8^+$ T cell distribution between good and poor clinical outcomes are not simply attributable to differences in the architecture of cancer cell islands or stroma.

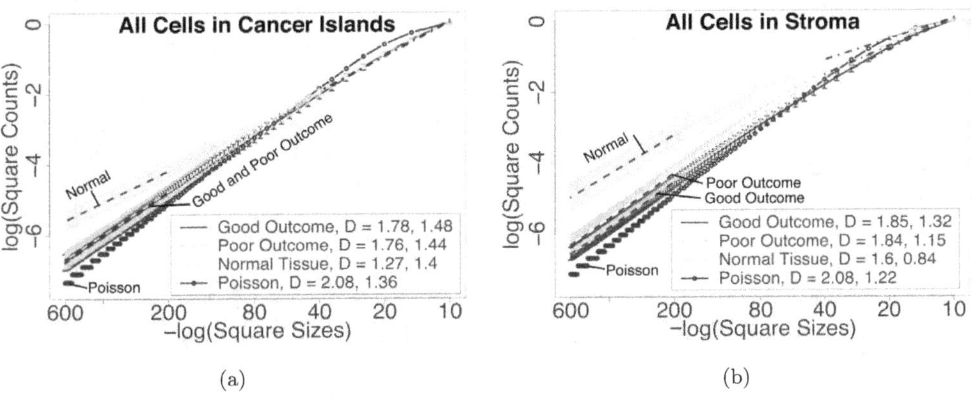

(a) (b)

Fig. 6. Log–log plot of the number of squares with at least one cell (of any type) that is in (a) a cancer cell island or epithelial tissue or (b) in the stroma versus the inverse box size in microns. (a) At long length scales (200–600 microns on the left side of plot), the mean fractal dimension x (slope) is 1.78 for good outcome (dark, solid), 1.76 for poor outcome (light, dashed), 1.27 for normal tissue (light, solid) and 2.08 for Poisson (black). The p-value for good versus poor outcome is 0.57 for long length scales. At short length scales (10–40 microns), the mean fractal dimension x (slope) is 1.48 for good outcome (dark, solid), 1.44 for poor outcome (light, dashed), 1.4 for normal tissue (light, solid) and 1.36 for Poisson (black). The p-value for good versus poor outcome is 0.49 at short length scales. (b) At long length scales (200–600 microns on the left side of plot), the mean fractal dimension x (slope) is 1.85 for good outcome (dark, solid), 1.84 for poor outcome (light, dashed), 1.6 for normal tissue (light, solid) and 2.08 for Poisson (black). The p-value for good versus poor outcome is 0.64 for long length scales. At short length scales (10–40 microns), the mean fractal dimension x (slope) is 1.32 for good outcome (dark, solid), 1.15 for poor outcome (light, dashed), 0.84 for normal tissue (light, solid) and 1.22 for Poisson (black). The p-value for good versus poor outcome is 0.013 for short length scales. Black dashed lines show the least squares linear regression fit at long and short length scales. Because different images had different numbers of squares with tissue (cells of any type), for each image, we normalized the number $n(L)$ of boxes with 1's by the total number $N(L)$ of boxes with cells in computing the fractal dimension. Thus, the y-axis values are negative. The error bars correspond to 95% confidence intervals.

Table 2. *P*-values and ROC AUC of cell density, fractal dimension (Frac Dim), fractal dimension difference, and the area under the occupancy curves (Occupancy AUC). The cell type is either CD8$^+$ T cells or all types of cells (All). The location is everywhere (All Tissue), cancer cell islands (Cancer), or Stroma. Long length scales correspond to 200–600 microns, short length scales correspond to 10–40 microns and all length scales correspond to 10–600 microns. "Mean Good" and "Mean Poor" are the mean values of the quantity averaged over the patients with good or poor clinical outcomes, respectively. The density is given in units of number of cells per square cm. The *p*-value corresponds to the null hypothesis that there is no difference in the quantity between good and poor clinical outcome. ROC AUC is the Receiver Operating Characteristic Area Under the Curve.

Quantity	Cell type	Location	Length scales	Mean good	Mean poor	*p*-value	ROC AUC
Cell Density	CD8$^+$	All Tissue		44,500	20,500	7.4×10^{-4}	0.79
Cell Density	CD8$^+$	Cancer		14,500	1830	0.0012	0.89
Cell Density	CD8$^+$	Stroma		40,200	21,300	0.021	0.79
Frac Dim	CD8$^+$	All Tissue	Long	1.7	1.5	2.8×10^{-4}	0.83
Frac Dim	CD8$^+$	All Tissue	Short	0.47	0.38	0.09	0.69
Frac Dim Difference	CD8$^+$	All Tissue		1.3	1.1	5.5×10^{-3}	0.78
Frac Dim	CD8$^+$	Cancer	Long	1.4	0.73	1.1×10^{-5}	0.89
Frac Dim	CD8$^+$	Cancer	Short	0.19	0.10	0.014	0.73
Frac Dim Difference	CD8$^+$	Cancer		1.2	0.62	1.5×10^{-5}	0.88
Frac Dim	CD8$^+$	Stroma	Long	1.7	1.4	0.0011	0.81
Frac Dim	CD8$^+$	Stroma	Short	0.49	0.40	0.066	0.69
Frac Dim Difference	CD8$^+$	Stroma		1.2	0.99	0.023	0.72
Frac Dim (thinned)	CD8$^+$	All Tissue	Long	0.85	0.90	0.41	0.57
Frac Dim (thinned)	CD8$^+$	All Tissue	Short	0.018	0.072	0.035	0.73
Frac Dim Difference (thinned)	CD8$^+$	All Tissue		0.83	0.83	0.98	0.47
Frac Dim	All	Cancer	Long	1.8	1.8	0.57	0.57
Frac Dim	All	Cancer	Short	1.5	1.4	0.49	0.61
Frac Dim	All	Stroma	Long	1.9	1.8	0.64	0.58
Frac Dim	All	Stroma	Short	1.3	1.2	0.013	0.73
Occupancy AUC	CD8$^+$	All Tissue	All	95.6	74.0	1.7×10^{-4}	0.85
Occupancy AUC	CD8$^+$	Cancer	All	65.4	29.8	1.4×10^{-5}	0.90
Occupancy AUC	CD8$^+$	Stroma	All	87.5	68.5	0.0011	0.83
Occupancy AUC (thinned)	CD8$^+$	All Tissue	All	39.5	35.5	0.28	0.64
Occupancy AUC (thinned)	CD8$^+$	Cancer	All	13.6	5.9	0.024	0.72
Occupancy AUC (thinned)	CD8$^+$	Stroma	All	58.4	54.7	0.33	0.69

Clinical significance of the occupancy, fractal dimension, and FD difference of CD8$^+$ T cells is comparable to that of CD8$^+$ T cell density.

Does the spatial distribution of CD8$^+$ T cells (as reflected in the occupancy AUC, fractal dimension and FD difference) differentiate between good and poor clinical outcome as well as that of cell density? Table 2 shows the p-values and ROC AUC of the occupancy, fractal dimension and FD difference of CD8$^+$ T cells in the various contexts described above. We compare these to the cell density (calculated by dividing the number of cells by the area) as well as to the fractal dimension of all types of cells in either the stroma or cancer cell islands. We see that in cancer cell islands and in all tissue, the clinical significance (as reflected in the p-values and ROC AUC) of the fractal dimension and FD difference of CD8$^+$ T cells at long length scales are comparable to that of CD8$^+$ T cell density, as is the area under the occupancy curves. Furthermore, for thinned CD8$^+$ T cells, the area under the occupancy curve in cancer cell islands and the fractal dimension (at short length scales) in all tissue has a small p-value, demonstrating that it is not just T cell density, but also the spatial distribution of T cells that matters clinically.

We investigated whether the clinical difference between good and poor outcome would increase by combining the CD8$^+$ T cell density and the fractal dimension of CD8$^+$ T cells. We tried various combinations of pairs where one member of the pair is CD8$^+$ T cell density and the other member is the CD8$^+$ T cell fractal dimension (all tissue, cancer cell islands, long and short length scales, thinned and un-thinned). We do not find any significant improvement in the p-values or ROC AUC, indicating that the CD8$^+$ T cell density is correlated with the CD8$^+$ T cell fractal dimension. We have confirmed this explicitly and find that the Pearson correlation coefficient (r) between the density and CD8$^+$ T cell fractal dimension varies between 0.5 and 0.9, depending on the length scales for the fractal dimension and on whether the cells are in all tissue, cancer cell islands or stroma. As expected, there is little correlation between cell density and the thinned fractal dimension.

5. Discussion

In this chapter, we have presented new techniques to characterize the spatial arrangements of cells (points in space). These include occupancy and novel applications of fractal dimensions that we used to quantify the spatial distribution of cells rather than morphology. Our results indicate that the spatial distribution of the CD8$^+$ T cells as reflected in their fractal dimension at short length scales is different between good and poor clinical outcomes, even if the cell density of various samples is normalized, i.e., made to be the same. In particular, the FD difference indicates that the CD8$^+$ T cells are more spatially dispersed in good outcome compared to poor outcome. This implies that the actual spatial distribution, and not just the density of CD8$^+$ T cells, matters clinically.

Our results raise the following questions: (1) Why is the spatial distribution of T cells fractal? (2) What picks out the intermediate length scale on the order of 100 microns where the fractal dimension changes between short and long length scales? (3) Why are the fractal dimension and occupancies of CD8$^+$ T cells different between good and poor clinical outcomes? (4) Why is the spatial distribution of CD8$^+$ T cells in excised tumor tissue correlated with whether cancer recurs? Our discussion of these questions follows:

(1) Why is the spatial distribution of T cells fractal? The self-similarity of the spatial distribution of CD8$^+$ T cells is reflected in their fractal dimension at short (10–40 microns) and long (200–600 microns) length scales. We speculate that this pattern may arise from the branching trajectories of the T cells as they patrol the tissue.[1,2] Branching structures such as trees and plant roots are self-similar, and hence, fractal, because they look the same over a range of length scales, i.e., over a range of magnifications. It may be that the paths followed by CD8$^+$ T cells have a branching structure because the T cells go around physical obstacles of various sizes such as other cells, blood vessels, and collagen fibers. Regions dense with collagen fibers can impede the motion of T cells.[31,32] In addition, T cells are known to travel along the outside of blood vessels[33] and loose collagen fibers[31,32] which can have a branching architecture.

(2) What picks out the length scale on the order of 100 microns where the fractal dimension changes between short and long length scales? It may be related to the fact that CD8$^+$ T cells tend to congregate around the outside of cancer cell islands which are typically hundreds of microns in size. A hundred microns is also comparable to the diffusion length of oxygen.[34–37]

(3) Why are the fractal dimension, FD difference and occupancies of CD8$^+$ T cells different between good and poor clinical outcomes? As we have seen, some of this is the result of the difference in CD8$^+$ T cell densities found in tumor tissue from good and poor outcome patients. But even after we correct for cell density, there is still a statistically significant difference between good and poor outcome. The reason for this is not clear. It appears to be related to the CD8$^+$ T cells being more spread out spatially (fractal dimension closer to 2) in good outcome compared to poor outcome. As we reduce the density by thinning the number of cells, the fractal dimension of the cells that are spread out will approach zero which is appropriate for individual points.

The more fundamental question is what determines the spatial distribution of CD8$^+$ T cells? Why do they go where they do? The difference in spatial distribution may reflect the difference in the spatial topography (obstacles) of the tumor microenvironment. One way to check this would be to look at the spatial distribution of other motile cell types. Another possibility is that the CD8$^+$ T cells in good outcome patients are more responsive to cytokines and more successful in finding their cognate antigen in the tumor microenvironment.

(4) Why is the spatial distribution of CD8$^+$ T cells in excised tumor tissue correlated with whether cancer recurs? It is well known that the density of T cells in excised tumor tissue can be used as a prognostic indicator.[6,7] We have found that the

spatial distribution as reflected in the fractal dimension and FD difference of $CD8^+$ T cells can also be associated with clinical outcome comparable in accuracy to cell density. The spatial distribution and density of T cells may be indicative of the responsiveness of the immune system to cancer. However, it is not understood why the spatial distribution and density of T cells in resected tumor tissue is associated with clinical outcome several years after the excision.

References

1. J. C. Wortman *et al.*, *bioRxiv* (2019). Available at: https://doi.org/10.1101/678607.
2. C. C. Yu *et al.*, arXiv:1911.11846 [physics.bio-ph] (2019).
3. W. H. Fridman, F. Pages, C. Sautes-Fridman and J. Galon, *Nat. Rev. Cancer.* **12**(4), 298 (2012). doi: 10.1038/nrc3245. PubMed PMID: 22419253.
4. B. Mlecnik *et al.*, *J. Clin. Oncol.* **29**(6), 610 (2011), doi: 10.1200/JCO.2010.30.5425. PubMed PMID: 21245428.
5. H. Angell and J. Galon, *Curr. Opin. Immunol.* **25**(2), 261 (2013), doi: 10.1016/j.coi.2013.03.004. PubMed PMID: 23579076.
6. A. Gabrielson *et al.*, *Cancer Immunol. Res.* **4**(5), 419 (2016), doi: 10.1158/2326-6066.CIR-15-0110. PubMed PMID: 26968206; PubMed Central PMCID: PMCPMC5303359.
7. P. C. Tumeh *et al.*, *Nature* **515**(7528), 568 (2014), doi: 10.1038/nature13954. PubMed PMID: 25428505; PubMed Central PMCID: PMCPMC4246418.
8. J. Saltz *et al.*, *Cell Rep.* **23**(1), 181 (2018), doi: 10.1016/j.celrep.2018.03.086. PubMed PMID: 29617659; PubMed Central PMCID: PMCPMC5943714.
9. A. Berthel *et al.*, *Oncoimmunology.* **6**(3), e1286436 (2017), doi: 10.1080/2162402X.2017.1286436. PubMed PMID: 28405518; PubMed Central PMCID: PMCPMC5384380.
10. Y. Yuan, *Cold Spring Harb. Perspect. Med.* **6**(8), 1 (2016), doi: 10.1101/cshperspect.a026583. PubMed PMID: 27481837; PubMed Central PMCID: PMCPMC4968167.
11. C. C. Maley, K. Koelble, R. Natrajan, A. Aktipis and Y. Yuan, *Breast Cancer Res.* **17**(1), 131 (2015), doi: 10.1186/s13058-015-0638-4. PubMed PMID: 26395345; PubMed Central PMCID: PMCPMC4579663.
12. A. Getis and J. K. Ord, *Geographical Analysis.* **24**(3), 189 (1992), doi: 10.1111/j.1538-4632.1992.tb00261.x. PubMed PMID: WOS:A1992JF93400001.
13. S. Nawaz, A. Heindl, K. Koelble and Y. Yuan, *Mod. Pathol.* **28**(6), 766 (2015), doi: 10.1038/modpathol.2015.37. PubMed PMID: 25720324.
14. Y. Yuan, *J. R. Soc. Interface.* **12**(103), 20141153 (2015), doi: 10.1098/rsif.2014.1153. PubMed PMID: 25505134; PubMed Central PMCID: PMCPMC4305416.
15. The Cancer Genome Atlas Program [Internet]. Available at: https://www.cancer.gov/about-nci/organization/ccg/research/structural-genomics/tcga.
16. R. Natrajan, H. Sailem, F. K. Mardakheh, M. Arias Garcia, C. J. Tape, M. Dowsett *et al.*, *PLoS Med.* **13**(2), e1001961 (2016), doi: 10.1371/journal.pmed.1001961. PubMed PMID: 26881778; PubMed Central PMCID: PMCPMC4755617.
17. B. B. Mandelbrot, *The Fractal Geometry of Nature* (W. H. Freeman and Co., New York, 1983).
18. M. Tambasco, M. Eliasziw and A. M. Magliocco, *J. Transl. Med.* **8**, 140 (2010), doi: 10.1186/1479-5876-8-140. PubMed PMID: 21194459; PubMed Central PMCID: PMCPMC3024250.
19. V. Velanovich, *Breast Cancer Res. Treat.* **49**(3), 245 (1998). PubMed PMID: 9776508.

20. A. Chan and J. A. Tuszynski, *R. Soc. Open Sci.* **3**(12), 160558 (2016), doi: 10.1098/rsos.160558. PubMed PMID: 28083100; PubMed Central PMCID: PMCPMC5210682.

21. J. W. Baish and R. K. Jain, *Nat. Med.* **4**(9), 984 (1998), doi: 10.1038/1952. PubMed PMID: 9734370.

22. J. W. Baish and R. K. Jain, *Cancer Res.* **60**(14), 3683 (2000). PubMed PMID: 10919633.

23. F. E. Lennon, G. C. Cianci, R. Kanteti, J. J. Riehm, Q. Arif, V. A. Poroyko *et al.*, *Sci. Rep.* **6**, 24578 (2016), doi: 10.1038/srep24578. PubMed PMID: 27080907; PubMed Central PMCID: PMCPMC4832330.

24. P. Bose, N. T. Brockton, K. Guggisberg, S. C. Nakoneshny, E. Kornaga, A. C. Klimowicz *et al.*, *BMC Cancer.* **15**, 409 (2015), doi: 10.1186/s12885-015-1380-0. PubMed PMID: 25976920; PubMed Central PMCID: PMCPMC4435912.

25. P. Waliszewski, F. Wagenlehner, S. Gattenlohner and W. Weidner, *Prostate.* **75**(4), 399 (2015), doi: 10.1002/pros.22926. PubMed PMID: WOS:000347811800007.

26. P. Waliszewski, *Front Physiol.* **7**, 34 (2016), doi: 10.3389/fphys.2016.00034. PubMed PMID: 26903883; PubMed Central PMCID: PMCPMC4749702.

27. H.-O. Peitgen, H. Jürgens and D. Saupe, *Chaos and Fractals: New Frontiers of Science* (Springer-Verlag, New York, 1992).

28. A. Baddeley and R. Turner, *J. Statist. Softw.* **12**(6), 1 (2005).

29. G. Pau, F. Fuchs, O. Sklyar, M. Boutros and W. Huber, *Bioinformatics.* **26**(7), 979 (2010).

30. J. A. Swets, *Science* **240**(4857), 1285 (1988).

31. H. Bougherara *et al.*, *Front Immunol.* **6**, 500 (2015), doi: 10.3389/fimmu.2015.00500. PubMed PMID: 26528284; PubMed Central PMCID: PMCPMC4600956.

32. H. Salmon *et al.*, *J. Clin. Invest.* **122**(3), 899 (2012), doi: 10.1172/JCI45817. PubMed PMID: 22293174; PubMed Central PMCID: PMCPMC3287213.

33. A. Boissonnas, L. Fetler, I. S. Zeelenberg, S. Hugues and S. Amigorena, *J. Exp. Med.* **204** (2), 345 (2007), doi: 10.1084/jem.20061890. PubMed PMID: 17261634; PubMed Central PMCID: PMCPMC2118741.

34. J. M. Brown and A. J. Giaccia, *Cancer Res.* **58**(7), 1408 (1998). PubMed PMID: 9537241.

35. I. J. Fidler, S. Yano, R. D. Zhang, T. Fujimaki and C. D. Bucana, *Lancet Oncol.* **3**(1), 53 (2002). Epub 2002/03/22. PubMed PMID: 11905606.

36. I. F. Tannock, *Br. J. Cancer.* **22**(2), 258 (1968). PubMed PMID: 5660132; PubMed Central PMCID: PMCPMC2008239.

37. L. H. Gray, A. D. Conger, M. Ebert, S. Hornsey and O. C. Scott, *Br. J. Radiol.* **26**(312), 638 (1953). Epub 1953/12/01. PubMed PMID: 13106296.

Chapter 4

Cooperation Among Tumor Cell Subpopulations
Leads to Intratumor Heterogeneity

Xin Li[*] and D. Thirumalai[†]

Department of Chemistry
University of Texas at Austin
Texas 78712, USA
xinlee0@gmail.com
†*dave.thirumalai@gmail.com*

Heterogeneity is a hallmark of all cancers. Tumor heterogeneity is found at different levels — interpatient, intrapatient, and intratumor heterogeneity. All of them pose challenges for clinical treatments. The latter two scenarios can also increase the risk of developing drug resistance. Although the existence of tumor heterogeneity has been known for two centuries, a clear understanding of its origin is still elusive, especially at the level of intratumor heterogeneity (ITH). The coexistence of different subpopulations within a single tumor has been shown to play crucial roles during all stages of carcinogenesis. Here, using concepts from evolutionary game theory and public goods game, often invoked in the context of the tragedy of commons, we explore how the interactions among subclone populations influence the establishment of ITH. By using an evolutionary model, which unifies several experimental results in distinct cancer types, we develop quantitative theoretical models for explaining data from *in vitro* experiments involving pancreatic cancer as well as *in vivo* data in glioblastoma multiforme. Such physical and mathematical models complement experimental studies, and could optimistically provide new ideas for the design of efficacious therapies for cancer patients.

Keywords: Cancer; complexity; heterogeneity; cooperation; public goods; unequal allocation.

1. Introduction

Cancer is frequently described as a genetic disease arising through clonal evolution of cells. It is well appreciated that cancer is not just one but a group of diseases. Based on the original cell types and organs, cancer has been classified into more than 100 different types.[1] From the latest statistics in the United States (see Table 1), it is estimated that there are more than 1.5 million people diagnosed as cancer patients, and over one third of them would die in 2019.[2] Cancer is the second leading cause of death worldwide, and it also leads to the highest economic loss, which is estimated at approximately \$1.16 trillion in 2010.[3,4] Therefore, cancer has become not only a major public health issue, but also imposes a great economic burden on the society.

However, we still do not have any effective cure or even the ability to control most cancer types, although significant breakthroughs have been achieved in the past few decades for cancer prevention, understanding and even treatment in certain cases.[1,5] The difficulty in the war against cancer is due to the complexity of this fatal disease, which exists in many different forms. Furthermore, the highly variable evolutionary properties also make it extremely difficult to treat.[1,6] In addition, cancer is a massively heterogeneous disease at different levels.[7,8] Even within the same tumor, it can contain many different subpopulations with either genetic or epigenetic variations.[9] Another aspect which makes the disease even more difficult to study is that cancer cells are not independent moieties but rather should be viewed as an evolving ecosystem.[10] Each cancer cell competes with others for limited resources and space subject to Darwinian evolution.[6,11] Also, it is frequently found that different cancer cell subpopulations cooperate with each other to overcome many biological constraints during their development.[10] Therefore, we might need new methods and models to understand the complexity of cancer at single cell level (μm size) to tumor (several milli meters), and finally at the scale of an individual (meters).

In this chapter, we first briefly review the complexity of cancer and discuss some progress in modeling cancer progression with particular emphasis on intratumor heterogeneity (ITH). Then, illustrate the use of the mathematical model to explain the phenomena of ITH observed *in vitro* experiments on pancreatic cancer and an *in vivo* study on glioblastoma multiforme (GBM). The theoretical study also provides some potential insights into cancer treatment methods.

2. Complexity of Cancer

2.1. *Cancer types*

A recent compilation shows that there are about 40 types of cancers based on the site of origin, such as lung, kidney, colon cancers and so on (see Table 1).[2] Due to the coexistence of different cell types in some organs, cancers can be further classified as carcinoma (epithelial cells), sarcoma (connective tissue), myeloma (plasma cells), leukemia (bone marrow), and lymphoma (cells of the lymph system), blastoma (precursor cells), etc. If we take kidney cancer as an example, there are renal cell carcinoma, transitional cell carcinoma, nephroblastomas, and renal sarcoma. The renal cell carcinoma (RCC) is the most common kidney cancer, contributing to 90% of the cases, and it can be classified into a few subgroups (clear cell RCC, Papillary RCC, Chromophobe RCC, duct RCC and others) according to the phenotype of cancer cells.[12,13] Indeed, it is quite astonishing that there are more than 100 different types of cancer known at present (see the full cancer list in *https : //www.cancer.gov/types#k*),[1,14] which explains the difficulty in determining broad principles that drive the origin and evolution of these diseases. Yes, we are facing not a single disease but hundreds of them. Although all of these diseases share some similar hallmarks as summarized in the landmark reviews,[1,15] there appears to

Table 1.   Cancer types, estimated number of new cancer cases and deaths in United States, 2019.[2]

	Estimated new cases	Estimated deaths
All sites	1,762,450	606,880
Tongue	17,060	3,020
Mouth	14,310	2,740
Pharynx	17,870	3,450
Other oral cavity	3,760	1,650
Esophagus	17,650	16,080
Stomach	27,510	11,140
Small intestine	10,590	1,590
Colon	101,420	51,020
Rectum	44,180	
Anus, anal canal, and anorectum	8,300	1,280
Liver and intrahepatic bile duct	42,030	31,780
Gallbladder and other biliary	12,360	3,960
Pancreas	56,770	45,750
Other digestive organs	7,220	2,860
Larynx	12,410	3,760
Lung and bronchus	228,150	142,670
Other respiratory organs	5,880	1,080
Bones and joints	3,500	1,660
Soft tissue (including heart)	12,750	5,270
Skin (excluding basal and squamous)	104,350	11,650
Breast	271,270	42,260
Uterine cervix	13,170	4,250
Uterine corpus	61,880	12,160
Ovary	22,530	13,980
Vulva	6,070	1,280
Vagina and other genital, female	5,350	1,430
Prostate	174,650	31,620
Testis	9,560	410
Penis and other genital, male	2,080	410
Urinary bladder	80,470	17,670
Kidney and renal pelvis	73,820	14,770
Ureter and other urinary organs	3,930	980
Eye and orbit	3,360	370
Brain and other nervous system	23,820	17,760
Endocrine system	54,740	3,210
blood cancer	176,200	56,770
Other and unspecified primary sites	31,480	45,140

be no universal treatment (nor will there be in the near future) for these diseases due to the lack of understanding of the underlying mechanisms of cancer evolution.

2.2. *Evolution of cancer*

In addition to the many faces of cancer that appear in different organs, another challenge for cancer treatment comes from tumor evolution. As proposed by Nowell in 1976, cancer is now widely viewed as a clonal evolutionary process.[6,11] Due to the imperfect genome replication process, environment exposures and heredity, a normal cell can transform into a cancer cell by acquiring genetic mutations, which occur on the time scale of a few decades.[16,11] During cancer progression, tumor cells constantly face the selective pressures derived from their complex microenvironment, such as competition from the surrounding healthy cells for nutrients and space for growth.[17] In addition, hypoxia, pH change, immune surveillance and other potential factors all threaten the survival and growth of tumor cells.[18–20] Therefore, tumor cells generate new traits through continuous evolution, which is just one of the major reasons for the failure of chemotherapy, radiotherapy and other widely applied methods used to treat cancer patients.[6,11]

We have gained considerable knowledge from studies of tumor evolution through animal models.[21,22] The advent of many modern techniques, such as next generation sequencing, has helped us get a deeper understanding for the genetic basis of cancer.[16] However, it is still very difficult to study tumor evolution in humans due to the inaccessibility of tumor biopsies from patients at different time points. Instead, the tumor evolutionary history obtained in most studies is derived from patient samples at a single time point based on assumptions, which are frequently violated as the tumor evolves.[23] Hence, a clear and complete picture about tumor evolution has not emerged. Many models have been proposed to describe the tumor evolution process, such as the linear sequential model, branched, neutral and punctuated evolution of tumors (see Fig. 1).[23,24] One common feature in all these models is that different subpopulations could appear and coexist in a tumor, which is the cause of pervasive cancer heterogeneity.

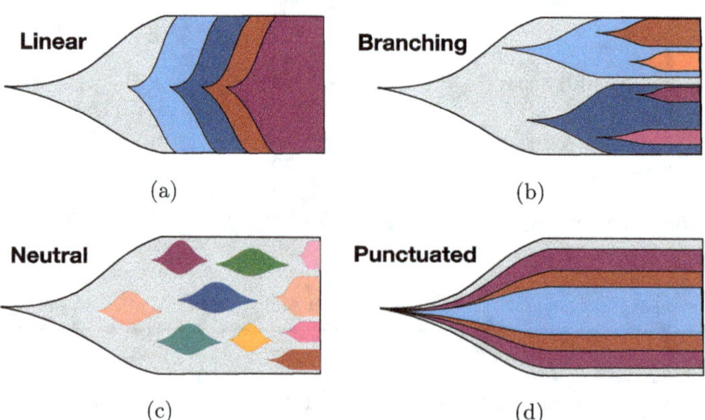

Fig. 1. **Four tumor evolution models.** (a) Linear evolution: Cells accumulate driver mutations (with selective advantage for the cancer cells) sequentially during which selective sweeps occur. (b) Branching evolution: New mutations appear before selective sweeps are completed and divergent subclones emerge from the same ancestor. (c) Neutral evolution: Only passenger mutations (without any selective advantage) accumulate stochastically in tumor cells. (d) Punctuated evolution: Most of the detectable subclonal mutations occur in short bursts of time at early stages of cancer evolution. Different colors represent subclones with different mutations.

2.3. *Cancer heterogeneities*

Cancer is driven by accumulation of genetic mutations especially the "driver mutations," which convey selective advantage to cancer cells. With the advent of sequencing technology, genome-wide association studies (GWAS) is affordable, thus giving us a powerful tool to search for cancer driver genes. The list of cancer driver genes is continuously growing and the number has reached around 300 recently.[25,26] A few cancer genes such as TP53 appear in many cancers while each cancer type usually has its own specific driver mutations. Therefore, different cancer patients have distinct tumor evolutionary processes, which leads to the interpatient heterogeneity (see Fig. 2).[7,8] Personalized cancer medicine has been proposed, and is necessary due to this type of heterogeneity.[27]

As the cancer cells escape the primary site and seed other sites of the body, they finish the transition into the last and fatal stage, referred to as cancer metastasis — responsible for 90% of cancer patient death.[28] Whether new driver mutations are required for the cancer metastasis is unresolved.[29–32] However, distinct new mutations can appear both at the primary and metastatic sites after the spatial isolation is established among cells at different sites.[33] It leads to the next level of heterogeneity, intrapatient heterogeneity which is one of the reasons for cancer recurrence after treatment (see Fig. 2).

In addition to the two types of cancer heterogeneities described above, the intratumor heterogeneity (ITH, see Fig. 2), which refers to the coexistence of different subpopulations in a single tumor, has been found in many cancers.[9,34–40] ITH plays a crucial role in almost all stages of cancer such as tumor progression, metastasis, drug resistance and recurrence of cancer.[41–43] Therefore, understanding the ITH is a major step in investigating cancer heterogeneity because it can help design better therapy to avoid drug resistance and cancer recurrence.

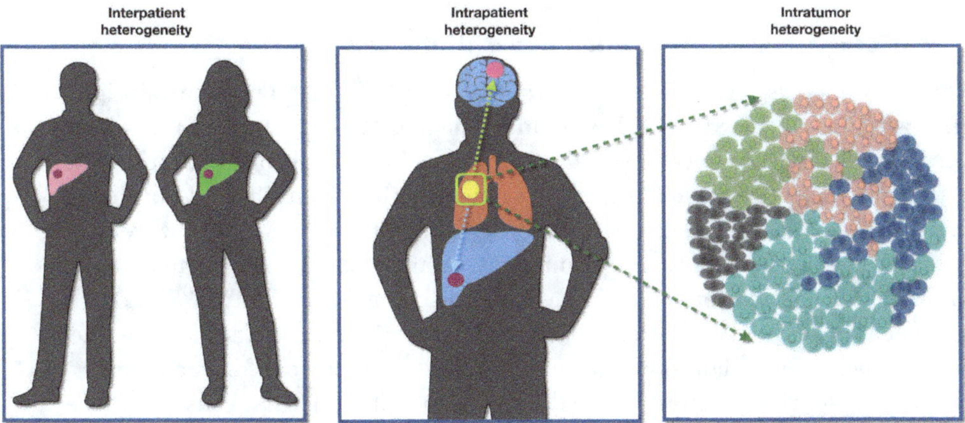

Fig. 2.   **Cancer heterogeneity.** Three levels of cancer heterogeneity: interpatient, intrapatient, and intratumor heterogeneity. Different color dots represent varied tumors or tumor cells with different genetic mutations.

The evolutionary models mentioned above all point to the possibility that different subpopulations (with distinct mutations) can coexist in a single tumor. In these models, the subpopulations interact with each other mainly through cell competition irrespective of the selective action under linear and branched models or the drift effect in neutral and punctuated models. However, increasing experimental evidence found that the cooperation among distinct cell subpopulations in a tumor is essential for tumor maintenance,[44] enhanced tumor growth,[45] and even cancer metastasis.[43,46,47] Surprisingly, a minor subpopulation is sufficient to support the whole tumor growth and determine the clinical course.[48–50] Hence, the different types of cell–cell interactions have to be considered to better understand the mechanisms of ITH, which is often neglected in many theoretical models. The lack of understanding of ITH greatly impairs the progress of developing more effective therapies for cancer patients.

3. Mathematical Models of ITH

3.1. *The multistage model of cancer evolution*

In the past few decades, many innovative mathematical models have been proposed to study the cancer evolutionary process, including cancer initiation, progression, and metastases.[51] A simple multistage model that cancer is driven by a number of driver mutation events was proposed to explain the age-dependent cancer incidence rate more than 60 years ago.[52] The Moran process accounts for the cell division, apoptosis, and mutation process in tumor evolution for a fixed cell population size. It has been widely used to investigate the accumulation of mutations and cancer initiation.[53–57] Other models which further include the spatial structure of tumors and its related microenvironment such as the oxygen and nutrient concentrations are considered to investigate the growth dynamics of tumor by using partial differential equations or an agent-based model.[58–62]

3.2. *ITH under competition and cooperation*

Evolutionary models have also been used to study ITH,[63–66] and many insightful results have been obtained from these studies, such as the extent of ITH and factors influencing ITH. There are two crucial assumptions in these theoretical studies. The genetic mutations only confer fitness advantage to the cell itself (cell-autonomous effect), and tumor cells interact with each other through competition. However, there is increasing evidence that tumors cannot overcome the microenvironmental constraints only through autonomous increase of the cell growth rate.[67–70] In contrast, enhancement of survival and proliferation rates through non-cell-autonomous effects by factors such as metalloproteinases and cytokines are critical for tumor progression.[46] Therefore, these factors secreted by certain tumor/normal cells ('producer') can bring cooperation among different cell subpopulations instead of competition. Such a cooperation can in principle accelerate tumor progression. Instead of one population accumulating all the mutations required for cancer development,[15] a few partially transformed subpopulations with each containing one or two mutations can realize this procedure through their cooperation.[10,71]

3.3. *Evolutionary game theory*

The evolutionary game theory (EGT) provides a novel and unique avenue for investigating ITH accounting for the interactions among subpopulations of tumor or between cancer and normal cells.[72–77] The EGT is a subfield of game theory (GT), which provides mathematical models for studying the strategic interaction among individuals.[78,79] An individual ('player') receives a payoff during the game depending on both the player and also the behavior ('strategy') of others. For EGT in cancer research, the players are cancer or normal cells, and the payoff is their fitness while the strategies are phenotypes adopted by players.[75] The advantage of EGT is that it can describe the time-dependent evolution of the relative abundance of each cell type, determine the equilibrium conditions and the stability of phenotype that coexist.[51] Here, we briefly describe the EGT. The evolutionary games for two cells of different types A, and B are frequently represented by the pay-off matrix

$$
\begin{array}{c|cc}
 & A & B \\
\hline
A & W_{AA} & W_{AB} \\
B & W_{BA} & W_{BB}
\end{array}
$$

where $W_{IJ}(I, J \equiv A, B)$ represents the fitness of the cell type I interacting with the cell type J. The cell–cell interaction can be direct or indirect and its effect can also be competition or cooperation due to the influences of space, nutrients, information, growth factors and other microenvironmental factors.[17,75,80] Therefore, the fitness function W_{IJ} of the cells can also be expressed by very different mathematical formula.[75] As the two different cell populations, instead of just two cells, are well-mixed with each other,[80] the average fitness (w_A, w_B) of the two cell types may be described as

$$w_A = f_A W_{AA} + (1 - f_A)W_{AB}, \tag{1}$$

$$w_B = (1 - f_B)W_{BA} + f_B W_{BB}, \tag{2}$$

where f_A, and f_B are the fractions of the cell type A, and B in the population, respectively. From the fitness functions of the cells, the time-(in)dependent properties of the cell population as mentioned above can be calculated.

In a few latest studies,[81,82] the EGT has been adopted to explain ITH in pancreatic cancer starting from the simplest case with only two different types of cancer cells. One cell type can produce a growth factor while the other does not. The growth factor (often called "public goods") can be consumed by both cell types and promote their proliferation. Qualitative conclusions are obtained for the coexistence of two cell subpopulations which are consistent with experimental observations.

4. Public Goods Game

In a recent study,[74] we investigated the ITH through the "public goods game" among cancer cells quantitatively. Instead of assuming a constant population size, and ideal fitness functions, we consider a growing cell population, and took cell growth rate as the fitness function. Both these quantities can be measured from experiments directly. In the following, we will briefly introduce the model and discuss some of the most important results that we discovered.

The public goods game is an economic model, which has extensive applications in many areas, such as microbial colonies, and insect communities, and even cancer research.[83–88,75] There are two players in this model based on whether they produce the public goods (called 'producer') or not ('non-producer'). Both of them derive benefits from the public goods while the producer usually has to pay a cost for the public goods production. We investigate the underlying mechanism of ITH during cancer progression within the framework of public goods game. We will give two examples to show how cooperation among cancer cell subpopulations can lead to the establishment of a stable ITH, which could be applicable to many other cancers.

4.1. *Models*

We consider two types of cancer cells with one of them producing a public good (see Fig. 3). In the insightful *in vitro* experiment,[81] the public good is the insulin-like growth factor II (IGF-II). First, we focus on this experiment and then, we apply the same idea to describe the ITH observed in an *in vivo* experiment on glioblastoma

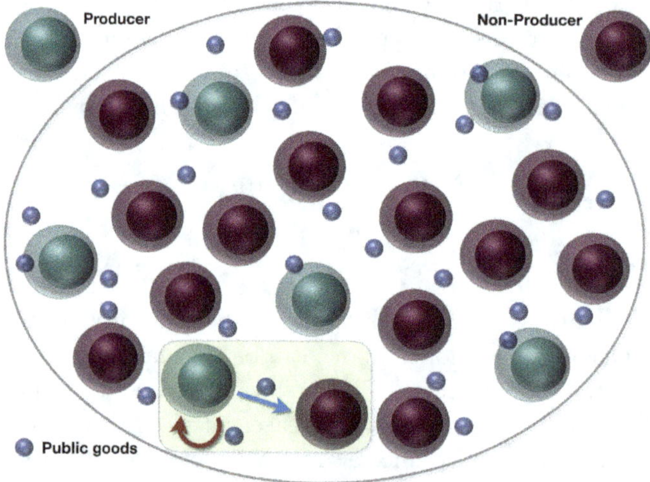

Fig. 3. **Schematic figure for the public goods game of two cell types.** The producers (cyan cells) generate public goods (small blue spheres) which are shared among producer and non-producer cells (dark red cells). The public goods promote the proliferation of both cell types. The inset rectangle: A detailed picture shows the circuit of public goods production and the feedback for the proliferation of producer and non-producer.

multiforme.[89] In the first experiment, both cell types are taken from cancer cells of mice with insulinomas (a type of pancreatic cancer). One cell type can produce IGF-II (the producer $(+/+)$) while the non-producer $(-/-)$ cells have IGF-II gene deleted, which means they can no longer produce IGF-II but can still consume it from the environment.

One critical element in using the concept of the public goods game is how to define the payoff function. Often a rather complex, frequency-dependent function is assigned to both players.[84,81] Here, we use the growth rate as the fitness functions for both the cell types as they can be measured in experiments directly. A nonlinear growth function[81] is found for the growth rate (w_-) of the non-producer cells as a function of IGF-II concentration (c), which is well described by the Hill-like function

$$w_- = a_1 + \lambda_1 c^\alpha / (a_2^\alpha + c^\alpha), \tag{3}$$

where $a_1 = 2.0$, $\lambda_1 = 18.9$, $\alpha = 0.7$, and $a_2 = 3.2$.[74] Although it has not been measured, we expect that the producer cells also have a similar functional form as Eq. (3) for their fitness (w_+), which leads to

$$w_+ = g(c) - p_0, \tag{4}$$

where $g(c)$ is the same Hill-like function as in Eq. (3), and p_0 is the cost paid by producer cells due to the IGF-II production. The IGF-II concentration (c) has two sources once two cell types are mixed with each other. One is from the production of the producer cells, and the other is supplied exogenously (in the culture medium). The IGF-II produced by the producer will depend on the fraction f_+ in the population. Therefore, the IGF-II concentration (c_+) available for the producer can be written as

$$c_+ = a f_+ + c_0, \tag{5}$$

where a is the coefficient for the allocation of IGF-II produced by $+/+$ cells, and c_0 represents the exogenous supply of IGF-II. Similarly, the available IGF-II (c_-) for the non-producer is given by

$$c_- = b f_+ + c_0. \tag{6}$$

The parameter b is the coefficient of allocation of IGF-II produced by $+/+$ cells. We use different parameters a, b to indicate that the producer and non-producer can get different fraction of IGF-II that are produced by the former cells.

The evolution of the fraction f_+ (f_-) of producers (non-producers) can be derived from the replicator equations

$$\frac{\partial f_+}{\partial t} = (w_+ - \langle w \rangle) f_+, \tag{7}$$

and,

$$\frac{\partial f_-}{\partial t} = (w_- - \langle w \rangle) f_-, \tag{8}$$

where the average fitness $\langle w \rangle$ is

$$\langle w \rangle = w_+ f_+ + w_- f_-, \tag{9}$$

with $f_+ + f_- = 1$ being the normalization condition. Let the number of producers and non-producers be N_+, and N_-, respectively. The whole population size, N, is the sum of N_+ and N_-. The system size, N, is a time-dependent quantity, which is often neglected in other models for simplicity.[81,90] Here, we describe the time-dependent changes in the system size, N, through

$$\frac{\partial N}{\partial t} = w_+ N_+ + w_- N_- = \langle w \rangle N. \tag{10}$$

Then, we can study the conditions for the coexistence of producers and non-producers from the equations above.

4.2. *Unequal allocation of public goods results in stable ITH*

The first important result derived using our model is that the emergence of a stable ITH requires an unequal allocation of public goods between producer and non-producer cells. The allocation of the public goods produced by the producers is determined by the ratio, b/a, in Eqs. (5) and (6). A unity for the value of b/a means that the two cell types share the public goods equally. Therefore, the two cell types will have the same fitness function, except for a constant shift p_0 due to the cost paid by the producer (see Eqs. (3) and (4) and Fig. 4(a)). There is only one stable state under this condition (see the filled blue dot in Figs. 4(a) and 4(d)), a homogeneous state consisting of only non-producer would be expected as long as the initial fraction $f_+(0) < 1$. On the other hand, given $b/a = 0$ which means producers do not share any public good with the non-producer. Then, the fitness of the non-producer does not depend on the fraction of the producer any more. There is a heterogeneous state but it is unstable under this condition (see the open circle in Fig. 4(a) and the evolution of f_+ in Fig. 4(e)). However, we find a stable heterogeneous state if $0 < b/a < 1$ (see the filled blue dots in Figs. 4(c) and 4(f)), which indicates that the producer maintains a higher fraction of the public goods produced. There is a negative feedback mechanism close to this stable coexisting state, which can be observed from Fig. 4(c).

To validate our conclusion that it is the unequal allocation of public goods leading to the establishment of a stable coexist state, we calculated the internal equilibrium fractions ($0 < f_+^i < 1$) of +/+ cells at different concentrations of serum (c_0) which has been detected in experiments. The +/+ cell fraction is measured after five days co-culture of +/+ and −/− cells under eleven different initial fractions of cells (with $f_+(t = 0)$ almost evenly distributed between 0 and 1) and different amounts of serum.[81] There are three free parameters in our model, a, b, and p_0. From the equal fitness (≈ 14.4) of producer and non-producer cells and the +/+ cell fraction approaching 1 at the stable internal state under $c_0 = 0$ observed in the experiments,[81] the parameter b ≈ 8 is obtained from Eqs. (3) and (6). Similarly, we can derive the values of a and p_0 based on the experimental results of internal equilibrium states.

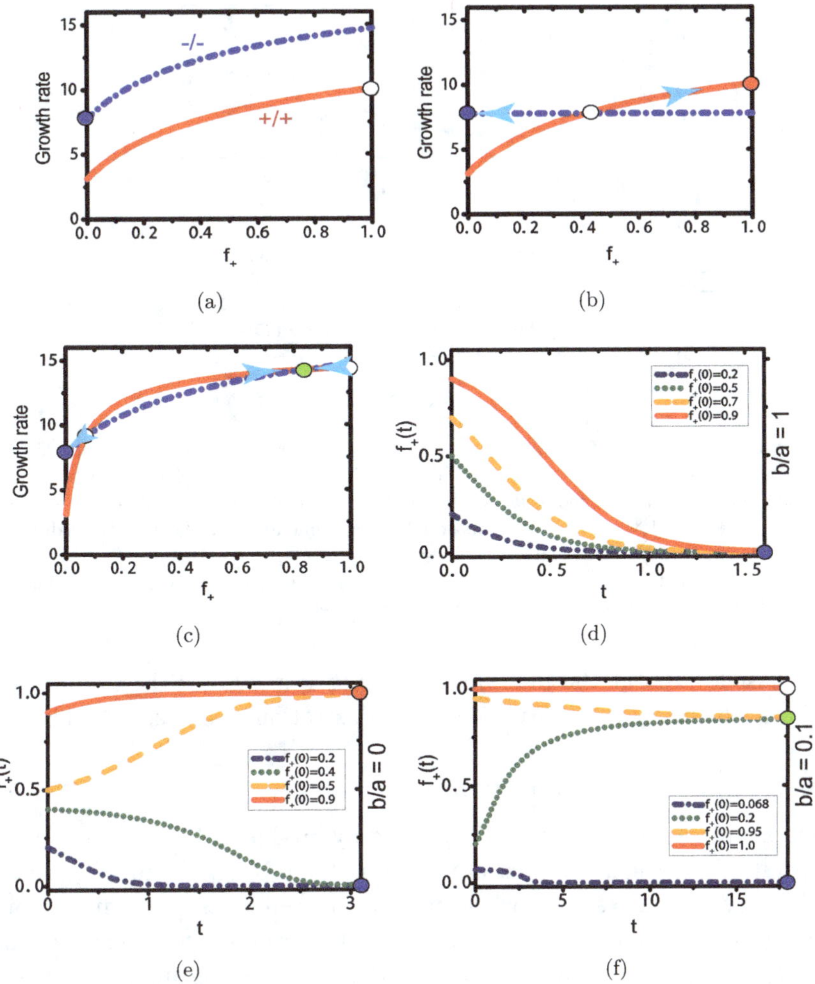

Fig. 4.   (a)–(c): The growth rates of producer $(+/+)$ and non-producer $(-/-)$ cells as a function of the $+/+$ cell fraction (f_+) at different allocation strategies of IGF-II generated by the $+/+$ cells. (a) The IGF-II are shared equally between two cell types $(b = a = 8)$, (b) $+/+$ cells do not share any IGF-II with $-/-$ cells $(b = 0$, and $a = 8)$, (c) $+/+$ cells only share a small fraction of IGF-II with $-/-$ cells $(b = 8$, and $a = 80)$. The parameter value $c_0 = 1$ and $p_0 = 4.65$. The growth rate of $+/+(-/-)$ cells are shown in solid red (dash-dotted blue) lines. The stable or unstable fixed points are represented by filled or empty circles, respectively. (d)–(f): The fraction $f_+(t)$ of $+/+$ cells as a function of time at different initial conditions corresponding to figures (a)–(c) on the left, accordingly. The growth rate is derived from the cell's relative density change at the log phase.[81] The time unit is in days. Figure is adopted from Ref. 74.

We find that our theoretical model captures the experimental observations very well (see the upper panel in Fig. 5). No quantitative explanations have been reached for similar experiments from other models at present. Interestingly, the ratio of b/a obtained through the comparison between our theoretical results and experiments is

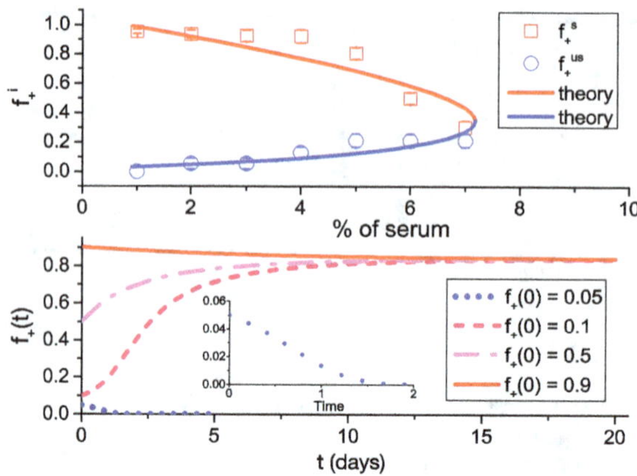

Fig. 5. Upper panel: The producer fraction f_+^i, ($i \equiv s$ or us, with s for stable and us for unstable state) at internal equilibrium states (observed on day 5 under different initial fractions of $f_+(t = 0)$ in experiments) as a function the levels of serum. The red squares (blue circles) represent stable (unstable) states. The error bars indicate the upper and lower boundaries observed in experiments. Our theoretical predictions are illustrated by the solid lines. The parameter values: $a = 80$, and $p_0 = 4.65$. Lower panel: The time dependence of $f_+(t)$ (the fraction of $+/+$ cells) at 3% of serum under various initial conditions. The inset is a zoom in for $f_+(t)$ with $f_+(0) = 0.05$. Figure is adopted from Ref. 74.

0.1 which is just located at the interval, we proposed above for the formation of a stable ITH (see the low panel of Fig. 5 for the illustration of one stable ITH state).

4.3. Cost p_0 influence cell cooperation

Another important result is that the cost p_0 paid by producers has a strong influence on cell cooperation. As the fitness of producers is reduced by p_0 due to the production of the public goods, we investigate whether this parameter has any influence on the cooperation of different subpopulations. From the diagram of the internal equilibrium fraction f_+^i of the producer as a function of the exogenous resource level at varied values of p_0 (see Fig. 6), we found a critical concentration for exogenous resources above which there is only one stable homogeneous state consisted of non-producers. And this critical concentration increases as the value of p_0 decreases. A second critical value (see the star symbols in Fig. 6(a)) is observed for exogenous resources as p_0 takes small values. There is no stable heterogeneous states after the exogenous resource level is lower than this critical value.

To better understand the influence of p_0 on cell cooperation, we calculated a phase diagram in terms of the initial fraction $f_+(0)$ and percentage of serum (see Figs. 6(b) and 6(c)). There are three distinct phases represented in different colors (pink, blue, and purple). Two stable homogeneous states consist of either producers (pink region) or non-producers (blue region). The third phase corresponds to a stable heterogeneous state consisted of both subpopulations (purple region). The area of the three regions is strongly influenced by the value of p_0 (see Figs. 6(b) and 6(c)). A heterogeneous phase is established at modest levels of exogenous resources and higher

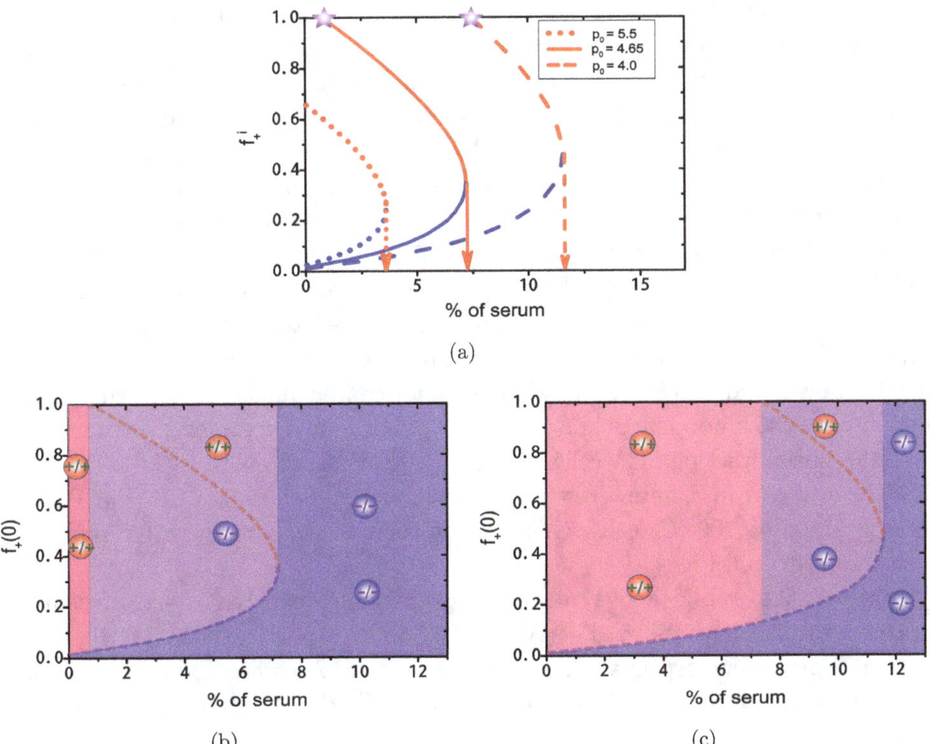

Fig. 6. (a) The producer fraction f_+^i, ($i \equiv s, us$) at internal equilibrium states as a function of the serum level at different values of p_0. The red (blue) lines indicate stable (unstable) equilibrium fractions f_+. Arrows and purple stars indicate the two critical serum concentrations. (b) and (c) Phase diagrams in terms of the initial fraction $f_+(0)$ and % of serum. The values of p_0 in (b) and (c) are 4.65 and 4.00, respectively. There are three stable phases in these two figures. (i) A homogeneous phase with producers only (see the region in pink color). (ii) A heterogeneous phase with both producers and non-producers (purple color). The red dashed line indicates the stable equilibrium fraction of producers. (iii) A homogeneous phase with non-producers only (see the blue color). The red and blue spheres labeled by $+/+$ and $-/-$ represent the producer and non-producer cells, respectively. Figure is adopted from Ref. 74.

$f_+(0)$. We also find that a higher cost paid by the producers actually helps the establishment of a heterogeneous system (see the change of the purple region in Figs. 6(b) and 6(c)).

4.4. *Cooperation among cancer subpopulations in Glioblastoma multiforme*

Cooperation among different cell types has been found in several types of cancer such as breast cancer,[44] prostate cancer[91] and Glioblastoma multiforme (GBM).[89] Here, we take GBM, which is the most aggressive and fatal brain cancer[92] as an example, to show how the same model based on public goods explains equally well the mechanism of ITH in GBM. Two types of cancer cells are frequently found in GBM with different expressions of epidermal growth factor receptor (EGFR).[93] Apparently, the coexistence of the two cell types in GBM promotes the cancer progression and leads to a worse prognosis of the patients.[94,95] Several studies found that the GBM cells with rearrangement of EGFR gene (Δ cells) can secrete factors (Interleukin-6 and Leukemia inhibitory factor) to enhance cell proliferation and inhibit its apoptosis.[89,96]

A recent study investigated the interactions between GBM cells with amplified levels of EGFR (referred to as WT cells) and Δ cells systematically.[89] A fixed amount of cancer cells (2×10^5 cells) with different initial fractions of the two cell types are injected into athymic nude mice (four to five weeks old). Then, the size of the established tumors was measured at different times (see the different symbols in Fig. 7). It is difficult for the WT cells alone to induce a new tumor in the mice (see the pink upside down-triangles in Fig. 7) while a tumor can be quickly generated by the Δ cell alone (see the blue squares in Fig. 7). Meanwhile, a faster tumor growth rate is found as the initial fraction of Δ cells increases until it reaches 90% (see the inset of Fig. 7). However, the maximum growth rate is found at the mixture of 10% WT and 90%Δ cells but not 100%Δ cells which seems to indicate a cooperative relation between these two cell types.

We applied the public goods model to explain the non-trivial experimental observations summarized above. The WT and Δ cells are considered to be the non-producer and producer, respectively. The IL-6 and/or LIF secreted by Δ cells is the public goods. We used a simple fitness function (w_-) for the WT cell as above which is given by

$$w_- = bf_+/(1 + bf_+). \tag{11}$$

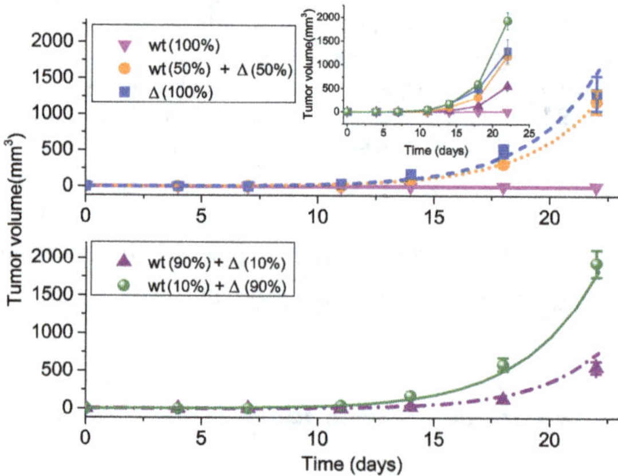

Fig. 7. The evolution of the tumor volume. The GBM tumor volume as a function of time at different initial fractions of WT and Δ cells. The experimental data are represented by symbols. The data from the upper panel are used to obtain the three free parameter values ($a = 68.4, b = 0.946$ and $p_0 = 0.651$) in the model. The purple and green curves at the lower panel are theoretical predictions. Error bars indicate the standard error. The inset gives the complete experimental data. Figure is adopted from Ref. 74.

The fitness function (w_+) for Δ cells is

$$w_+ = af_+/(1 + af_+) - p_0. \tag{12}$$

The different parameters a and b indicate that the public goods (produced by the Δ cells) can be shared unequally between the two cell types. There are only three free parameters in our model, which can be calculated in the following way. The average fitness of the population is given by $\langle w \rangle = w_+$ as it contains only producer cells with $f_+ = 1$ (see Eq. (9)). Therefore, the population size, N, can be calculated from Eq. (10) which increases exponentially, $N = N_0 e^{w_+ t}$. From the growth curve (see the blue square in the upper panel of Fig. 6) of the Δ cells measured in experiments, we obtain

$$\frac{a}{1+a} - p_0 \approx 0.335, \tag{13}$$

where 0.335 is obtained from the exponential fit. It is found that the two types of cells grow at the same rate given $f_+ = f_-$,[89] which leads to

$$\frac{0.5a}{1+0.5a} - p_0 = \frac{0.5b}{1+0.5b}. \tag{14}$$

In addition, the evolution of $N(t)$ can be described by $N(t) = N_0 e^{w_- t}$ given $f_+ = f_-$, and $w_+ = w_-$ as the average fitness $\langle w \rangle = 0.5(w_+ + w_-) = w_-$. Thus, we obtain the following expression:

$$\frac{0.5b}{1+0.5b} \approx 0.321, \tag{15}$$

from the tumor growth curve (orange dots) in the upper panel of Fig. 7. The constant 0.321 is obtained from the exponential fit (see the orange dotted line in Fig. 7). Therefore, the three parameters in our model can be derived using Eqs. (13)–(15), which leads to, $a = 68.4$, $b = 0.946$, and $p_0 = 0.651$.

To test our model, we predicted the tumor growth behavior at 10% and 90% of Δ cells, and found that our predictions agree very well with experimental observations (see the lower panel of Fig. 7). We also calculated the tumor growth rate as a function of the fraction of Δ cells[74] and found that there is a maximum tumor growth rate at an intermediate value (≈ 0.77) of the Δ cells fraction which is consistent with experimental observations and can be test in future experiments.

Our theory also provides some hints for the frequent observations of both WT cells and Δ cells in GBM patient and its poor prognosis. From the evolution of the fraction of Δ cells under different initial conditions (Fig. 8(a)), we found that the heterogeneous state is stable even $f_+(0)$ varies from 0.1 to 0.9. Therefore, a quick recurrence of GBM is possible if some Δ cells still exist in the tumor. Cooperation between different cancer types greatly enhanced their survival and progression. Hence, new drugs and methods should be developed to eliminate such a relation. Our theory predicts that a rich culture medium can encourage cell competition instead of cooperation, which

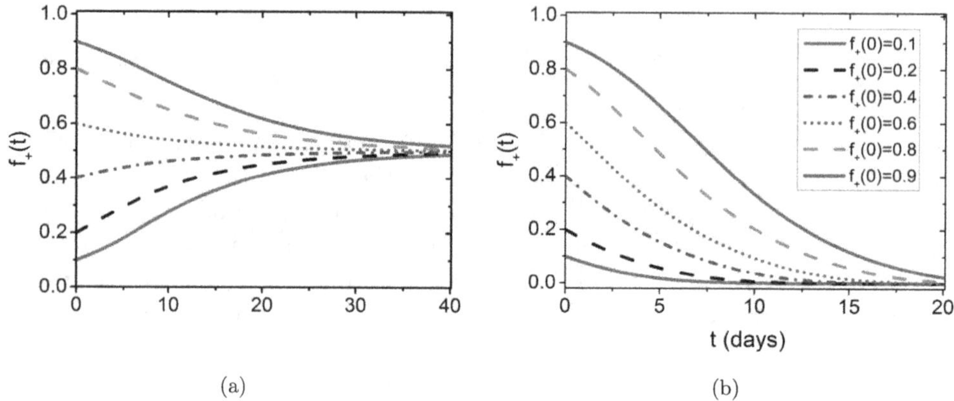

(a)   (b)

Fig. 8. Prediction of the fraction $f_+(t)$ of Δ cells as a function of time under distinct initial conditions $(f_+(t=0))$ in GBM. (a) Without external supply of public goods $(c_0 = 0.0)$; (b) With external supply of public goods, and the total public goods becomes $af_+ + c_0$ $(bf_+ + c_0)$ for the producer (non-producer) with $c_0 = 1.0$. Other parameters are the same as in Fig. 7. The labels in (a) and (b) are the same. Figure is adopted from Ref. 74.

might provide potential treatment strategies for GBM and other similar cancer types (see Fig. 8(b)).

5. Discussion

In this chapter, we briefly reviewed different aspects of cancer complexity from the surprisingly long list of cancers. How to deal with such a complex evolving system poses a big challenge for biologists, clinicians and presents new opportunities for physicists. The ten hallmarks of cancer summarized by Hanahan and Weinberg in 2011 provide a guideline for understanding the progression of neoplastic disease.[15] Due to the ubiquitous heterogeneity present in cancer, there is no universal treatment method for all cancers although the latest immunotherapy brings great hope in this direction.[97,98] A deeper understanding for the cancer evolution and the heterogeneity at all levels is a potentially reachable goal that would enable control or even devise effective treatment of the fatal disease in the future. One of the obstacles slowing down progress comes from the complex interactions among cancer cells, which leads to the formation of a complex cancer cell society, and to the emergence of unexpected phenomena.[10] This could serve as a platform for physicists to give detailed and quantitative understandings of the underlying mechanisms of complex systems using mathematical models. We have provided just one example to illustrate the possibility of applying physical models in studying important questions in cancer research. It is quite encouraging that a simple physical model provides a unified description of different phenomena in pancreatic cancer and GBM in a quantitative manner. In addition, the model can also be used to explain the origin and maintenance of ITH in other cancers in which similar cell–cell interactions are present.[44,91] These ideas could be useful in microbial colonies, insect communities, human society, and other systems.[99–101] We expect that similar physical models and ideas from very different fields[51,102] could help us better understand the disease, and might also provide new ideas and methods for more efficacious cancer therapy.

Acknowledgments

This work is supported by the National Science Foundation (PHY 17-08128 and CHE 16-32756) and the Collie-Welch Chair through the Welch Foundation (F-0019).

References

1. D. Hanahan and R. A. Weinberg, The hallmarks of cancer, *Cell* **100**(1), 57–70 (2000).
2. R. L. Siegel, K. D. Miller and A. Jemal, Cancer statistics, *CA: A Cancer J. Clin.* **69**(1), 7–34 (2019).
3. H. Wang *et al.*, Global, regional, and national life expectancy, all-cause mortality, and cause-specific mortality for 249 causes of death, 1980–2015: A systematic analysis for the Global Burden of Disease Study 2015, *Lancet* **388**(10053), 1459–1544 (2016).
4. B. Steward and C. P. Wild, *World Cancer Report 2014* (International Agency for Research on Cancer, Lyon, 2014), pp. 16–69.
5. J. Heymach *et al.*, Clinical cancer advances 2018: Annual report on progress against cancer from the American Society of Clinical Oncology, *J. Clin. Oncol.* **36**(10), 1020–1044 (2018).
6. P. C. Nowell, The clonal evolution of tumor cell populations, *Science* **194**(4260), 23–28 (1976).
7. P. L. Bedard, A. R. Hansen, M. J. Ratain and L. L. Siu, Tumour heterogeneity in the clinic, *Nature* **501**(7467), 355–364 (2013).
8. E. J. Fox and L. A. Loeb, Cancer: One cell at a time, *Nature* **512**(7513), 143–144 (2014).
9. M. Gerlinger *et al.*, Intratumor heterogeneity and branched evolution revealed by multiregion sequencing, *New Engl. J. Med.* **366**(10), 883–892 (2012).
10. D. P. Tabassum and K. Polyak, Tumorigenesis: It takes a village, *Nature Reviews Cancer* **15**(8), 473 (2015).
11. M. Greaves and C. C. Maley, Clonal evolution in cancer, *Nature* **481**(7381), 306–313 (2012).
12. J. Ferlay, H. R. Shin, F. Bray, D. Forman, C. Mathers and D. M. Parkin, Estimates of worldwide burden of cancer in 2008: GLOBOCAN 2008, *International Journal of Cancer* **127**(12), 2893–2917 (2010).
13. J. C. Cheville, C. M. Lohse, H. Zincke, A. L. Weaver and M. L. Blute, Comparisons of outcome and prognostic features among histologic subtypes of renal cell carcinoma, *Am. J. Surg. Pathol.* **27**(5), 612–624 (2003).
14. A-Z list of cancer types, https://www.cancer.gov/types#k.
15. D. Hanahan and R. A. Weinberg, Hallmarks of cancer: The next generation, *Cell* **144**(5), 646–674 (2011).
16. B. Vogelstein, N. Papadopoulos, V. E. Velculescu, S. Zhou, L. A. Diaz and K. W. Kinzler, Cancer genome landscapes, *Science* **339**(6127), 1546–1558 (2013).
17. E. Moreno, Is cell competition relevant to cancer? *Nat. Rev. Cancer* **8**(2), 141–147 (2008).

18. D. Ackerman and M. C. Simon, Hypoxia, lipids, and cancer: Surviving the harsh tumor microenvironment, *Trends Cell Biol.* **24**(8), 472–478 (2014).

19. L. A. Liotta and E. C. Kohn, The microenvironment of the tumour–host interface, *Nature* **411**(6835), 375–379 (2001).

20. F. R. Balkwill, M. Capasso and T. Hagemann, The tumor microenvironment at a glance, *J. Cell Sci.* **125**(23), 5591–5596 (2012).

21. T. Van Dyke and T. Jacks, Cancer modeling in the modern era: Progress and challenges, *Cell* **108**(2), 135–144 (2002).

22. K. K. Frese and D. A. Tuveson, Maximizing mouse cancer models, *Nat. Rev. Cancer* **7**(9), 654–658 (2007).

23. A. Davis, R. Gao and N. Navin, Tumor evolution: Linear, branching, neutral or punctuated? *Biochim. Biophys. Acta (BBA), Rev. Cancer* **1867**(2), 151–161 (2013).

24. R. G. Gupta and R. A. Somer, Intratumor heterogeneity: Novel approaches for resolving genomic architecture and clonal evolution, *Mol. Cancer Res.* **15**(9), 1127–1137 (2017).

25. P. A. Futreal *et al.*, A census of human cancer genes, *Nat. Rev. Cancer* **4**(3), 177 (2004).

26. M. H. Bailey *et al.*, Comprehensive characterization of cancer driver genes and mutations, *Cell* **173**(2), 371–385 (2018).

27. L. Chin, J. N. Andersen and P. A. Futreal, Cancer genomics: From discovery science to personalized medicine, *Nat. Med.* **17**(3), 297 (2011).

28. C. L. Chaffer and R. A. Weinberg, A perspective on cancer cell metastasis, *Science* **331** (6024), 1559–1564 (2011).

29. D. R. Robinson *et al.*, Integrative clinical genomics of metastatic cancer, *Nature* **548** (7667), 297–303 (2017).

30. L. R. Yates *et al.*, Genomic evolution of breast cancer metastasis and relapse, *Cancer Cell* **32**(2), 169–184 (2017).

31. P. Priestley *et al.*, Pan-cancer whole-genome analyses of metastatic solid tumours, *Nature* **575**(7781), 210–216 (2019).

32. N. J. Birkbak and N. McGranahan, Cancer genome evolutionary trajectories in metastasis, *Cancer Cell* **37**(1), 8–19 (2020).

33. S. Turajlic and C. Swanton, Metastasis as an evolutionary process, *Science* **352**(6282), 169–175 (2016).

34. A. Sottoriva *et al.*, Intratumor heterogeneity in human glioblastoma reflects cancer evolutionary dynamics, *Proc. Nat. Acad. Sci.* **110**(10), 4009–4014 (2013).

35. A. Bashashati *et al.*, Distinct evolutionary trajectories of primary high-grade serous ovarian cancers revealed through spatial mutational profiling, *J. Pathol.* **231**(1), 21–34 (2013).

36. M. Gerlinger *et al.*, Genomic architecture and evolution of clear cell renal cell carcinomas defined by multiregion sequencing, *Nat. Genet.* **46**(3), 225–233 (2014).

37. E. C. de Bruin *et al.*, Spatial and temporal diversity in genomic instability processes defines lung cancer evolution, *Science* **346**(6206), 251–256 (2014).

38. L. R. Yates *et al.*, Subclonal diversification of primary breast cancer revealed by multiregion sequencing, *Nat. Med.* **21**(7), 751–759 (2015).

39. P. C. Boutros *et al.*, Spatial genomic heterogeneity within localized, multifocal prostate cancer, *Nat. Genet.* **47**(7), 736–745 (2015).

40. S. Ling *et al.*, Extremely high genetic diversity in a single tumor points to prevalence of non-Darwinian cell evolution, *Proc. Nat. Acad. Sci.* **112**(47), E6496–E6505 (2015).

41. R. A. Burrell, N. McGranahan, J. Bartek and C. Swanton, The causes and consequences of genetic heterogeneity in cancer evolution, *Nature* **501**(7467), 338 (2013).

42. D. A. Lawson, K. Kessenbrock, R. T. Davis, N. Pervolarakis and Z. Werb, Tumour heterogeneity and metastasis at single-cell resolution, *Nat. Cell Biol.* **20**(12), 1349 (2018).

43. M. Janiszewska *et al.*, Subclonal cooperation drives metastasis by modulating local and systemic immune microenvironments, *Nat. Cell Biol.* **21**(7), 879–888 (2019).

44. A. S. Cleary, T. L. Leonard, S. A. Gestl and E. J. Gunther, Tumour cell heterogeneity maintained by cooperating subclones in WNT-driven mammary cancers, *Nature* **508**(7494), 113–117 (2014).

45. A. Marusyk, D. P. Tabassum, P. M. Altrock, V. Almendro, F. Michor and K. Polyak, Non-cell-autonomous driving of tumour growth supports sub-clonal heterogeneity, *Nature* **514**(7520), 54–58 (2014).

46. A. Chapman, L. F. del Ama, J. Ferguson, J. Kamarashev, C. Wellbrock and A. Hurlstone, Heterogeneous tumor subpopulations cooperate to drive invasion, *Cell Rep.* **8**(3), 688–695 (2014).

47. N. Aceto *et al.*, Circulating tumor cell clusters are oligoclonal precursors of breast cancer metastasis, *Cell* **158**(5), 1110–1122 (2014).

48. C. G. Mullighan *et al.*, Genomic analysis of the clonal origins of relapsed acute lymphoblastic leukemia, *Science* **322**(5906), 1377–1380 (2008).

49. B. E. Johnson *et al.*, Mutational analysis reveals the origin and therapy-driven evolution of recurrent glioma, *Science* **343**(6167), 189–193 (2014).

50. A. S. Morrissy *et al.*, Divergent clonal selection dominates medulloblastoma at recurrence, *Nature* **529**(7586), 351–357 (2016).

51. P. M. Altrock, L. L. Liu and F. Michor, The mathematics of cancer: Integrating quantitative models, *Nat. Rev. Cancer* **15**(12), 730 (2015).

52. P. Armitage and R. Doll, The age distribution of cancer and a multi-stage theory of carcinogenesis, *Br. J. Cancer* **8**(1), 1 (1954).

53. F. Michor, Y. Iwasa and M. A. Nowak, Dynamics of cancer progression, *Nat. Rev. Cancer* **4**(3), 197 (2004).

54. N. Beerenwinkel *et al.*, Genetic progression and the waiting time to cancer, *PLoS Comput. Biol.* **3**(11), e225 (2007).

55. L. M. Merlo, J. W. Pepper, B. J. Reid and C. C. Maley, Cancer as an evolutionary and ecological process, *Nat. Rev. Cancer* **6**(12), 924 (2006).

56. C. Tomasetti, B. Vogelstein and G. Parmigiani, Half or more of the somatic mutations in cancers of self-renewing tissues originate prior to tumor initiation, *Proc. Nat. Acad. Sci.* **110**(6), 1999–2004 (2013).

57. H. Teimouri, M. P. Kochugaeva and A. B. Kolomeisky, Elucidating the correlations between cancer initiation times and lifetime cancer risks, *Sci. Rep.* **9**(1), 1–8 (2019).

58. J. A. Sherratt and M. A. Chaplain, A new mathematical model for avascular tumour growth, *J. Math. Biol.* **43**(4), 291–312 (2001).

59. M. A. Nowak, F. Michor and Y. Iwasa, The linear process of somatic evolution, *Proc. Natl. Acad. Sci.* **100**(25), 14966–14969 (2003).

60. N. Beerenwinkel, R. F. Schwarz, M. Gerstung and F. Markowetz, Cancer evolution: Mathematical models and computational inference, *Syst. Biol.* **64**(1), e1–e25 (2014).

61. A. N. Malmi-Kakkada, X. Li, H. S. Samanta, S. Sinha and D. Thirumalai, Cell growth rate dictates the onset of glass to fluidlike transition and long time superdiffusion in an evolving cell colony, *Phys. Rev. X* **8**(2), 021025 (2014).

62. S. Sinha, A. N. Malmi-Kakkada, X. Li, H. S. Samanta and D. Thirumalai, Spatially heterogeneous dynamics of cells in a growing tumor spheroid: Comparison between theory and experiments, *Soft Matter* **16**, 5294–5304 (2020).

63. A. Kansal, S. Torquato, E. Chiocca and T. Deisboeck, Emergence of a subpopulation in a computational model of tumor growth, *J. Theor. Biol.* **207**(3), 431–441 (2000).

64. I. Bozic *et al.*, Accumulation of driver and passenger mutations during tumor progression, *Proc. Natl. Acad. Sci.* **107**(43), 18545–18550 (2010).

65. R. Durrett, J. Foo, K. Leder, J. Mayberry and F. Michor, Intratumor heterogeneity in evolutionary models of tumor progression, *Genetics* **188**(2), 461–477 (2011).

66. B. Waclaw, I. Bozic, M. E. Pittman, R. H. Hruban, B. Vogelstein and M. A. Nowak, A spatial model predicts that dispersal and cell turnover limit intratumour heterogeneity, *Nature* **525**(7568), 261 (2015).

67. R. A. Gatenby and R. J. Gillies, A microenvironmental model of carcinogenesis, *Nat. Rev. Cancer* **8**(1), 56 (2008).

68. M. J. Bissell and W. C. Hines, Why don't we get more cancer? A proposed role of the microenvironment in restraining cancer progression, *Nat. Med.* **17**(3), 320 (2011).

69. D. F. Quail and J. A. Joyce, Microenvironmental regulation of tumor progression and metastasis, *Nat. Med.* **19**(11), 1423 (2013).

70. K. Curtius, N. A. Wright and T. A. Graham, An evolutionary perspective on field cancerization, *Nat. Rev. Cancer* **18**(1), 19 (2018).

71. R. Axelrod, D. E. Axelrod and K. J. Pienta, Evolution of cooperation among tumor cells, *Proc. Natl. Acad. Sci.* **103**(36), 13474–13479 (2006).

72. D. Dingli, Chalub FAdCC, F. Santos, S. Van Segbroeck and J. M. Pacheco, Cancer phenotype as the outcome of an evolutionary game between normal and malignant cells, *Br. J. Cancer* **101**(7), 1130 (2009).

73. D. Basanta, J. G. Scott, M. N. Fishman, G. Ayala, S. W. Hayward and A. R. Anderson, Investigating prostate cancer tumour–stroma interactions: Clinical and biological insights from an evolutionary game, *Br. J. Cancer* **106**(1), 174 (2012).

74. X. Li and D. Thirumalai, Share, but unequally: A plausible mechanism for emergence and maintenance of intratumour heterogeneity, *J. Roy. Soc. Interface* **16**(150), 20180820 (2019).

75. M. Archetti and K. J. Pienta, Cooperation among cancer cells: Applying game theory to cancer, *Nat. Rev. Cancer* **19**(2), 110–117 (2019).

76. M. Gluzman, J. G. Scott and A. Vladimirsky, Optimizing adaptive cancer therapy: Dynamic programming and evolutionary game theory, preprint (2018), arXiv:181201805.

77. K. Staňková, J. S. Brown, W. S. Dalton and R. A. Gatenby, Optimizing cancer treatment using game theory: A review, *JAMA Oncol.* **5**(1), 96–103 (2019).

78. J. M. Smith, *Evolution and the Theory of Games* (Cambridge University Press, 1982).

79. M. J. Osborne *et al.*, *An Introduction to Game Theory* (Oxford university Press, New York, 2004).

80. J. M. Pacheco, F. C. Santos and D. Dingli, The ecology of cancer from an evolutionary game theory perspective, *Interface Focus* **4**(4), 20140019 (2014).

81. M. Archetti, D. A. Ferraro and G. Christofori, Heterogeneity for IGF-II production maintained by public goods dynamics in neuroendocrine pancreatic cancer, *Proc. Natl. Acad. Sci.* **112**(6), 1833–1838 (2015).

82. G. J. Kimmel, P. Gerlee, J. S. Brown and P. M. Altrock, Neighborhood size-effects shape growing population dynamics in evolutionary public goods games, *Commun. Biol.* **2**(1), 53 (2019).

83. J. Hofbauer and K. Sigmund, *Evolutionary Games and Population Dynamics* (Cambridge University Press, 1998).

84. C. Hauert, M. Holmes and M. Doebeli, Evolutionary games and population dynamics: Maintenance of cooperation in public goods games, *Proc. Roy. Soc. Lond. B, Biol. Sci.* **273**(1600), 2565–2570 (2006).

85. M. A. Nowak, Five rules for the evolution of cooperation, *Science* **314**(5805), 1560–1563 (2006).

86. B. Allen, J. Gore and M. A. Nowak, Spatial dilemmas of diffusible public goods, *Elife* **2**, e01169 (2013).

87. M. Nanda and R. Durrett, Spatial evolutionary games with weak selection, *Proc. Natl. Acad. Sci.* **114**(23), 6046–6051 (2017).

88. M. Bauer and E. Frey, Multiple scales in metapopulations of public goods producers, *Phys. Rev. E* **97**(4), 042307 (2018).

89. I. Maria-del Mar *et al.*, Tumor heterogeneity is an active process maintained by a mutant EGFR-induced cytokine circuit in glioblastoma, *Genes Dev.* **24**(16), 1731–1745 (2010).

90. R. A. Blythe and A. J. McKane, Stochastic models of evolution in genetics, ecology and linguistics, *J. Stat. Mech. Theory Exp.* **2007**(7), P07018 (2007).

91. F. Mateo *et al.*, SPARC mediates metastatic cooperation between CSC and non-CSC prostate cancer cell subpopulations, *Mol. Cancer* **13**(1), 237 (2014).

92. O. Gallego, Nonsurgical treatment of recurrent glioblastoma, *Curr. Oncol.* **22**(4), e273–281 (2015).

93. R. Nishikawa *et al.*, A mutant epidermal growth factor receptor common in human glioma confers enhanced tumorigenicity, *Proc. Natl. Acad. Sci.* **91**(16), 7727–7731 (1994).

94. N. Shinojima *et al.*, Prognostic value of epidermal growth factor receptor in patients with glioblastoma multiforme, *Cancer Res.* **63**(20), 6962–6970 (2003).

95. A. B. Heimberger *et al.*, Prognostic effect of epidermal growth factor receptor and EGFRvIII in glioblastoma multiforme patients, *Clin. Cancer Res.* **11**(4), 1462–1466 (2005).

96. P. C. Heinrich, I. Behrmann, H. Serge, H. M. Hermanns, G. Müller-Newen and F. Schaper, Principles of interleukin (IL)-6-type cytokine signalling and its regulation, *Biochem. J.* **374**(1), 1–20 (2003).

97. I. Mellman, G. Coukos and G. Dranoff, Cancer immunotherapy comes of age, *Nature* **480**(7378), 480–489 (2011).

98. A. Ribas and J. D. Wolchok, Cancer immunotherapy using checkpoint blockade, *Science* **359**(6382), 1350–1355 (2018).

99. S. Dobata and K. Tsuji, Public goods dilemma in asexual ant societies, *Proc. Natl. Acad. Sci.* **110**(40), 16056–16060 (2013).

100. K. Drescher, C. D. Nadell, H. A. Stone, N. S. Wingreen and B. L. Bassler, Solutions to the public goods dilemma in bacterial biofilms, *Curr. Biol.* **24**(1), 50–55 (2014).

101. I. Kaul, I. Grungberg and M. A. Stern, *Global Public Goods: International Cooperation in the 21st Century* (Oxford University Press, New York, 1999).

102. T. Kirkpatrick and D. Thirumalai, Colloquium: Random first order transition theory concepts in biology and physics, *Rev. Mod. Phys.* **87**(1), 183 (2015).

Chapter 5

Cell Mechanics Drives Migration Modes

Claudia Tanja Mierke

Faculty of Physics and Earth Science
Peter Debye Institute of Soft Matter Physics
Biological Physics Division
University of Leipzig, Linnéstr. 5
Leipzig 04103, Germany
claudia.mierke@uni-leipzig.de

The classical migration modes, such as mesenchymal or amoeboid migration modes, are essentially determined by molecular, morphological or biochemical properties of the cells. These specific properties facilitate the cell migration and invasion through artificial extracellular matrices mimicking the environmental conditions of connective tissues. However, during the migration of cells through narrow extracellular matrix constrictions, the specific extracellular matrix environments can either support or impair the invasion of cells. Beyond the classical molecular or biochemical properties, the migration and invasion of cells depends on intracellular cell mechanical characteristics and extracellular matrix mechanical features. The switch between cell states, such as epithelial, mesenchymal or amoeboid states, seems to be mainly based on epigenetic changes and environmental cues that induce the reversible transition of cells toward another state and thereby promote a specific migration mode. However, the exact number of migration modes is not yet clear. Moreover, it is also unclear whether every individual cell, independent of the type, can undergo a transition between all different migration modes in general. A newer theory states that the transition from the jamming to unjamming phase of clustered cells enables cells to migrate as single cells through extracellular matrix confinements. This review will highlight the mechanical features of cells and their matrix environment that regulate and subsequently determine individual migration modes. It is discussed whether each migration mode in each cell type is detectable or whether some migration modes are limited to artificially engineered matrices *in vitro* and can therefore not or only rarely be detected *in vivo*. It is specifically pointed out how the intracellular architecture and its contribution to cellular stiffness or contractility favors the employment of a distinct migration mode. Finally, this review envisions a connection between mechanical properties of cells and matrices and the choice of a distinct migration mode in confined 3D microenvironments.

Keywords: Contractile forces; cell invasion; stiffness; migration modes; extracellular matrix.

Abbreviations

AFM: Atomic force microscopy
MMP: Matrix metalloproteinase
UV: Ultraviolet
MCAM: Melanoma cell adhesion molecule
DexMa: dextran methacrylate
PDMS: Polydimethylsiloxane
MLCK: myosin-light chain kinase
WRAMP: Wnt5a-receptor-actomyosin polarity
MCAM: melanoma cell adhesion molecule
CaM: calmodulin
FAK: focal adhesion kinase

1. Introduction

The number of migration modes increases slowly from year to year. But still, all these different migration modes that enable a cell to move through an artificial matrix, a channel or native tissue can be found under specific conditions in an individual cell or in a specific cell type. The focus of this review lies on the single cell migration. In general, from a biophysical point of view, the migration modes rely on the distribution and dynamics of protrusion exertion, the generation of contraction and the adhesive state of a migrating cell. When extending the migration scenario to a broader view, the mechanical properties of the surrounding matrix environment need to be taken into account. The matrix properties themselves are altered by the composition, the topology and mechanics (Fig. 1). More precisely, the spatial distribution and/or remodeling dynamics of actin-driven protrusions, the contractile

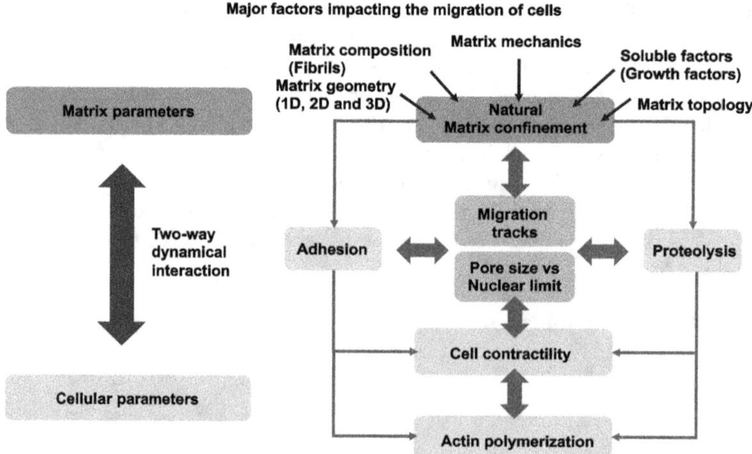

Fig. 1. Tissue environmental and cell intrinsic factors regulating the cell migration capacity and velocity.

force generating interaction between actin and myosin and the adhesive state of the cells determine the migration mode and speed and subsequently the overall migratory capacity of cells (Fig. 1). However, both cell mechanics and matrix mechanics determine the migratory capacity of cells. Based on these criteria, several migration modes have been identified that can be employed by normal healthy cells and pathological cells, such as cancer cells.[1–4,59] Since actin filaments are not the only cytoskeletal components in living cells, intermediate filaments and microtubules are supposed to play a role in regulating the migration mode or the switching between individual migration modes.

Previously, many cell mechanical properties, such as cell stiffness or inverse cell deformability, have been determined under static conditions suggesting that these mechanical properties do not change during the rather "static" migration and invasion of cells when they face different environmental conditions on their track through the connective tissue microenvironment. However, it has been suggested that the migration mode might be a continuous one. This continuous migration mode can additionally change in order to adapt to varying environmental conditions. Since the mechanical properties of a cell are regulated by environmental cues, the cell mechanics can be altered during their migration and invasion accordingly due to varying matrix mechanical properties.[5,6] Subsequently, the continuous variation of the cellular microenvironment can be employed to differentiate stem cells under normal physiological conditions[7] or to modulate the stemness of cancer cells under diseased conditions.[8]

In turn, specific stem cell types, including tumor stem cells, can be selected based on mechanical properties of the microenvironment.[6,9] The continuous interaction of stem cells with their local surroundings, which contain cellular and acellular constituents, leads to the hypothesis that a niche for stem cells exists helping them to keep their self-renewing capacity and their undifferentiated state including their cell multipotency.[10] The idea of a niche for stem cells can be transferred to cancer stem cells. Similarly, cancer stem cells represent a rare subpopulation of cancer cells within the primary tumor that possess the capacity to promote the initiation of a tumor, lead to the heterogeneity among cancer cells, cause the occurrence of drug resistance and subsequently can reemerge a tumor after a remission period.[11,12]

Besides the well-identified chemical players, physical properties, such as stiffness and tension of the local microenvironment, are moving into the focus of current and future cancer research, as these properties facilitate the further development and the malignant progression of cancer.[13,14] Moreover, the development of cancer seems to be pronouncedly supported by increased stiffness of the tissue microenvironment that is altered by both, the increased stromal collagen production and the presence of fibrosis.[15–17] Hence, it can be proposed that the cell mechanical properties are altered by matrix mechanics that subsequently alter the migratory and invasive behavior of cells.[4,14,18] Based on this knowledge, the mechanical properties of cells have been investigated directly when analyzing their migration and invasion through the 3D extracellular matrix environment.[19,20,169] Finally, migration and invasion studies focus on the relationship between the mechanical properties of cells, the regulation of biological signaling pathways and, more recently, matrix mechanics. Hence, the focus of this review is on the cell mechanics defining the specific migration modes. The migration modes are compared in terms of similarities and transition options between modes.

2. The Cell Mechanical Phenotype: Static Versus Dynamic Properties

The mechanical properties of cells seem to determine their migration mode in 3D confined extracellular matrices. It has been reported that there is a correlation between cell mechanics and the invasive capacity of distinct cell types.[21,22,20,23–27,169] Moreover, for specific cancer cell types, the mechanical phenotype has been used to identify the invasive capacity and hence the malignant potential of certain cancer types including oral cancer[28,29] and head and neck cancer.[30] In principle, there are two major hypotheses as to how the mechanical properties of cells must be to promote cell migration and invasion.[31] On the one hand, it has been reported that invasive cells need to be softer in order to migrate and invade increasingly.[28–30] This leads to the first hypothesis that increasingly invasive cells are more compliable and hence softer than less invasive controls. On the other hand, it has been seen that the highly migrating and invasive cells display an increased stiffness compared to weakly migrating and invasive cells.[23,19,22,20] These results formed the foundation for the second opposing hypothesis that invasive cells are stiffer compared to control cells.

Besides these obvious contradictory hypotheses, there is another issue in determining the cell mechanics that requires to be enlightened in this review. The cell mechanical properties are usually assessed, when the cells have undergone a specific treatment in order to be measured and the measurement is performed at a specific time point, usually without synchronization of the cells. A major weakness is the static analysis of cell mechanics.

The behavior of cells is rather not static, instead, it is rather dynamic. This dynamic behavior of cells generally relies on a tightly regulated balance between forces that contract the cell to cause a compact cell shape and forces that extend the cell boundaries to facilitate an enlarged and less compact cell shape.[32] Both forces are required for the migration of cells and the migration-driven remodeling of the extracellular matrix environment that occurs in many physiological processes, such as tissue development and wound healing after tissue injury. In detail, contractile forces of cells are able even to shift the cell's center of gravity. This may either lead to a full or partly deadhesion of the cells from its substrate or matrix compression including its remodeling. Protrusive forces evoke the spreading and adhesion of cells or lead to the extension of lamellipodia and filopodia that play a major role in cell migration and invasion. This maximizes the cell contact surface with the environment such as a 3D extracellular matrix or other substrates to which the cells adhere. Therefore, the protrusive and the contractile forces of cells are tightly balanced to fulfill specific functions that are required for the normal and healthy physiological behavior.

Since the cell itself is not a static living matter, the mechanical properties of cells were simultaneously analyzed during their migration and invasion through confined 3D extracellular matrix. In detail, the forces that cells exert during their migration through a 3D matrix were determined and the flow fields around these cells can be revealed.[19,20,169] Thereby, it is important not to include fiducial bead markers in the 3D matrices,[33,34] since migrating and invading cells can phagocytize them and no longer move on through the matrix.[35] The dynamic alterations of the mechanical properties of migrating cells can be seen in protrusion waves, which are initiated by a mechano-chemical crosstalk between focal cell adhesions, the tension of the membrane and the actin-driven protrusion of the cell at its leading edge.[36] These mechanical alterations are coordinated with alterations in cellular morphology (shape changes) and are based on the wave-like dynamics of the actin cytoskeleton that can be activated, for example, by cell mechanical changes or matrix environmental changes.[36]

3. Influence of Environmental Conditions on Cell Mechanics

There are several research studies on the direction in which environmental properties affect the mechanical properties of cells.[4,37–39] Many sophisticated migration and invasion assays in special environments have been developed to explore the effect of the cellular surroundings on cell mechanics. All these techniques and assays are aimed at overcoming or improving the limitations of a flat 2D environment that has been used for migration and cell mechanical analysis in recent decades. In order to show how cells react to environmental disturbances, five major different approaches can be carried out in which the cellular microenvironments are controlled to a certain degree.

First, the patterning of ligands on defined substrates, such as multiprotein 2D or 3D patterning or precisely regulated molecular density techniques, can be utilized that controls the process of cell adhesion and hence force generation.[40–44] Second, the regulation of the rheological properties of the environment of cells, including supported bilayers, mechanical transducers, such as flexible micropillars[45] or deformable hydrogels in 2D using fiducial tracers, can be employed.[23,46–50] Moreover, in 3D environments without any embedded fiducial makers[19,20,169] the matrix rheological properties are adaptable and their effect on the migration and invasion can be examined.

These techniques help to quantify cellular force exertion under certain constrains. Third, facing cells to different topologies ranging from nanotopography via fibril substrates (cellulose) to geometric confinements alters cell mechanics and mechanotransduction processes and impacts their cell functions and gene or protein expression.[51,52] Fourth, the regulation of the spatial confinement and variation of the properties (as mentioned in one to three) over time, such as soluble or matrix embedded gradients or time-varying environmental properties, seem to impact the mechanical phenotype of cells and subsequently cellular functions.[53–55] Fifth, mechano-invasion assays can be performed to figure out how mechanical stimulation by fiber alignment and stiffness alteration can impact the migration and invasion of cells. In parallel the gene and protein expression of invasive and adhesive cells can be analyzed.[56]

3.1. *Amoeboid and blebbing migration mode*

The amoeboid and the blebbing migration modes are classical modes that have been identified and well characterized.[57,58] In this specific migration mode, the cells regulate at least two major parameters: the proteolytic activity is abolished and the cell adhesiveness to the 3D environment is relatively low. The amoeboid migration mode can be simulated, when the proteolytic activity is impaired and a low detachment rate of $k_d = 0.02\,\mathrm{s}^{-1}$ is employed. Under these conditions, the simulated cell exerts random pseudopodia-like protrusions toward nearby pores of the 3D extracellular matrix environment.[2] Based on the polymerization of actin at the cell front, the cell expands at its front and the cell protrusions are still enlarged in their size and protrude into the pores of the 3D extracellular matrix.[56] In detail, one protrusion can dominate over the other and subsequently pull the entire cell into the migration direction, because the cell tries to conserve its volume. This migration mode seems to be similar to the experimentally identified amoeboid migration mode.[57] Moreover, when the formation of adhesion sites is fully blocked, the simulated cell performs a migration that is still highly similar to the adhesion-based cell. In detail, the extracellular matrix interactions represent only steric hindrances for the cell to migrate and invade through the extracellular matrix. The cells move in the amoeboid mode at lower speed, because they can generate traction forces solely relatively inefficient. In line with other reports, this migration mode is highly similar to the blebbing migration mode (Fig. 2).[59,60] It has been stated that the experimentally observed blebbing migration mode can be explained by the membrane detachment from the cell's actin cortex subsequently leading to an extended plasma membrane that is pushing outwards by the hydrostatic pressure inside the cell (Fig. 2).[61,58] There exist models to simulate the experimentally identified mechanism of blebbing motion. However, these models simulate the morphodynamic behavior of the cells similar to cells migrating in a blebbing migration mode. In fact, the morphological

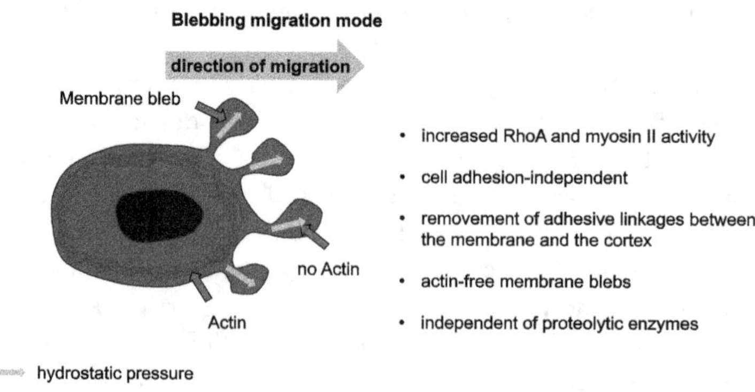

Fig. 2.   Blebbing migration mode and the relevant parameters.

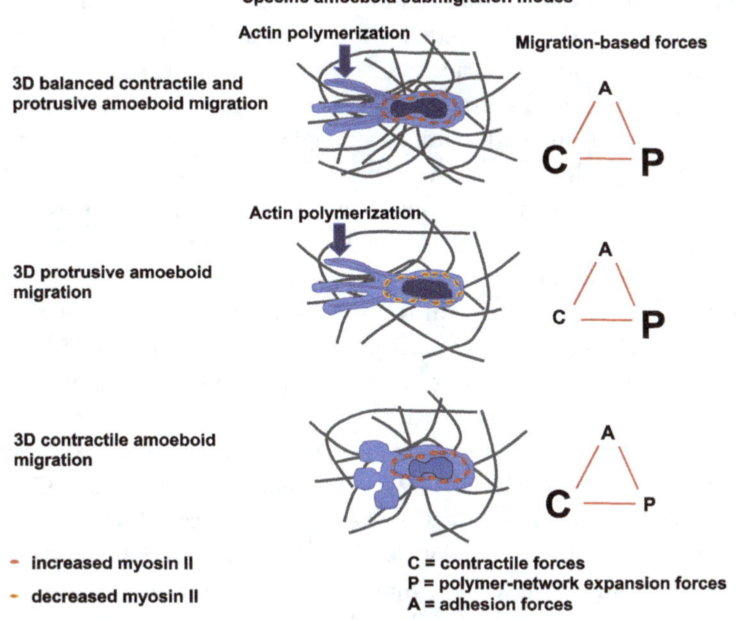

Fig. 3. The three amoeboid migration types and their associated forces. The increase of a specific force or forces favor the choice of a specific amoeboid migration mode.

description "amoeboid" migration summarizes several submigration types of rather different biophysical modes of cellular migration and invasion that cover a broad range of modes such as the blebbing migration mode and an entirely actin polymerization-based gliding migration mode.[57,61]

These submigration types of the amoeboid migration are based on different principles of the generation and transduction of forces which ultimately lie at the root of these different amoeboid phenotypes (Fig. 3). When the balance between actin protrusion, actomyosin contraction and cell–matrix adhesion is altered, these different amoeboid modes can be explained. In detail, the blebbing and gliding migration modes are hardly extreme variants of a single, common unique migration strategy. Due to variations of physiological and experimental conditions and the choice of a specific cell type, the submode of amoeboid migrating cells can switch between at least three distinct mechanical submodes. More precisely, different modes of amoeboid migration can be described by analyzing the cell-generated forces, such as protrusive or contractile force generation and force transduction that determine the overall cellular adhesiveness. By varying the balance between these three force types, the different submodes of amoeboid migration and invasion can be defined. In fact, the force-relationship between adhesion, contraction and polymer-network expansion finally leads to the overall amoeboid phenotype that is presented in multiple experimental approaches. These three major forces acting in cell migration are the adhesion force, F_A, contraction force, F_C, and polymer-network expansion force, F_P (Fig. 3).

In conclusion, the amoeboid migration mode is usually reported to be well defined in terms of mechanism, however, these definitions are mainly based on morphological descriptions of the amoeboid migration and do not refer to the exact mechanism. Under amoeboid migration a broad spectrum of heterogenous types of cell mechanical driven migration modes are described. These submodes of amoeboid migration are rather ill-defined and cannot be easily distinguished from each other. The general description of the amoeboid migration mode usually covers roundish cells that migrate and invade through 3D confined extracellular matrices without containing any mechanical characterization of the cells including cytoskeletal remodeling dynamics. Keratocytes are such an example for cells migrating in a gliding migration mode that does not belong to the overall amoeboid migration mode[62] since they exhibit a constant roundish shape. Thus, keratocytes represent no candidate cells for the amoeboid migration mode.[63] The consideration of actomyosin-based mechanics is still mainly based on treating it as a static event. However, the regulatory and hence dynamical mechanics including the actomyosin-dependent or other cytoskeletal element-dependent mechanics, such as actin-crosslinkers, microtubules, vimentin (or other intermediate filaments), and septins should be considered as well[64–67] and should be analyzed in migrating cells. Finally, the contribution of these factors in well-defined amoeboid migrating cells will help to uncover various migration mechanics and lead to mechanically well-defined migration modes.

3.2. *Chimneying migration mode*

The chimneying migration mode has been detected in mesenchymal cells that can even migrate without cell adhesion (Fig. 4).[68] In more detail, these mesenchymal cells do not stop migration, when the cell adhesion is impaired. As far as I understood, the chimneying migration mode is similar to a blebbing migration mode, but seems to be rather restricted to mesenchymal cells. Whether leukocytes can also use this mode of migration remains still not yet clearly understood. The chimneying migration mode can be seen as a coupled friction-poroelasticity migration mode demonstrating that

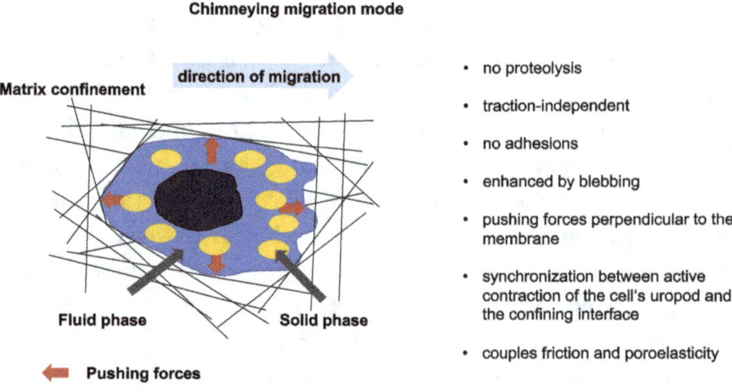

Fig. 4.   Chimneying migration mode and the relevant parameters.

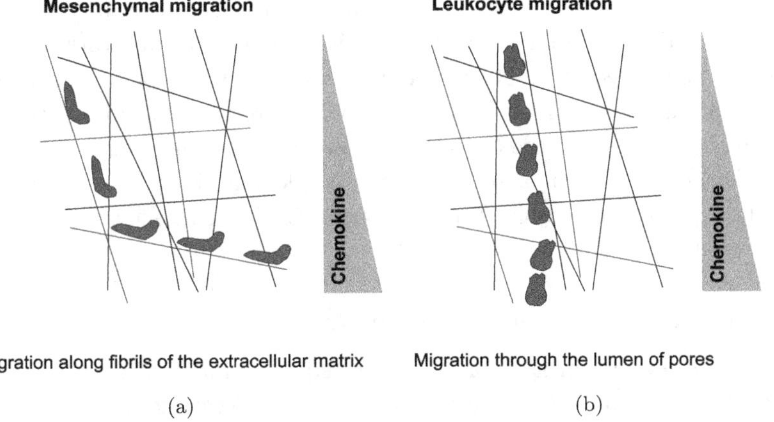

Fig. 5. Migratory patterns impact the migration path of mesenchymal migrating cells. (a) in the presence of a gradient, whereas leukocytes, (b) are not affected.

confined cells are able to migrate and invade mechanically without cell adhesions (Fig. 4). Thereby, the cell employs the chimneying climbing technique and exerts pushing forces perpendicular to the membrane. The cell itself actively contracts its uropod, which is continuously synchronized to the passive friction force between the cell's surface and the confinement surface (Fig. 4). The difference between a mesenchymal cell and leukocyte that migrate through confined 3D extracellular matrices is that the mesenchymal cell migrates along fibers of the matrix scaffold, whereas the leukocyte migrates directly through the center of the pores in these matrix scaffolds (Fig. 5).[69]

The adhesion-driven cell motility is currently the most investigated and well-known migration mode. However, when cells migrate in an adhesion-free confined environment, such as simple 1D channels, or when cells migrate with a higher migration speed, they are able to switch to a blebbing-driven migration mode, termed chimneying migration. This chimneying migration mode requires a precise synchronization of the active contraction of the cell's uropod and the passive friction force that is present between the cell and the surface of its confinement, such as the inner side of the channel surface (Fig. 4). Hence, a one-dimensional poroelastic model of chimneying can be used to reveal the active cell strains. However, a major improvement is the identification of the synchronization between these internal cell strains and the external friction forces that is coordinated by the cell itself. Subsequently, this synchronization is required for the forward movement of cells to be self-determined. Simulations of these migration modes display that the blockage of cell adhesion cannot fully abolish the migration and invasion of mesenchymal cells. Mesenchymal cells, which generally possess proteolytic activity, but which is impaired, are still able to migrate and invade through a densely confined 3D extracellular matrix by employing a chimneying migration mode,[68] in which small protrusions are exerted that extend into the pores of the 3D extracellular matrix on the cellular side (Fig. 5).

These protrusions into the pores of the surrounding confinement are composed of actin networks that are integral components of the entire cellular actin network and thereby are mechanically connected to it. The retrograde actin flow within the cell can generate traction forces through the steric effect between the cell protrusions and the surrounding extracellular matrix. However, the cell's movement in this chimneying migration mode seems to be less efficient compared with cells employing the proteolysis-dependent mesenchymal migration mode, because the cells possess a low efficiency in the generation of traction forces.

3.3. *Finger-like protrusive migration mode*

The formation of cellular protrusions is a prerequisite for a distinct type of cell migration. Cells migrating in 3D environments and *in vivo* can exert a large variety of different protrusion types that range from actin polymerization-based lamellipodia to contractility-based cellular blebs. The possibility for cells to switch between different protrusions and hence different migration modes is supposed to drive the migration of cells in complex environments and subsequently lead to the progression of the cancer cell dissemination in targeted tissues. However, the plasticity of the formation of cell protrusions has been precisely analyzed solely during the transition between amoeboid and mesenchymal migration modes that is tightly associated with an alteration in cell morphology and shape. Thus, even the minimal prerequisites for the blebs to lamellipodia transitions, including the time scales on which they appear, stay unknown. In order to address this point, the switching between the two types of cell protrusions during the migration and invasion of individual cells has been determined. When cells are used that can switch between blebs or lamellipodia driven migration states, the mechanical properties for each migration mode, including the dynamics of the switching process itself, can be revealed. Moreover, it has been convincingly shown that the shifting of the balance between the actin protrusivity and actomyosin contractility finally uncovers immediate transitions in migrating and invading cells. In specific detail, the switching between the migration modes takes place without alteration in the overall cell shape, cell polarity or cell adhesion. In addition, the rapid transition between blebs and lamellipodia has been shown to be initiated by alterations in the cell-substrate adhesion during their migration on micropatterns. Finally, the type of the protrusion exerted by the migrating and invading cells is regulated dynamically and independently of the cell's overall morphology. These findings lead to the hypothesis that the formation of protrusions is an autonomous module within the regulatory cellular network that facilitates the cell plasticity during their migration.

The finger-like protrusive migration mode can also be termed simply protrusive migration mode (Fig. 6). Multiple cell types can migrate by the exertion of long, finger-like protrusions originating from the cell body, extending in the 3D extracellular matrix environment and generating traction forces at the protrusive tips.[70] It can be proposed that the cell shape of both migration types can be simulated

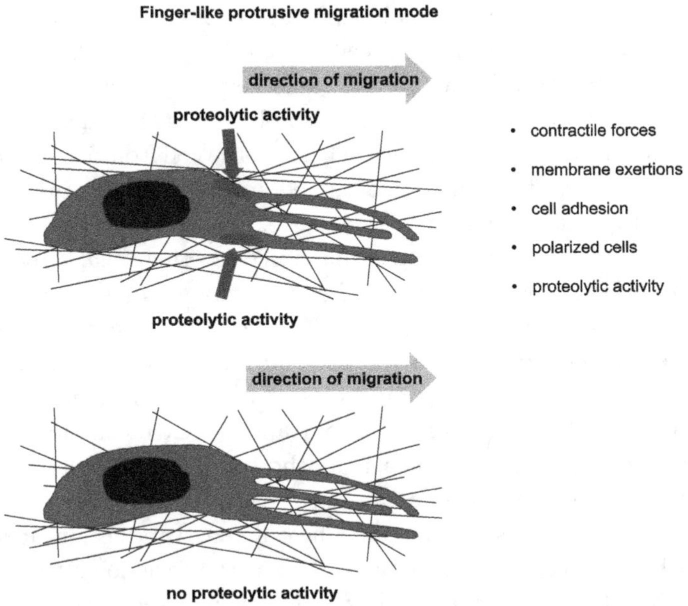

Finger-like protrusive migration mode

direction of migration

proteolytic activity

proteolytic activity

- contractile forces
- membrane exertions
- cell adhesion
- polarized cells
- proteolytic activity

direction of migration

no proteolytic activity

Fig. 6. Finger-like protrusive migration mode and the relevant parameters.

by actively adapting the actin polymerization and the cell adhesion formation in the specific protrusion. In order to limit the simulated protrusion to a length of 12 mm and a width of 2 mm, the radius of the cell was set at about 6 mm. Combined with a strong myosin-driven contraction, the simulated cytoskeletal actin network can create a high contraction force that acts between the cell protrusion and the remaining cell body. Since the focal adhesions are highly concentrated in protrusions, the entire cell body can be pulled toward and over the protrusion. At the leading edge of the cell protrusion, the polymerization of actin extends the plasma membrane outwards and thereby pushes the entire plasma membrane into the pores of the 3D extracellular matrix scaffold and thereby overcomes the resisting adhesion force at local focal adhesions. Subsequently, a continuous cell migration and invasion is achieved. During this finger-like protrusive migration mode, the cell needs to de-adhere and hence rupture the existing cell–matrix adhesion bonds at its neck region. The local extracellular matrix proteolysis can even support the rupture of bonds, but it is unnecessary for cell movement. Even without the proteolytic activity, a simulated cell can continue to migrate, if the cell first exerts contractile forces within the protrusion that are powerful, and second if there are strong focal adhesions that can be dispersed over the entire protrusion and third if the extracellular matrix scaffold can easily be broken.

3.4. *Mesenchymal migration*

The classical mesenchymal migration mode is commonly utilized by cancer cells that possess a strong proteolytic activity and pronounced focal adhesions. The experimental findings can easily be modeled. More precisely, when the proteolytic activity is present and a relatively low cell de-adhesion rate of $k_d = 0.02\,\mathrm{s}^{-1}$ is employed, the simulated cell exhibits an elongated cell shape, where the nucleus is positioned at the rear end of the cell. All of this is in line with the experimental shape of cells migrating in a mesenchymal manner. Thereby, the cell front is pushed forward in the direction of motion through the actin polymerization toward the plasma membrane and subsequently the actin network is expanded. At the same time, the cell's rear retracts by the cell-extracellular matrix connecting focal adhesions. The proteolytic activity of the cells is critical for the mesenchymal migration mode, since cells can overcome steric hinderances, such as the extracellular matrix scaffold, that represents a barrier for cell migration. Thus, efficient migration seems to be possible. When cells move in mesenchymal migration, they create transient adhesion connections between the intracellular actin network and the extracellular surrounding matrix at the sides of the cell. In contrast, at the cell's leading edge, cell–matrix adhesions generate traction forces, since the local actin network flow rate is faster than the overall translocation rate of the cell. The cell–matrix adhesion sites are then moved to the back of the cell, reducing the flow velocity of the local actin network. When the local flow velocity of actin becomes slower than the cell translocation rate, the cell–matrix adhesions represent resistant forces on the cell. However, the cells can move and penetrate continuously, as the cell adhesions gradually mature and ultimately detach from the cell when they are located on the back of the cell.

3.5. *Lobopodial migration mode*

The lobopodial migration mode represents an intermediate migration mode, since it possesses features of two other major migration modes, such as the protrusive migration mode and the blebbing migration mode, that are combined (Fig. 7). Initially, this lobopodial migration mode has been detected and initially defined in fibroblasts.[71–73] More precisely, the lobopodial migration mode can be observed in fibroblasts that migrate and invade through linearly elastic matrices. In fact, the rheological properties of a specific extracellular matrix scaffold seem to be critically affecting the type of migration mode chosen by a migrating and invading cell.

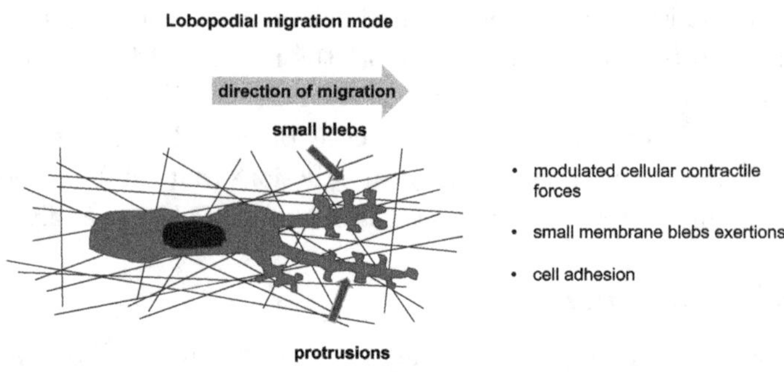

Fig. 7.   Lobopodial migration mode and the relevant parameters.

In highly crosslinked 3D extracellular matrices that display a linearly elastic behavior and where the matrix rigidity does not increase with an applied force, fibroblasts are able to employ a pressure-based lobopodial migration mode, which depends on RhoA-ROCK-myosin II-facilitated cell contractility[71,74] and paxillin and vinculin containing focal adhesions.[71] Dissimilar to the lamellipodial-based migration, fibroblasts exhibiting the lobopodial migration mode display first a blunt, cylindrical leading-edge protrusion and intracellular blebs that are expanding by intracellular pressure, but they are second not containing increased cortactin, and third these cells migrate and invade without the polarization of the two Rho GTPases Rac1 and Cdc42.[71] In contrast, a vimentin-facilitated coupling occurs between the nucleoskeleton and the plasma membrane, which leads to the establishment of cell compartments with different pressures, whereby the nucleus functions as a piston and thus enables the formation of pressure-driven blebs at the front edge of the cell.[74] In 3D cell-derived matrices the knockdown of RhoA or the ROCK inhibition can lead to a transition between the fibroblast migration modes from lobopodial migration to lamellipodial migration, which has no effect on the velocity of the cell migration.[71] In addition, there has been some evidence that cancer cells cannot employ the lobopodial migration mode. However, this hypothesis needs more investigations in order to confirm it. There is some support from a specific fibrosarcoma cell line, the HT1080 cells, that cannot switch to a lobopodial migration mode in linearly elastic cell-derived matrices, but instead they adopt amoeboidal or mesenchymal–lamellipodial morphologies and seem to migrate using one of these migration modes.[71] Hence, we do not know whether cancer cells can generally employ a lobopodial migration mode to move themselves through connective tissues containing extracellular matrix scaffolds or whether they use indeed a lobopodial migration mode under distinct conditions such as when a certain combination of microenvironmental and cell signaling features induces such an intermediate switch between the classical amoeboid migration modes and the mesenchymal migration mode.[73]

3.6. *Nuclear piston migration mode*

The nuclear piston migration mode represents a relatively novel migration mode and seems to be definitely based on the mechanical properties of cells and their nuclear compartment, the cell nucleus (Fig. 8).[75] In more detail, adherent primary human

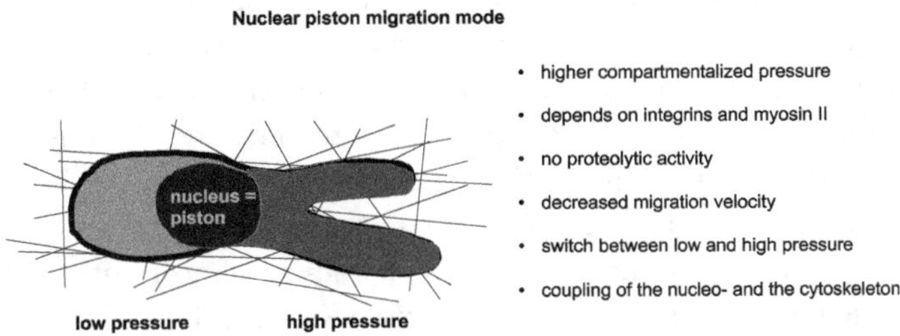

Fig. 8.   Nuclear piston migration mode and the relevant parameters.

fibroblasts can change from a low-pressure lamellipodia to a high-pressure lobopodial protrusions-based migration mode, when they migrate through a highly crosslinked 3D extracellular matrix confinement, such the extracellular matrix of mammalian dermis or cell-derived matrix.[71] Moreover, nonadherent fibroblasts can employ a third specific migration mode in 3D extracellular matrix environments, termed the A1 amoeboid migration mode.[76] More precisely, in lobopodial fibroblasts, the acto-myosin contractility pulls the cell nucleus in the direction of migration similar to a piston in a cylinder to elevate the cytoplasmic hydraulic pressure in front of the cell nucleus and in the direction of movement.[74] Hence, this compartmentalized pressure facilitates the forward movement of the lobopodial membrane rather than the "simple" actin polymerization-based Brownian ratchet that is commonly linked with the lamellipodial cell protrusion. Subsequently, this nuclear piston migration mechanism is employed for the efficient migration and invasion of primary fibroblasts in cross-linked 3D extracellular matrices. It is clearly known that primary fibroblasts and cancer cells can switch between distinct migration modes, however it is still not well understood whether these cells can switch solely between the same migration modes or whether the plasticity in cell migration and invasion is controlled by similar mechanisms independent of the specific migration modes. In order, to analyze whether the hypothesis holds true that the migratory plasticity is cell type specific and hence primary fibroblasts and their malignant counterparts differ in their migration mode, it has been investigated whether polarized human HT1080 fibrosarcoma cells can perform a nuclear piston mechanism on their migration and invasion in 3D cell-derived matrices. In detail, the intracellular pressures in front of and behind the cell nucleus of HT1080 have been determined. As proposed, the nuclear piston mechanism is usually inactive in the malignant fibrosarcoma cells, however, the mechanisms can be switched in elongated, polarized cancer cells, when the proteolytic matrix metalloproteinase (MMP) activity of the cells is impaired by adding, for instance, a specific pharmacological inhibitor, such as GM6001.[75]

3.7. *Slingshot migration mode (nonproteolytic mechanism)*

The slingshot migration mode represents a relatively new migration mode.[77] It is based on synthetic matrices that can be tuned in the alignment of their fibers and consequently in the matrix stiffness. To reveal how the alignment of matrix fibers and the stiffness of the 3D extracellular matrix affect the migration and invasion of distinct cell types, a designed and well-characterized synthetic biomimetic extracellular matrix that is composed of electrospun dextran methacrylate (DexMA) fibers that enable the regulation of the orthogonal fiber alignment and the bulk stiffness of the matrix.[78] In specific detail, cell-perceived extracellular matrix mechanics can be regulated by utilizing electrospinning fibrous matrices, which are placed on an array of microfabricated polydimethylsiloxane (PDMS) wells to cancel out an effect from the underlying mechanically rigid support layer (Fig. 9). In order to alter the alignment of the fibers, the shape of the electric field at the collecting surface needs to

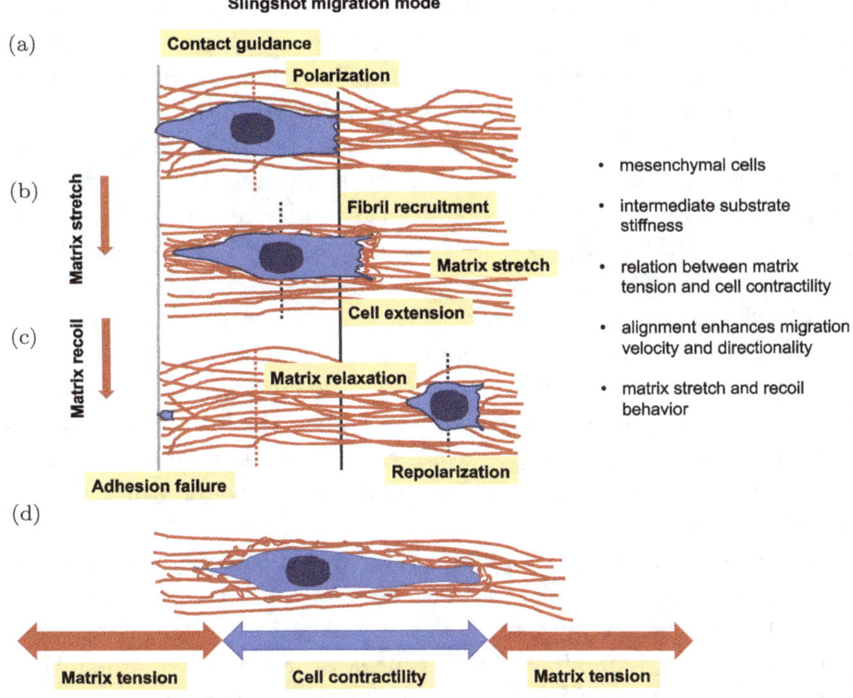

Fig. 9.   The tugging force migration mode and the relevant parameters.

be changed during the electrospinning procedure. Therefore, the separation distance between two parallel collecting electrodes requires to be altered.[79] When the distance is increased, the fibers are increasingly aligned. In specific detail, the photo-initiated DexMA crosslinking is regulated through the duration of the ultraviolet (UV) exposure time that controls the stiffness at the single fiber level. The bending stiffness of the fiber can be determined by employing a three-point bending using atomic force microscopy (AFM). In order to reveal the overall stiffness of the substrate, the bulk substrate stiffness can be analyzed by tension measurements, in which a 1 mm cylindrical indenter is pressed in the suspended matrices. A crosslinking can be chosen to cover the full spectrum of matrix deformation ranging from maximal to undetectable displacements due to cell-generated forces. Indeed, the values of the Young's modulus of single fibers are the same as those that have been determined for single fibrin fibers, elastin fibers and intermediately stretched fibronectin fibrils.[80] The range of the bulk stiffness values lies between 1 kPa and 30 kPa, which is within the broad variety of tissues[18] and in the stiffness range where the transition from normal to cancerous mammary tissue can be observed.[13] Matrix properties, such as the density of ligands, the diameter of the fibers and the thickness of the scaffold need to be kept constant by using a specific cRGD concentration, a specific polymer weight and a specific duration of the fiber generation. When all these properties of the 3D extracellular matrices are kept precisely controlled in naturally derived extracellular matrix proteins such as collagen and fibrin hydrogels, it can be investigated that these matrix properties are basically coupled to protein concentrations and the conditions for the gelation process.[81–83] 3D extracellular matrices with free-radical polymerized DexMA fibers can be used to obtain proteolytically uncleavable meth-acrylate–methacrylate crosslinks and hence the effect of cell-driven degradation of these matrices over time is impaired.[84–86] Therefore, the focus of these experiments is on the analyses of nonproteolytic migration modes in order to reveal the mechanisms of proteolytic-independent 3D cell migration and invasion.[87] Moreover, it has been addressed that the cell mechanics are altered in the human MDA-MB-231 breast cancer cells and the human HT1080 fibrosarcoma cells, when the proteolytic degradation is inhibited by adding a broad spectrum MMP inhibitor, termed GM6001.[88] By its addition, the two cell types do not only round up, they additionally lose their actomyosin-dependent contractility, possess decreased stress fibers and subsequently, their nuclear mechanics are disturbed. In specific detail, the nucleus is softer compared to untreated controls, which is based on an elevated phosphorylation of the nuclear envelope protein laminA/C.[88]

How realistic is such an artificial migration mode with a matrix that can be tuned in terms of stiffness and pore-size, but which cannot be cleaved by the migrating and invading cells? The mechanical invasion mode is performed as follows: After seeding the cells, they were cultured for 6 h before the time-lapse imaging is recorded. During that time most of the cells migrated and invaded into the matrix and consequently, the cells embedded themselves into the mechanically tunable matrix. At each stiff-ness level, the alignment of the matrices causes the migration of cells at higher velocities and in a more directional manner, which is similar to the migration of cells on single tracks. The effect of the fiber alignment on the directionality of the cell migration, which can be revealed by the deviation of the position of the cell from a linear fit for its overall linear migration track, has been reported to be consistent over all matrix stiffnesses indicating a strong influence of contact guidance on cell migration and invasion[89,90] and on the directionality that seem to be rather stiffness-independent under these conditions. This result supports previous studies employing 3D collagen fiber hydrogels.[83]

For cell migration of multiple cells types, a biphasic migration behavior has been revealed that can be found in 2D and 3D migration and invasion matrices.[91] The migration and invasion of cells in 3D extracellular matrix networks with pore sizes above 5 μm is elevated with increasing stiffness of the matrices, whereas the cell invasion in matrices with smaller pores (below 5 μm) is decreased with increasing stiffness of the matrices. Finally, these results demonstrate that the 3D cell invasion is increased by higher matrix stiffness, which contrasts the cell migration behavior on 2D substrates, when the pore size of these 3D matrices is not below a critical value rendering the matrices a barrier to cell migration due to excessive steric obstacles.[91]

On 2D substrates, a classic biphasic dependence of the migration speed on matrix stiffness can be predicted by simulations. Hence, it can be predicted in general that cell types exhibiting a higher force-generating capacity do not need to reduce their migration speed on very stiff matrices, instead, they just disable the biphasic response behavior.[92,93] A lower number of membrane receptors on the cell surface seem to restrict the biphasic response phenomenon, since those cells need stiffer matrices in order to migrate efficiently. A cell type employing the robust biphasic migration on flat 2D substrates is predicted from the simulation to acquire a faster migration speed in channel-like confined microenvironments with different widths and heights in increasingly confined environments.[92] When the simulations are performed in flatter microchannels, the biphasic mechanosensitive cell migration response is more stable on 2D micropatterns than in channel-like 3D confinements.[92] Moreover, the dimensionality alterations of the matrix confinement affect the migration path that cells sense and how they respond to the specific stiffness of the matrix. When these conditions are simulated new phenotypes for stiffness- and topography-sensitive cell migration can be obtained that depend crucially on both, the cell-intrinsic properties, such as cell mechanics, and the matrix properties such as matrix mechanics. These theoretical results can help to understand the multiple mechanosensitive migration modes that allow the migration and invasion of cancer cells through topographically very heterogeneous microenvironments. In line with these results, the alignment of the 3D extracellular matrix fibers enables the migrating cells to move along these fibers more easily through the confinement, which can be observed *in vivo* in tissues, where cancer cells migrate out of the primary tumor.

Finally, a major question can still be raised of whether the specific slingshot migration mode has been seen in *in vivo* tissues or does it seem to be connected to the artificial 3D engineered matrix environment?

3.8. *Tugging force (guided) migration mode*

The tugging force migration mode is based on dynamically fluctuating ("tugging") traction of the cells (Fig. 10). It is well documented that cells generate contractile forces toward their substrate to which they adhere to and toward neighboring cells primarily through myosin-generated forces based on the tension of the actomyosin cytoskeleton.[94,95] This actomyosin activity is crucial for the morphogenesis of entire tissues and the dynamic regulation of cell–cell or cell–matrix adhesions. In addition, myosin II is required for stereotypical cell shape alterations and the remodeling of cell–cell adhesions associated with the extension (or expansion) of the tissue by the intercalation of cells in both Drosophila[96,97] and Xenopus models.[98,99] In line with these results, the inhibition of the Rho kinase, myosin-light chain kinase (MLCK) or the activity of the myosin ATPase can interfere with the adherence junctions' growth and their maintenance of in several systems.[100–104] E-cadherin and N-cadherin cell surface receptors cluster into focal adhesion-like structures when cells are seeded on cadherin-coated planar substrates. More precisely, the clustering of the cadherins requires actomyosin-driven contractility.[103,105,106] Dissimilar to mechanosensitivity of focal adhesions,[107] the force-dependent assembly of the cadherins in cell–cell adherence junctions is not well understood. However, focal adhesions and adherence

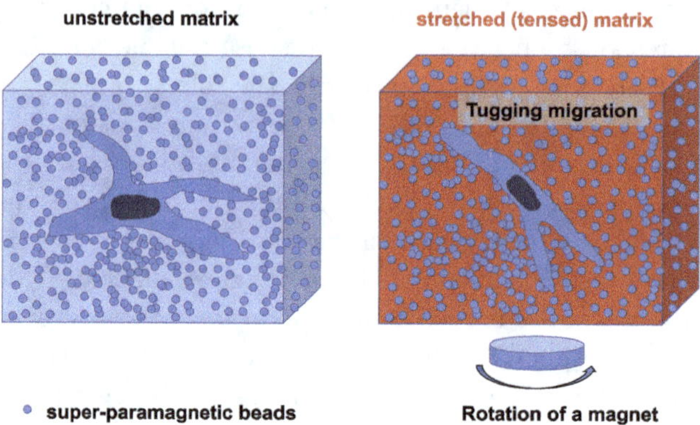

Fig. 10.   The slingshot migration mode is chosen by contractile cells in a matrix-stiffness-dependent manner. At an intermediate stiffness, the velocity and the directionality of migrating cells exhibits highest values. Schematic presentation of the key steps of the slingshot migration in aligned fibrous extracellular matrices. (a) Contact guidance-driven polarization of the cell. (b) Protrusion extension occurs simultaneously with the recruitment of fibrils and the stretch of the extracellular matrix environment. (c) The adhesion fails (rupture of thethers) and hence the fast recoiling of the cell is induced that subsequently leads to the translocation of the entire cell. (d) Force balance between the tension of the extracellular matrix and the cell-generated contractile forces.

junctions are closely linked to the actin cytoskeleton that connect the entire cell to the extracellular matrix environment or to neighboring cells through integrin cell–matrix receptors or cadherin cell–cell receptors, respectively, and subsequently support the cells to withstand substantial forces from the outside.[108] An experimental approach of microfabricated force sensors can be used to quantitatively determine both in parallel, the force applied to the cell–cell contact and the size of the adherence junction. It has been reported that certain cell types, such as endothelial cells, generate considerable forces that can be up to about 120 nN and normally pull to the axis of the cell–cell adhesions, known as the intercellular tugging force. Moreover, there exists a robust correlation between the size of adherence junctions and the tugging force. Both junctional parameters, the size of adherence junctions and the tugging force, respond directly to the activation of myosin. Moreover, this actomyosin contractility works together with a Rac-driven signaling pathway to set the ultimate size of the adherence junctions. In addition, the acute stimulation with RhoA or the exertion of external pulling forces leads to a fast adherence junction growth. Subsequently, these results show that adherence junctions can begin to grow due to mechanical stress, leading to the hypothesis that cell-generated forces impact the architecture and structure of tissues by self-regulating the strength of cell–cell adhesions. Tugging forces regulate the size of cell–cell adhesions[109] and hence determine the migration and invasion of cells. In general, tugging forces play an important role in morphogenesis, providing mechanical integrity, and generating mechanical stress gradients for the pattern regulation of cell functions.[110–112] Previous research has been focused on the role of externally applied forces that are required to break-down or remodel cell–cell adhesions.[113–115] In addition, the endogenous stresses generated between cells have been measured directly and it has been shown that these tugging forces initiate the growth of the adherence junction. In more detail, tugging forces can be determined using bowtie patterns to limit the cells to a single contacting interface.

In general, the capacity of the cells that can migrate in a directional manner toward a stiffer region within the 3D extracellular matrix scaffold is termed durotaxis. This process seems to be crucial for fundamental tissue functions, such as development and wound healing, whereas the durotaxis can also facilitate the malignant progression of cancer including cancer metastasis. Cell migration is regulated by integrin-associated focal adhesions, the assemblies of proteins that connect the contractile actomyosin cytoskeleton with the plasma membrane, which transmit forces that are generated and transmitted by the cytoskeleton toward the 3D extracellular matrix environment and subsequently, and in turn translate the microenvironmental mechanics in cellular biochemical signals. To sense the stiffness of the surrounding extracellular matrix, migrating fibroblasts adapt the focal adhesion mechanics on the nanoscale. Thereby they exert forces reminiscent of repeated tugging on the 3D extracellular matrix scaffold. All focal adhesions tug autonomously in an individual cell and hence act as small local individual sensors for rigidity. This behavior allows a cell to discriminate between the rigidity differences within the local 3D extracellular matrix with high spatial resolution on the level of single focal adhesions.[116] Finally, cell motility assays based on the tugging force migration mode have been developed. Thereby, the 3D extracellular matrix scaffold contains fiber-associated super-paramagnetic beads with 0.5–1.5 μm diameters that are tensioned by a magnet that is slightly rotated underneath the 3D matrix scaffold (Fig. 10). Consequently, the fibers of the scaffold are tensioned and aligned due to the amount of magnet rotation. In specific detail, the migration and invasion of distinct cells can be triggered by mechanical transient stimulation of the cells.[56]

3.9. *Rear-squeezing migration mode*

In order to mimic the nucleus-at-front migration type,[117] the radius of the simulated cell is limited to 6 μm, which means that the nucleus occupies then a higher fraction of the entire cell volume. Moreover, 90% of the actin nucleation events are located at the rear half of the migrating cell. Thus, the actin network is mainly concentrated behind the cell nucleus and the actin scaffold contraction can apply a squeezing force on the cell nucleus that finally pushes the entire cell forward in the migration direction (Fig. 11). The cell-extracellular matrix adhesion is necessary for this specific rear-squeezing migration mode, since steric connections between the cell and the 3D extracellular matrix scaffold are largely diminished due to the small interacting region between the cell's rear and the 3D extracellular matrix scaffold.

It is clearly known how cells employ the polymerization at their leading edge to move themselves forward on 2D substrates, whereas the migration modes of cells in 3D matrices relevant to *in vivo* microenvironments remain still rather elusive and

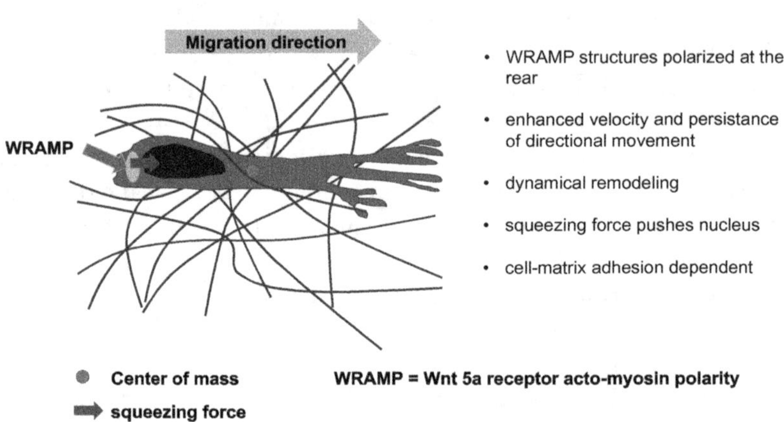

Fig. 11. The rear-squeezing migration mode and the relevant parameters.

poorly understood. Myosin II activity-dependent intracellular tension drives the regulation of cell migration. Moreover, there is evidence from theoretical and experimental studies that myosin II is highly concentrated at the rear of the migrating cell that is either isoform dependent or isoform independent and hence supports the 3D migration and invasion by myosin II tension at the cell's posterior end.

This scenario of rear-end myosin II tension-based migration is not restricted to amoeboid migration modes. Additionally, it has been identified for the mesenchymal migration mode, where a 2D-like migration mode facilitated by front protrusions at the cell's leading edge is common, suggesting a universal mechanism that underlies this migration type. From a biophysical point-of-view, it needs to be enlightened how the anisotropic localization of myosin II causes the 3D extracellular matrix migration and invasion of cells. In fact, a mechanism has been revealed where an interplay between the mechanical recruitment of myosin II and its biochemical activation leads to the support of directional cell migration in 3D extracellular matrix confinements. For amoeboid 3D extracellular matrix migration, myosin II first localizes mainly to the rear of the cell, which causes a slight polarization of the cell. Under the action of a tension-driven mechanism, the cell exhibits a uropod-like structure. Underlying biochemical signaling pathways trigger the actomyosin contractility, which either generates traction forces on the adhesion system or evokes prominent migration forces by causing blebbing activity on the cell membrane that leads to the movement of the cells. In the mesenchymal 3D extracellular matrix migration, cells employ the elastic properties of 3D extracellular matrices to migrate in and invade through them. The minor isoform of myosin is myosin IIB that is held back by relatively stiff 3D extracellular matrices at the rear (posterior side) of the cells. Myosin IIB is then activated through signal transduction processes, whereby prominent cell polarization is facilitated by determining the cell polarity from front to back and by subsequently generating the cell's back. More precisely, myosin IIB induces the polarization of the entire cell and regulates together with the major myosin IIA isoform the coordinated assembly of myosin IIA-depending stress fibers, in order to promote the directional movement of cells in the 3D confinements.[118,119]

The mechanisms at the leading edge of the cell during the migration are well known, whereas the polarized mechanisms at the cell's back that regulates the directional migration are not clearly understood. A new intracellular complex, the Wnt5a-receptor-actomyosin polarity (WRAMP) structure, has been identified that controls the regulation of the polarized localization of actin, myosin IIB and melanoma cell adhesion molecule (MCAM, CD146, Muc-18) upon induction by Wnt5a.[120]

During the locomotion, cells need to regulate the exertion of membrane protrusions and the adhesion of cells to the extracellular matrix at the leading edge of the cell in a coordinated manner. At the same time, the cell needs to retract the rear membrane and subsequently rear focal adhesions are disassembled. At the cell's leading edge, lamellipodia are created through the polymerization of actin via a Rac, Cdc42, WASP/WAVE, and Arp2/3-dependent mechanism supporting the cell adhesion to the extracellular matrix by focal adhesion formation. Finally the forward migration of cells is promoted.[121–123] Moreover, these proteins are involved in the maintenance and directional migration.[122,124] In contrast, the rear end mechanisms in migrating and invading cells are less investigated and hence still largely elusive. At the back of the cell, localized F-actin and nonmuscle myosin provide a mechanical force for moving the entire cell body forward.[125] The regulation of the directional movement of cells is still unclear. It is known that the directional movement of cells is driven by many processes, such as development, angiogenesis, inflammation and cancer. Therefore, the identification of novel signaling processes regulating the migration and invasion of cells will lead to new therapeutic targets for multiple human diseases. More precisely, Wnt5a triggers a new cell entity and polarizes the rear end of the cell.[126–129]

Wnt5a functions as a signaling-ligand that controls the spatial organization in cells, since it translocates proteins to specific intracellular locations in an anisotropical manner.[130–132] During metazoan development, Wnt5a regulates the control of cell polarity responses, including the formation of the cell body axis and the orientation of cell division via the planar cell polarity (synonymously convergent extension) pathways.[133] In addition, Wnt5a elevates the cytosolic Ca^{2+} levels, although the mechanisms are not yet clearly revealed.[134,135] However, Wnt5a supports the cell migration and invasion in melanoma, lung, gastric, pancreatic, and ovarian cancer cells.[136–141] In fact, Wnt5a has been found to be present at elevated levels in aggressive, metastatic melanomas and in primary tumors at their invasive edge, where the cancer cells can metastasize toward the surrounding lymph nodes.[136,137] However, also the mechanisms regulating the Wnt5a-facilitated migration and invasion of these cancer cells are not well understood.

In melanoma cells, it has been revealed that Wnt5a triggers the formation of a novel protein scaffold that induces the polarized actomyosin establishment at the cell's rear and Ca^{2+} signaling.[128] This structure is termed the Wnt5a-receptor-actomyosin polarity (WRAMP) structure and is associated with an anisotropic localization of the cell surface receptor MCAM (Fig. 11). A proposed mechanism seems to utilize the internalization of transmembrane receptors, such as MCAM, in endosomes, which can then interact with the actomyosin-dependent cytoskeleton to achieve the dynamic and transient WRAMP structure. It is important that the WRAMP structure entity is temporally linked to the F-actin fiber recruitment and nonmuscle myosin-IIB and thereby locally increases cytoplasmic Ca^{2+} ion levels. Thus, the WRAMP structure plays an important role, since it facilitates a mechanism for the direction of a highly localized second messenger signaling event, which is required for actomyosin contractility, membrane lifting, and retraction.

Moreover, the characteristics of the WRAMP structure lead to the suggestion that it functions in an event at the cell's rear promoting cell migration and invasion. Isotropic Wnt5a and a gradient of the chemokine CXCL12/SDF1 can cause a distal localization of WRAMP structures from the Golgi apparatus, which is associated with the cell's rear polarity. In detail, WRAMP structures are concentrated in the cell's rear due to the direction of nucleokinesis.[127] However, the behavior of these WRAMP structures in the extended cell migration and its precise function in directional movement have not yet been clearly understood. Using live-cell imaging, it has been figured out what role the WRAMP structures play in the directional migration of cells. These results clearly show that WRAMP structures are kept at the cell's rear in migrating melanoma cells and healthy control cells. In a large timeframe, individual cells can fluctuate between missing and existing WRAMP structures. In the presence of WRAMP structures cells can translocate (and hence move) over a longer distance with highspeed and display a larger persistence of the directional cell movement. When these WRAMP structures are absent, the cells can alter their direction of migration. This process is driven by a relocalization of intracellular WRAMP structures to define a new position for the cell's rear end.

Finally, these results suggest that WRAMP structures are polarized at the cell's rear end during cell movement and they facilitate the velocity and persistence of directional migration and invasion. All those are fundamental processes that are needed for cell locomotion. Hence, through the characterization of these WRAMP structures a deeper understanding of the cell's rear polarity mechanisms regulating cell migration is provided and new candidates for targeting cancer cell invasion can be identified and tested.[120]

3.10. *Connection between cell mechanics and migration modes*

Indeed, there seems to be a strong connection between the cell mechanics and the migration modes a cell can utilize in order to migrate into a distinct tissue type and invade through it.[142] There have never been fewer migration modes precisely defined than today. Hence, it is not surprisingly that the growing number of migration modes does not help to provide a clear picture of how and when a distinct migration can be utilized by a specific type of cells. Moreover, even the environment of the cells plays an important role. More precisely, the matrix mechanical properties can additionally, regulate the mechanical properties of cells and finally, they can also impair or trigger migration and invasion. Besides this, the proteolytic digestion of the surrounding microenvironment of cells may guide their way through a complex extracellular environment and hence interfere with both the mechanical properties of the matrix environment and the cells. These enzymes can either directly degrade their surroundings or shed membrane receptors directly from their own cells.[143,87] Thereby, in the latter case, the membrane's mechanical properties may be impacted, which additionally affects the migration and invasion of these cells. Additionally, also the adhesive state of the cells may be affected by the shedding of membrane receptors.

3.11. *Switch between migration modes*

It is not yet fully clear how the switch between migration modes can be performed and what role the microenvironment plays in this transition. Under confinement and low adhesion cells seem to adapt an amoeboid migration mode, if they cannot degrade enzymatically the surrounding confinement of extracellular matrix.[76] The experimental analysis of how cells migrate in and invade through 3D extracellular matrix scaffolds has identified a multiple number of cell migration mechanisms and hence led to the establishment of distinct migration modes.[1,144,145] Thereby, these migration modes are not exclusively associated to a specific cell type and hence many cell types can even switch between two or more distinct migration modes due to environmental disturbances.[146,71,76,147,148] In specific detail, metastatic cancer cells that migrate and invade through a 3D extracellular matrix can even transit between distinct migration modes by using specific mechanisms and this transition is usually reversible.[149,146,147] For example, adherent, elongated and hence mesenchymal cancer cells employ MMPs to overcome the limitation of the pore size confinement of 3D collagen matrices and subsequently, migrate through them with their bulky cell nucleus.[150–154] However, when the proteolytic activity of the cells is reduced, they display increased actomyosin contractility and thereby exhibit a round (amoeboid) cell shape and become less adherent to the surrounding environment.[146,155,147] Moreover, this increase in the actomyosin contractility causes a switch from the mesenchymal migration mode to the initiation of a blebbing migration mode in a 3D microenvironment and hence enables the rounded cells to utilize a rapid and adhesion-independent migration mode to invade through intact 3D extracellular matrices.[156,76,148] Taken together, this amoeboid–mesenchymal transition of the cells has been reported for the first time for HT1080 cells that stably express MT1-MMP, termed HT1080/MT1 cells.[146] However, later it has also been observed in a variety of other cell types.[157,148]

4. Role of Rho-GTPases in 2D and 3D Migration and Migration Mode Selection

The regulation of force seems to be crucial for the determination or the switch between migration modes. An example are fibroblasts that are initially spherical in a nonadhesive state (suspended) and spread when they are in contact with a cell substrate to exhibit a flattened cell shape. Hence the adhesion process seems to be analogous to the lamellipodia and filopodia extension at the cell's leading edge and thereby these highly polarized cells move over the substrate. The spreading of the cell is regulated by the polymerization of actin, which is enhanced by the two Rho-family GTPases Rac1 and Cdc42. More precisely the Rac1 and Cdc42 regulate the polymerization start through different signal transduction pathways.[158] Cdc42 causes the activation of the Arp2/3 complex that is essential for the nucleation of new actin filaments in order to build a dendritic actin filament scaffold.[159] In contrast, Rac1 can evoke the uncapping of pre-exisitng actin filaments at their barbed ends in order to enable their further growth.[160] After the cell spreading, specialized intracellular structures, such as actin stress fibers and focal adhesions, can facilitate both the adhesion of the cells to their substrate. In Fig. 12, the adhesion of a fibroblast to a flat laminin-coated substrate is presented. There are two different cell types analyzed, such as fibroblasts in which Arp3 is knocked down and control fibroblasts expressing Arp3. It can be seen that these two cell types exhibit a different morphology, since Arp3 knock-down cells display no lamellipodia, whereas the control cells can still form lamellipodia (Fig. 12). Moreover, the migration and invasion of fibroblasts expressing Arp3 is increased compared to Arp3 knock down cells. These two processes, the spreading and decapping, demand a myosin-based contractile force triggered by RhoA GTPase- and calcium/calmodulin (CaM)-dependent signal transduction processes, that stimulate the phosphorylation of myosin II in its regulatory light chain in nonmuscle cells.[161,162] Subsequently, the cell extension and adhesion processes are closely coordinated. In detail, in the initial phase of cell

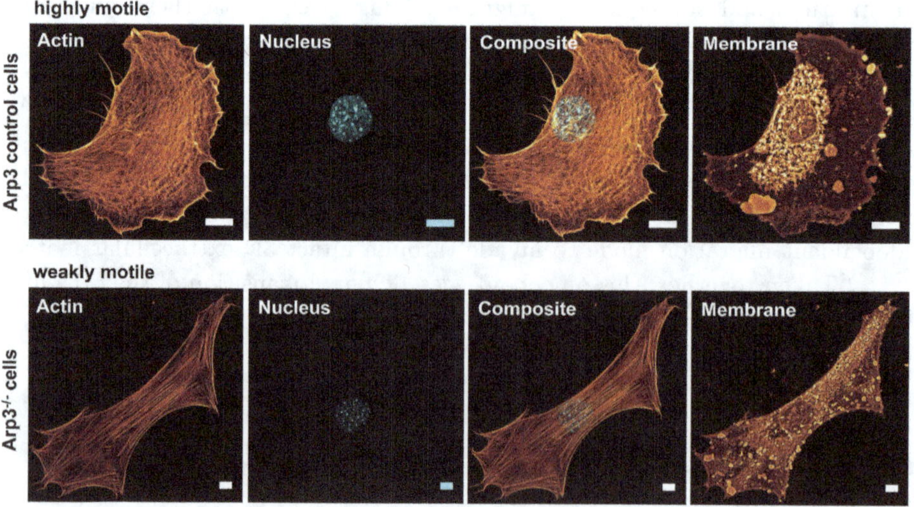

Fig. 12.   Fibroblasts expressing Arp3 display lamellipodia, whereas in Arp3 knock-down fibroblast the formation of lamellipodia is impaired. All scale bars are 10 μm.

spreading, the binding of cell–matrix adhesion receptors, such as integrins, to extracellular matrix ligands abolishes the activity of RhoA through the activation of p190RhoGAP by c-Src and focal adhesion kinase (FAK).[163–165] This observation leads to the formulation of the following hypothesis stating that the deactivation of RhoA impairs the myosin-dependent contractile force generation in order to enable the cell spreading process. Moreover, this hypothesis leads to the prediction that the myosin activity is inversely correlated with the cell spreading rate. Hence this hypothesis has been tested experimentally by determining the myosin activity, such as the phosphorylation level of the regulatory light chain and the spreading rate in chicken embryo fibroblasts.

Subsequently, current migration models propose that the cell extension is driven by the actin filament growth, which is regulated by the incorporation of actin monomers at the barbed actin filament ends. In specific detail, this hypothesis also assumes that the barbed ends of these actin filaments are not impaired by capping proteins binding to these barbed ends. This hypothesis can be proven or disproven by the analysis of the binding capacity of the actin filaments to the pharmacological drug cytochalasin D. More precisely, when the actin filaments are free, they bind cytochalasin D at close to nanomolar concentrations with the known binding constant of free actin filaments.[166] At sub-nanomolar concentrations of cytochalasin D, the extension of the cell edges should be impaired, whereas at higher, micromolar concentrations the cell shape and cell deformability are affected, since cytochalasin D competes with the barbed end capping proteins for the binding of the actin filaments.[167] Indeed, these suggestions have been confirmed experimentally.[5]

Another hypothesis states that the myosin activation stimulates the interaction of myosin with actin filaments, which are anchored to the cell membrane and hence elevates the cell stiffness.[168] As a consequence, the resistance of the cell to extend itself by actin polymerization is increased. Moreover, this hypothesis predicts that an increase in myosin activation causes elevated cell stiffness levels. Indeed, this hypothesis has been confirmed by analysis of the cell stiffness with a cell indentation assay. In summary, the mechanotransduction signaling is involved in cell migration and invasion and can even lead to a change in the migration mode, since cells respond to environmental cues.[5]

5. Concluding Remarks

In conclusion, the choice of the individual migration mode depends mainly on the mechanical phenotype of the cells, such as their cell mechanics. The selection of a distinct migration mode depends rather less on the cell shape and morphology, which has been used frequently in previous studies for determining the individual migration mode. In line with this, the switch of the cells from a rather epithelial morphology to a more fibroblast-like mesenchymal morphology, which enabled the cells to migrate in 3D matrix confinements, has been based on the specific alterations in the cell shape and morphology rather than on cell mechanical features. Many mechanical properties can still only be determined when the cells are outside a 3D microenvironment and therefore much research effort is needed to develop new or improved biophysical techniques for determining the mechanical properties of cells in a 3D microenvironment. Instead of measuring the static (mechanical) properties of cells, future approaches will have to address the dynamic remodeling of these mechanical properties using new biophysical techniques.

Acknowledgments

I thank Thomas M. L. Mierke for his helpful discussions and proof reading.

References

1. P. Friedl and K. Wolf, Plasticity of cell migration: A multiscale tuning model, *J. Cell Biol.* **188**(1), 11 (2009).
2. J. Zhu and A. Mogilner, Comparison of cell migration mechanical strategies in three-dimensional matrices: A computational study, *Interface Focus* **6**(5), 20160040 (2016).
3. C. T. Mierke, Physical view on migration modes, *Cell Adhesion Migr.* **9**(5), 367 (2015).
4. C. T. Mierke, The matrix environmental and cell mechanical properties regulate cell migration and contribute to the invasive phenotype of cancer cells, *Rep. Prog. Phys.* **82**(6), 064602 (2019).
5. V. Ruprecht, P. Monzo, A. Ravasio, Z. Yue, E. Makhija, P. O. Strale, N. Gauthier, G. V. Shivashankar, V. Studer, C. Albiges-Rizo and V. Virgile Viasnoff, How cells respond to environmental cues – insights from bio-functionalized substrates, *J. Cell Sci.* **130**, 51 (2017).
6. R. D. González-Cruz, V. C. Fonseca and E. M. Darling, Cellular mechanical properties reflect the differentiation potential of adipose-derived mesenchymal stem cells, *PNAS* **109**(24), E1523 (2012).
7. R. Chen and D. Dean, Mechanical properties of stem cells from different sources during vascular smooth muscle cell differentiation, *Mol. Cell Biomech.* **14**(3), 153 (2017).
8. D. J. Flanagan, M. C. Hodder and O. J. Sansom, Microenvironmental cues in cancer stemness, *Nat. Cell Biol.* **20**, 1102 (2018).
9. A. R. Choudhury, S. Gupta, P. K. Chaturvedi, N. Kumar and D. Pandey, Mechanobiology of Cancer Stem Cells and Their Niche, *Cancer Microenviron.* **12**, 17 (2019).
10. J. Zhang and L. Li, Stem cell niche: Microenvironment and beyond, *J. Biol. Chem.* **283**(15), 9499 (2008).
11. M. F. Clarke *et al.*, Cancer stem cells-perspectives on current status and future directions: AACR workshop on cancer stem cells, *Cancer Res.* **66**, 9339 (2006).
12. V. Plaks, N. Kong and Z. Werb, The cancer stem cell niche: How essential is the niche in regulating stemness of tumor cells? *Cell Stem Cell* **16**(3), 225 (2015).
13. K. R. Levental *et al.*, Matrix crosslinking forces tumor progression by enhancing integrin signaling, *Cell* **3**, 891 (2009).
14. M. J. Paszek *et al.*, Tensional homeostasis and the malignant phenotype, *Cancer Cell* **8**(3), 241 (2005).
15. N. F. Boyd *et al.*, Mammographic breast density as an intermediate phenotype for breast cancer, *Lancet Oncol.* **6**(10), 798 (2005).
16. C. E. Daniels and J. R. Jett, Does interstitial lung disease predispose to lung cancer? *Curr. Opin. Pulm. Med.* **11**(5), 431 (2005).

17. C. T. Mierke, F. Sauer, S. Grosser, S. Puder, T. Fischer and J. A. Käs, The two faces of enhanced stroma: Stroma acts as a tumor promoter and a steric obstacle, *NMR Biomed.* **31**(10), e3831 (2018).
18. A. J. Engler, S. Sen, H. L. Sweeney and D. E. Discher, Matrix elasticity directs stem cell lineage specification, *Cell* **126**, 677 (2006).
19. C. T. Mierke, T. Fischer, S. Puder, T. Kunschmann, B. Soetje and W. H. Ziegler, Focal adhesion kinase activity is required for actomyosin contractility-based invasion of cells into dense 3D matrices, *Sci. Rep.* **7**, 42780 (2017).
20. T. Kunschmann, S. Puder, T. Fischer, A. Steffen, K. Rottner and C. T. Mierke, The small GTPase Rac1 increases cell surface stiffness and enhances 3D migration into extracellular matrices, *Sci. Rep.* **9**(1), 7675 (2019).
21. J. Guck, S. Schinkinger, B. Lincoln, F. Wottawah, S. Ebert, M. Romeyke, D. Lenz, H. M. Erickson, R. Ananthakrishnan, D. Mitchell, J. Käs, S. Ulvick and C. Bilby, Optical deformability as an inherent cell marker for testing malignant Transformation and metastatic competence, *Biophys. J.* **88**, 3689 (2005).
22. T. Kunschmann, S. Puder, T. Fischer, J. Perez, N. Wilharm and C. T. Mierke, Integrin-linked kinase regulates cellular mechanics facilitating the motility in 3D extracellular matrices, *Biochim. Biophys. Acta-Mol. Cell Res.* **1864**(3), 580 (2017).
23. C. T. Mierke, B. Frey, M. Fellner, M. Herrmann and B. Fabry, Integrin $\alpha5\beta1$ facilitates cancer cell invasion through enhanced contractile forces, *J. Cell Sci.* **124**(3), 369 (2011).
24. S. Puder, T. Fischer and C. T. Mierke, The transmembrane protein fibrocystin/polyductin regulates cell mechanics and cell motility, *Phys. Biol.* **16**, 066006 (2019).
25. M. Lekka, Discrimination between normal and cancerous cells using AFM, *BioNaNoScience* **6**(1) 65 (2016).
26. M. Lekka and J. Pabijan, Measuring elastic properties of single cancer cells by AFM, *Methods Mol. Biol.* **1886**, 315 (2019).
27. E. A. Ekpenyong, G. Whyte, K. Chalut, S. Pagliara, F. Lautenschlä, C. Fiddler, S. Paschke, U. F. Keyser, E. R. Chilvers and J. Guck, Viscoelastic properties of differentiating blood cells are fate- and function-dependent, *PLoS One* **7**, e45237 (2012).
28. T. W. Remmerbach, F. Wottawah, J. Dietrich, B. Lincoln, C. Wittekind and J. Guck, Oral cancer diagnosis by mechanical phenotyping, *Cancer Res.* **69**, 1728 (2009).
29. J. Runge, T. E. Reichert, A. Fritsch, J. Käs, J. Bertolini and T. W. Remmerbach, Evaluation of single-cell biomechanics as potential marker for oral squamous cell carcinomas: A pilot study, *Oral Dis.* **20**, e120 (2014).
30. F. Meinhövel, R. Stange, J. Schnauß, M. Sauer, J. A. Käs and T. W. Remmerbach, Changing cell mechanics — a precondition for malignant transformation of oral squamous carcinoma cells, *Converg. Sci. Phys. Oncol.* **4**, 034001 (2018).
31. C. T. Mierke, The role of the optical stretcher is crucial in the investigation of cell mechanics regulating cell adhesion and motility, *Front. Cell Dev. Biol.* **7**, 184 (2019).
32. T. Wakatsuki, R. B. Wysolmerski and E. L. Elson, Mechanics of cell spreading: Role of myosin II, *J. Cell Sci.* **116**(8), 1617 (2003).
33. T. M. Koch, S. Münster, N. Bonakdar, J. P. Butler and B. Fabry, 3D Traction forces in cancer cell invasion, *PLoS One* **7**(3), e33476 (2012).
34. J. Steinwachs, C. Metzner, K. Skodzek, N. Lang, I. Thievessen, C. Mark, S. Münster, K. E. Aifantis and B. Fabry, Three-dimensional force microscopy of cells in biopolymer networks, *Nat. Methods* **13**(2), 171 (2016).
35. C. T. Mierke, Phagocytized beads reduce the $\alpha5\beta1$ integrin facilitated invasiveness of cancer cells by regulating cellular stiffness, *Cell Biochem. Biophys.* **66**(3), 599 (2013).
36. J. Müller and M. Sixt, Cell Migration: Making the Waves. *Curr. Biol.* **27**, R19 (2017).

37. X. Guo, K. Bonin, K. Scarpinato and M. Guthold, The effect of neighboring cells on the stiffness of cancerous and non-cancerous human mammary epithelial cells, *New J. Phys.* **16**, 105002 (2014).

38. C. T. Mierke, The biomechanical properties of 3d extracellular matrices and embedded cells regulate the invasiveness of cancer cells, *Cell Biochem. Biophys.* **61**(2), 217 (2011).

39. E. L. Baker, J. Lu, D. Yu, R. T. Bonnecaze and M. H. Zaman, Cancer cell stiffness: Integrated roles of three-dimensional matrix stiffness and transforming potential, *Biophys. J.* **99**, 2048 (2010).

40. P. Kanchanawong, G. Shtengel, A. M. Pasapera, E. B. Ramko, M. W. Davidson, H. F. Hess and C. M. Waterman, Nanoscale architecture of integrin-based cell adhesions, *Nature* **468**, 580 (2010).

41. I. Patla, T. Volberg, N. Elad, V. Hirschfeld-Warneken, C. Grashoff, R. Fässler, J. P. Spatz, B. Geiger and O. Medalia, Dissecting the molecular architecture of integrin adhesion sites by cryo-electron tomography, *Nat. Cell Biol.* **12**, 909 (2010).

42. S. R. Coyer, A. Singh, D. W. Dumbauld, D. A. Calderwood, S. W. Craig, E. Delamarche and A. J. Garcia, Nanopatterning reveals an ECM area threshold for focal adhesion assembly and force transmission that is regulated by integrin activation and cytoskeleton tension, *J. Cell Sci.* **125**, 5110 (2012).

43. E. A. Cavalcanti-Adam and J. P. Spatz, Receptor clustering control and associated force sensing by surface patterning: When force matters, *Nanomedicine* **10**, 681 (2015).

44. M. Arnold, E. A. Cavalcanti-Adam, R. Glass, J. Blümmel, W. Eck, M. Kantlehner, H. Kessler and J. P. Spatz, Activation of integrin function by nanopatterned adhesive interfaces, *Chemphyschem* **5**, 383 (2004).

45. S. Rahmouni, A. Lindner, F. Rechenmacher, S. Neubauer, T. R. A. Sobahi, H. Kessler, E. A. Cavalcanti-Adam and J. P. Spatz, Hydrogel micropillars with integrin selective peptidomimetic functionalized nanopatterned tops: A new tool for the measurement of cell traction forces transmitted through alphavbeta3- or alpha5beta1-integrins, *Adv. Mater.* **25**, 5869 (2013).

46. J. Fu, Y. Wang, M. Yang, R. Desai, X. Yu, Z. Liu and C. Chen, Mechanical regulation of cell function with geometrically modulated elastomeric substrates, *Nat. Methods* **7**(9), 733 (2010).

47. J. Tan, J. Tien, D. Pirone, D. Gray and K. Bhadriraju and C. Chen, Cells lying on a bed of microneedles: An approach to isolate mechanical force, *Proc. Natl. Acad. Sci. USA* **100**(4), 1484 (2003).

48. P. W. Oakes, Y. Beckham, J. Stricker and M. L. Gardel, Tension is required but not sufficient for focal adhesion maturation without a stress fiber template, *J. Cell Biol.* **196**, 363 (2012).

49. S. V. Plotnikov, B. Sabass, U. S. Schwarz and C. M. Waterman, Highresolution traction force microscopy, *Methods Cell Biol.* **123**, 367 (2014).

50. J. R. Soine, C. A. Brand, J. Stricker, P. W. Oakes, M. L. Gardel and U. S. Schwarz, Model-based traction force microscopy reveals differential tension in cellular actin bundles, *PLoS Comput. Biol.* **11**, e1004076 (2015).

51. C. J. Bettinger, R. Langer and J. T. Borenstein, Engineering substrate topography at the micro- and nanoscale to control cell function, *Angew. Chem. Int. Ed. Engl.* **48**, 5406 (2009).

52. C.-H. Yu, J. B. K. Law, M. Suryana, H. Y. Low and M. P. Sheetz, Early integrin binding to Arg-Gly-Asp peptide activates actin polymerization and contractile movement that stimulates outward translocation, *Proc. Natl. Acad. Sci. USA* **108**, 20585 (2011).

53. I. Dupin, M. Dahan and V. Studer, Investigating axonal guidance with microdevice-based approaches, *J. Neurosci.* **33**, 17647 (2013).

54. A. del Rio, R. Perez-Jimenez, R. Liu, P. Roca-Cusachs, J. M. Fernandez and M. P. Sheetz, Stretching single talin rod molecules activates vinculin binding, *Science* **23**, 638 (2009).

55. X. Hu, C. Jing, X. Xu, N. Nakazawa, V. W. Cornish, F. M. Margadant and M. P. Sheetz, Cooperative vinculin binding to talin mapped by time resolved super resolution microscopy, *Nano Lett.* **16**, 4062 (2016).

56. A. N. Gasparski, S. Ozarkar and K. A. Beningo, Transient mechanical strain promotes the maturation of invadopodia and enhances cancer cell invasion *in vitro*, *J. Cell Sci.* **130**, 1965 (2017).

57. T. Lämmermann and M. Sixt, Mechanical modes of 'amoeboid' cell migration, *Curr. Opin. Cell Biol.* **21**, 636 (2009).

58. E. K. Paluch and E. Raz, The role and regulation of blebs in cell migration, *Curr. Opin. Cell Biol.* **25**(5), 582 (2013).

59. M. Bergert, S. D. Chandradoss, R. A. Desai and E. Paluch, Cell mechanics control rapid transitions between blebs and lamellipodia during migration, *Proc. Natl. Acad. Sci. USA* **109**, 14434 (2012).

60. W. Strychalski and R. D. Guy, A computational model of bleb formation, *Math. Med. Biol.* **30**, 115 (2013).

61. M. Goudarzi *et al.*, Bleb expansion in migrating cells depends on supply of membrane from cell surface invaginations, *Dev. Cell* **43**(5), 577 (2017).

62. Y. Asano, T. Mizuno, T. Kon, A. Nagasaki, K. Sutoh and T. Q. Uyeda, Keratocyte-like locomotion in amiB-null Dictyostelium cells, *Cell Motil. Cytoskeleton* **59**, 17 (2004).

63. K. Keren, Z. Pincus, G. M. Allen, E. L. Barnhart, G. Marriott, A. Mogilner and J. A. Theriot, Mechanism of shape determination in motile cells, *Nature* **453**, 475 (2008).

64. M. J. Brown, J. A. Hallam, E. Colucci-Guyon and S. Shaw, Rigidity of circulating lymphocytes is primarily conferred by vimentin intermediate filaments, *J. Immunol.* **166**, 6640 (2001).

65. G. Laevsky and D. A. Knecht, Cross-linking of actin filaments by myosin II is a major contributor to cortical integrity and cell motility in restrictive environments, *J. Cell Sci.* **116**, 3761 (2003).

66. A. J. Tooley, J. Gilden, J. Jacobelli, P. Beemiller, W. S. Trimble, M. Kinoshita and M. F. Krummel, Amoeboid T lymphocytes require the septin cytoskeleton for cortical integrity and persistent motility, *Nat. Cell Biol.* **11**, 17 (2008).

67. J. Xu, F. Wang, A. Van Keymeulen, M. Rentel and H. R. Bourne, Neutrophil microtubules suppress polarity and enhance directional migration, *Proc. Natl. Acad. Sci. USA* **102**, 6884 (2005).

68. R. J. Hawkins, M. Piel, G. Faure-Andre, A. M. Lennon-Dumenil, J. F. Joanny, J. Prost, R. Voituriez, Pushing off the walls: A mechanism of cell motility in confinement, *Phys. Rev. Lett.* **102**, 058103 (2009).

69. J. Renkawitz and M. Sixt, Mechanisms of force generation and force transmission during interstitial leukocyte migration, *EMBO Rep.* **11**, 745 (2010).

70. W. R. Legant, J. S. Miller, B. L. Blakely, D. M. Cohen, G. M. Genin and C. S. Chen, Measurement of mechanical tractions exerted by cells in three-dimensional matrices, *Nat. Methods* **7**, 969 (2010).

71. R. J. Petrie, N. Gavara, R. S. Chadwick and K. M. Yamada, Nonpolarized signaling reveals two distinct modes of 3D cell migration, *J. Cell Biol.* **197**, 439 (2012).

72. C. T. Mierke, The fundamental role of mechanical properties in the progression of cancer disease and inflammation, *Rep. Prog. Phys.* **77**(7), 076602 (2014).

73. C. D. Paul, P. Mistriotis and K. Konstantopoulos, Cancer cell motility: Lessons from migration in confined spaces, *Nat. Rev. Cancer* **17**(2), 131 (2017).

74. R. J. Petrie, H. Koo and K. M. Yamada, Generation of compartmentalized pressure by a nuclear piston governs cell motility in a 3D matrix, *Science* **345**, 1062 (2014).

75. R. J. Petrie, H. M. Harlin, L. I. T. Korsak and K. M. Yamada, Activating the nuclear piston mechanism of 3D migration in tumor cells, *J. Cell Biol.* **216**(1), 93 (2017).

76. Y.-L. Liu, M. Le Berre, F. Lautenschlaeger, P. Maiuri, A. Callan-Jones, M. Heuze, T. Takaki, R. Voituriez and M. Piel, Confinement and low adhesion induce fast amoeboid migration of slow mesenchymal cells, *Cell* **160**, 659 (2015).

77. W. V. Wang, C. D. Davidson, D. Lin and B. M. Baker, Actomyosin contractility-dependent matrix stretch and recoil induces rapid cell migration, *Nat. Commun.* **10**, 1186 (2019).

78. B. M. Baker, B. Trappmann, W. Y. Wang, M. S. Sakar, I. L. Kim, V. B. Shenoy, J. A. Burdick and C. S. Chen, Cell-mediated fibre recruitment drives extracellular matrix mechanosensing in engineered fibrillar microenvironments, *Nat. Mater.* **14**, 1262 (2015).

79. D. Li, Y. Wang and Y. Xia, Electrospinning of polymeric and ceramic nanofibers as uniaxially aligned arrays, *Nano Lett.* **3**, 1167 (2003).

80. M. Guthold, W. Liu, E. A. Sparks, L. M. Jawerth, L. Peng, M. Falvo, R. Superfine, R. R. Hantgan and S. T. Lord, A comparison of the mechanical and structural properties of fibrin fibers with other protein fibers, *Cell Biochem. Biophys.* **49**, 165 (2007).

81. L. Li, J. Eyckmans and C. S. Chen, Designer biomaterials for mechanobiology, *Nat. Mater.* **16**, 1164 (2017).

82. K. Wolf, S. Alexander, V. Schacht, L. M. Coussens, U. H. Von Andrian, J. van Rheenen, E. Deryugina and P. Friedl, Collagen-based cell migration models *in vitro* and *in vivo*, *Semin. Cell Dev. Biol.* **20**, 931 (2009).

83. K. M. Riching *et al.*, 3D collagen alignment limits protrusions to enhance breast cancer cell persistence, *Biophys. J.* **107**, 2546 (2014).

84. B. Trappmann, B. M. Baker, W. J. Polacheck, C. K. Choi, J. A. Burdick and C. S. Chen, Matrix degradability controls multicellularity of 3D cell migration, *Nat. Commun.* **8**, 371 (2017).

85. J. L. West and J. A. Hubbell, Polymeric biomaterials with degradation sites for proteases involved in cell migration, *Macromolecules* **32**, 241 (1999).

86. M. P. Lutolf, J. L. Lauer-Fields, H. G. Schmoekel, A. T. Metters, F. E. Weber, G. B. Fields and J. A. Hubbell, Synthetic matrix metalloproteinase-sensitive hydrogels for the conduction of tissue regeneration: Engineering cell-invasion characteristics, *Proc. Natl Acad. Sci. USA* **100**, 5413 (2003).

87. K. Wolf and P. Friedl, Extracellular matrix determinants of proteolytic and non-proteolytic cell migration, *Trends Cell Biol.* **21**, 736 (2011).

88. A. Das, A. Barai, M. Monteiro, S. Kumar and S. Sen, Nuclear softening is essential for protease-independent migration, *Matrix Biol.* **82**, 4 (2019).

89. W. Y. Wang, A. T. Pearson, M. L. Kutys, C. K. Choi, M. A. Wozniak, B. M. Baker and C. S. Chen, Extracellular matrix alignment dictates the organization of focal adhesions and directs uniaxial cell migration, *APL Bioeng.* **2**, 046107 (2018).

90. A. Ray, O. Lee, Z. Win, RM. Edwards, P. W. Alford, D.-H. Kim and P. P. Provenzano, Anisotropic forces from spatially constrained focal adhesions mediate contact guidance directed cell migration, *Nat. Commun.* **8**, 14923 (2017).

91. N. R. Lang, K. Skodzek, S. Hurst, A. Mainka, J. Steinwachs, J. Schneider, K. E. Aifantis and B. Fabry, Biphasic response of cell invasion to matrix stiffness in three-dimensional biopolymer networks, *Acta Biomat.* **13**, 61 (2015).

92. A. Pathak, Modeling and predictions of biphasic mechanosensitive cell migration altered by cell-intrinsic properties and matrix confinement, *Phys. Biol.* **15**(6), 065001 (2018).

93. A. Pathak and S. Kumar, Independent regulation of tumor cell migration by matrix stiffness and confinement, *Proc. Natl. Acad. Sci. USA* **109**, 10334 (2012).

94. K. Burridge and M. Chrzanowska-Wodnicka, Focal adhesions, contractility, and signaling, *Annu. Rev. Cell Dev. Biol.* **12**, 463 (1996).

95. R. M. Mege, J. Gavard and M. Lambert, Regulation of cell–cell junctions by the cytoskeleton, *Curr. Opin. Cell Biol.* **18**(5), 541 (2006).

96. C. Bertet, L. Sulak and T. Lecuit, Myosin-dependent junction remodelling controls planar cell intercalation and axis elongation, *Nature* **429**(6992), 667 (2004).

97. M. Rauzi, P. Verant, T. Lecuit and P. F. Lenne, Nature and anisotropy of cortical forces orienting Drosophila tissue morphogenesis, *Nat. Cell Biol.* **10**(12), 1401 (2008).

98. A. Rolo, P. Skoglund and R. Keller, Morphogenetic movements driving neural tube closure in Xenopus require myosin IIB, *Dev. Biol.* **327**(2), 327 (2009).

99. P. Skoglund, A. Rolo, X. Chen, B. M. Gumbiner and R. Keller, Convergence and extension at gastrulation require a myosin IIB-dependent cortical actin network, *Development* **135**(14), 2435 (2008).

100. S. Abraham, M. Yeo, M. Montero-Balaguer, H. Paterson, E. Dejana, C. J. Marshall and G. Mavria, VE-Cadherin-mediated cell–cell interaction suppresses sprouting via signaling to MLC2 phosphorylation, *Curr. Biol.* **19**(8), 668 (2009).

101. J. de Rooij, A. Kerstens, G. Danuser, M. A. Schwartz and C. M. Waterman-Storer, Integrindependent actomyosin contraction regulates epithelial cell scattering, *J. Cell Biol.* **171**(1), 153 (2005).

102. A. I. Ivanov, M. Bachar, B. A. Babbin, R. S. Adelstein, A. Nusrat and C. A. Parkos, A unique role for nonmuscle myosin heavy chain IIA in regulation of epithelial apical junctions, *PLoS One* **2**(7), e658 (2007).

103. A. M. Shewan, M. Maddugoda, A. Kraemer, S. J. Stehbens, S. Verma, E. M. Kovacs and A. S. Yap, Myosin 2 is a key Rho kinase target necessary for the local concentration of E-cadherin at cell–cell contacts, *Mol. Biol. Cell* **16**(10), 4531 (2005).

104. S. Yamada and W. J. Nelson, Localized zones of Rho and Rac activities drive initiation and expansion of epithelial cell–cell adhesion, *J. Cell Biol.* **178**(3), 517 (2007).

105. J. Gavard, M. Lambert, I. Grosheva, V. Marthiens, T. Irinopoulou, J. F. Riou, A. Bershadsky and R. M. Mège, Lamellipodium extension and cadherin adhesion: Two cell responses to cadherin activation relying on distinct signalling pathways, *J. Cell Sci.* **117**(Pt 2), 257 (2004).

106. M. Lambert, O. Thoumine, J. Brevier, D. Choquet, D. Riveline and R. M. Mège, Nucleation and growth of cadherin adhesions, *Exp. Cell Res.* **313**(19), 4025 (2007).

107. A. D. Bershadsky, N. Q. Balaban and B. Geiger, Adhesion-dependent cell mechanosensitivity, *Annu. Rev. Cell Dev. Biol.* **19**, 677 (2003).

108. C. S. Chen, J. Tan and J. Tien, Mechanotransduction at cell–matrix and cell–cell contacts, *Annu. Rev. Biomed. Eng.* **6**, 275 (2004).

109. Z. Liu, J. L. Tan, D. M. Cohen, M. T. Yang, N. J. Sniadecki, S. A. Ruiz, C. N. Nelson and C. S. Chen, Mechanical tugging force regulates the size of cell–cell junctions, *PNAS* **107**(22), 9944 (2010).

110. C. M. Nelson, R. P. Jean, J. L. Tan, W. F. Liu, N. J. Sniadecki, A. A. Spector and C. S. Chen, Emergent patterns of growth controlled by multicellular form and mechanics, *Proc. Natl. Acad. Sci. USA* **102**(33), 11594 (2005).

111. I. C. Scott and D. Y. Stainier, Developmental biology: Twisting the body into shape, *Nature* **425**(6957), 461 (2003).

112. D. E. Ingber and I. Tensegrity, Cell structure and hierarchical systems biology, *J. Cell Sci.* **116**(Pt 7), 1157 (2003).

113. Y. S. Chu, W. A. Thomas, O. Eder, F. Pincet, E. Perez, J. P. Thiery and S. Dufour, Force measurements in E-cadherin-mediated cell doublets reveal rapid adhesion strengthened by actin cytoskeleton remodeling through Rac and Cdc42, *J. Cell Biol.* **167**(6), 1183 (2004).

114. K. S. Ko, P. D. Arora and C. A. McCulloch, Cadherins mediate intercellular mechanical signaling in fibroblasts by activation of stretch-sensitive calcium-permeable channels, *J. Biol. Chem.* **276**(38), 35967 (2001).

115. U. S. Potard, J. P. Butler and N. Wang, Cytoskeletal mechanics in confluent epithelial cells probed through integrins and E-cadherins, *Am. J. Physiol.* **272**(5 Pt 1), C1654 (1997).

116. S. V. Plotnikov and C. M. Waterman, Guiding cell migration by tugging, *Curr. Opin. Cell Biol.* **25**(5) (2013), doi: 10.1016/j.ceb.2013.06.003.

117. W. Shih and S. Yamada, Myosin IIA dependent retrograde flow drives 3D cell migration, *Biophys. J.* **98**, L29 (2010).

118. Q. Chi, T. Yin, H. Gregersen, X. Deng, Y. Fan, J. Zhao, D. Liao and G. Wang, Rear actomyosin contractility-driven directional cell migration in three-dimensional matrices: A mechano-chemical coupling mechanism, *J. R. Soc. Interface* **11**, 20131072 (2014).

119. T. Vallenius, Actin stress fibre subtypes in mesenchymal-migrating cells, *Open Biol.* **3**(6), 130001 (2013).

120. M. K. Connacher, J. W. Tay and N. G. Ahn, Rear-polarized Wnt5a-receptor-actin-myosin- polarity (WRAMP) structures promote the speed and persistence of directional cell migration, *Mol. Biol. Cell* **28**(14), 1924 (2017).

121. A. J. Ridley, M. A. Schwartz, K. Burridge, R. A. Firtel, M. H. Ginsberg, G. Borisy, J. T. Parsons and A. R. Horwitz, Cell migration: Integrating signals from front to back, *Science* **302**, 1704 (2003).

122. R. J. Petrie, A. D. Doyle and K. M. Yamada, Random versus directionally persistent cell migration, *Nat. Rev. Mol. Cell Biol.* **10**, 538 (2009).

123. J. T. Parsons, A. R. Horwitz and M. A. Schwartz, Cell adhesion: Integrating cytoskeletal dynamics and cellular tension, *Nat. Rev. Mol. Cell Biol.* **11**, 633 (2010).

124. M. Krause and A. Gautreau, Steering cell migration: Lamellipodium dynamics and the regulation of directional persistence, *Nat. Rev. Mol. Cell Biol.* **15**, 577 (2014).

125. L. P. Cramer, Mechanism of cell rear retraction in migrating cells, *Curr. Opin. Cell Biol.* **25**, 591 (2013).

126. B. Bowerman, Cell signaling. Wnt moves beyond the canon, *Science* **320**, 327 (2008).

127. E. S. Witze, E. S. Litman, G. M. Argast, R. T. Moon and N. G. Ahn, Wnt5a control of cell polarity and directional movement by polarized redistribution of adhesion receptors, *Science* **320**, 365 (2008).

128. E. S. Witze, M. K. Connacher, S. Houel, M. P. Schwartz, M. K. Morphew, L. Reid, D. B. Sacks, K. S. Anseth and N. G. Ahn, Wnt5a directs polarized calcium gradients by recruiting cortical endoplasmic reticulum to the cell trailing edge, *Dev. Cell* **26**, 645 (2013).

129. K. M. Mladinich and A. Huttenlocher, WRAMPing up calcium in migrating cells by localized ER transport, *Dev. Cell* **26**, 560 (2013).

130. M. Nishita, M. Enomoto, K. Yamagata and Y. Minami, Cell/tissue-tropic functions of Wnt5a signaling in normal and cancer cells, *Trends Cell Biol.* **20**, 346 (2010).

131. A. Kikuchi, H. Yamamoto, A. Sato and S. Matsumoto, Wnt5a: Its signalling, functions and implication in diseases, *Acta Physiol. (Oxford)* **204**, 17 (2012).

132. E. Gomez-Orte, B. Saenz-Narciso, S. Moreno and J. Cabello, Multiple functions of the noncanonical Wnt pathway, *Trends Genet.* **29**, 545 (2013).

133. B. Gao, Wnt regulation of planar cell polarity (PCP), *Curr. Topics Dev. Biol.* **101**, 263 (2012).

134. D. C. Slusarski, J. Yang-Snyder, W. B. Busa and R. T. Moon, Modulation of embryonic intracellular Ca2+ signaling by Wnt-5A, *Dev. Biol.* **182**, 114 (1997).

135. A. D. Kohn and R. T. Moon, Wnt and calcium signaling: Beta-catenin-independent pathways, *Cell Calcium* **38**, 439 (2005).

136. M. Bittner *et al.*, Molecular classification of cutaneous malignant melanoma by gene expression profiling, *Nature* **406**, 536 (2000).

137. A. T. Weeraratna, Y. Jiang, G. Hostetter, K. Rosenblatt, P. Duray, M. Bittner and J. M. Trent, Wnt5a signaling directly affects cell motility and invasion of metastatic melanoma, *Cancer Cell* **1**, 279 (2002).

138. M. Kurayoshi, N. Oue, H. Yamamoto, M. Kishida, A. Inoue, T. Asahara, W. Yasui and A. Kikuchi, Expression of Wnt-5a is correlated with aggressiveness of gastric cancer by stimulating cell migration and invasion, *Cancer Res.* **66**, 10439 (2006).

139. S. Ripka, A. Konig, M. Buchholz, M. Wagner, B. Sipos, G. Kloppel, J. Downward, T. Gress and P. Michl, WNT5A–target of CUTL1 and potent modulator of tumor cell migration and invasion in pancreatic cancer, *Carcinogenesis* **28**, 1178 (2007).

140. M. P. O'Connell *et al.*, The orphan tyrosine kinase receptor, ROR2, mediates Wnt5A signaling in metastatic melanoma, *Oncogene* **29**, 34 (2010).

141. C. Peng, X. Zhang, H. Yu, D. Wu and J. Zheng, Wnt5a as a predictor in poor clinical outcome of patients and a mediator in chemoresistance of ovarian cancer, *Int. J. Gynecol. Cancer* **21**, 280 (2011).

142. C. T. Mierke, Physical view on the interactions between cancer cells and the endothelial cell lining during cancer cell transmigration and invasion, *Biophys. Rev. Lett.* **10**(1), 1 (2015).

143. F. Sabeh *et al.*, Tumor cell traffic through the extracellular matrix is controlled by the membrane-anchored collagenase MT1-MMP, *J. Cell Biol.* **167**, 769 (2004).

144. R. J. Petrie and K. M. Yamada, At the leading edge of three-dimensional cell migration, *J. Cell Sci.* **125**, 5917 (2012).

145. G. Charras and E. Sahai, Physical influences of the extracellular environment on cell migration, *Nat. Rev. Mol. Cell Biol.* **15**, 813 (2014).

146. K. Wolf, I. Mazo, H. Leung, K. Engelke, U. H. von Andrian, E. I. Deryugina, A. Y. Strongin, E. B. Brocker and P. Friedl, Compensation mechanism in tumor cell migration: Mesenchymal-amoeboid transition after blocking of pericellular proteolysis, *J. Cell Biol.* **160**, 267 (2003).

147. C. D. Madsen, S. Hooper, M. Tozluoglu, A. Bruckbauer, G. Fletcher, J. T. Erler, P. A. Bates, B. Thompson and E. Sahai, STRIPAK components determine mode of cancer cell migration and metastasis, *Nat. Cell Biol.* **17**, 68 (2015).

148. V. Ruprecht *et al.*, Cortical contractility triggers a stochastic switch to fast amoeboid cell motility, *Cell* **160**, 673 (2015).

149. E. Sahai and C. J. Marshall, Differing modes of tumour cell invasion have distinct requirements for Rho/ROCK signalling and extracellular proteolysis, *Nat. Cell Biol.* **5**, 711 (2003).

150. X. Yu *et al.*, N-WASP coordinates the delivery and F-actin-mediated capture of MT1-MMP at invasive pseudopods, *J. Cell Biol.* **199**, 527 (2012).

151. K. Wolf, M. Te Lindert, M. Krause, S. Alexander, J. Te Riet, A. L. Willis, R. M. Hoffman, C. G. Figdor, S. J. Weiss and P. Friedl, Physical limits of cell migration: Control by ECM space and nuclear deformation and tuning by proteolysis and traction force, *J. Cell Biol.* **201**, 1069 (2013).

152. P. M. Davidson, C. Denais, M. C. Bakshi and J. Lammerding, Nuclear deformability constitutes a rate-limiting step during cell migration in 3-D environments, *Cell. Mol. Bioeng.* **7**, 293 (2014).

153. T. Harada *et al.*, Nuclear lamin stiffness is a barrier to 3D migration, but softness can limit survival, *J. Cell Biol.* **204**, 669 (2014).
154. C. M. Denais, R. M. Gilbert, P. Isermann, A. L. McGregor, M. te Lindert, B. Weigelin, P. M. Davidson, P. Friedl, K. Wolf and J. Lammerding, Nuclear envelope rupture and repair during cancer cell migration, *Science* **352**, 353 (2016).
155. M. Bergert, A. Erzberger, R. A. Desai, I. M. Aspalter, A. C. Oates, G. Charras, G. Salbreux and E. K. Paluch, Force transmission during adhesion-independent migration, *Nat. Cell Biol.* **17**, 524 (2015).
156. T. Lämmermann *et al.*, Rapid leukocyte migration by integrin-independent flowing and squeezing, *Nature* **453**, 51 (2008).
157. V. Sanz-Moreno, G. Gadea, J. Ahn, H. Paterson, P. Marra, S. Pinner, E. Sahai and C. J. Marshall, Rac activation and inactivation control plasticity of tumor cell movement, *Cell* **135**, 510 (2008).
158. R. A. Worthylake, S. Lemoine, J. M. Watson and K. Keith Burridge, RhoA is required for monocyte tail retraction during transendothelial migration, *J. Cell Biol.* **154**(1), 147 (2001).
159. G. G. Borisy and T. M. Svitkina, Actin machinery: Pushing the envelope, *Curr. Opin. Cell Biol.* **12**, 104 (2000).
160. J. H. Hartwig, G. M. Bokoch, C. L. Carpenter, P. A. Janmey, L. A. Taylor, A. Toker and T. P. Stossel, Thrombin receptor ligation and activated Rac uncap actin filament barbed ends through phosphoinositide synthesis in permeabilized human platelets, *Cell* **82**, 643 (1995).
161. M. Amano, M. Ito, K. Kimura, Y. Fukata, K. Chihara, T. Nakano, Y. Matsuura and K. Kaibuchi, Phosphorylation and activation of myosin by Rho-associated kinase (Rho-kinase), *J. Biol. Chem.* **271**, 20246 (1996).
162. K. Kimura *et al.*, Regulation of myosin phosphatase by Rho and Rho-associated kinase (Rho-kinase), *Science* **273**, 245 (1996).
163. W. T. Arthur and K. Burridge, RhoA inactivation by p190RhoGAP regulates cell spreading and migration by promoting membrane protrusion and polarity, *Mol. Biol. Cell.* **12**, 2711 (2001).
164. X. D. Ren, W. B. Kiosses and M. A. Schwartz, Regulation of the small GTP-binding protein Rho by cell adhesion and the cytoskeleton, *EMBO J.* **18**, 578 (1999).
165. X. D. Ren, W. B. Kiosses, D. J. Sieg, C. A. Otey, D. D. Schlaepfer and M. A. Schwartz, Focal adhesion kinase suppresses Rho activity to promote focal adhesion turnover, *J. Cell Sci.* **113**, 3673 (2000).
166. J. A. Cooper, Effects of cytochalasin and phalloidin on actin, *J. Cell Biol.* **105**, 1473 (1987).
167. T. Wakatsuki, B. Schwab, N. C. Thompson and E. L. Elson, Effects of cytochalasin D and latrunculin B on mechanical properties of cells, *J. Cell Sci.* **114**, 1025 (2001).
168. Y. Miyake, N. Inoue, K. Nishimura, N. Kinoshita, H. Hosoya and S. Yonemura, Actomyosin tension is required for correct recruitment of adherens junction components and zonula occludens formation, *Exp. Cell Res.* **312**(9), 1637 (2006).
169. T. Fischer, N. Wilharm, A. Hayn and C. Tanja, Mierke Matrix and cellular mechanical properties are the driving factors for facilitating human cancer cell motility into 3D engineered matrices, *Converg. Sci. Phys. Oncol.* **3**, 044003, https://doi.org/10.1088/2057-1739/aa8bbb.

Chapter 6

Physics of the Extracellular Matrix and Biology of Tumors — A Close Relationship

Hanna Engelke

Department of Chemistry and Center for NanoScience
Ludwig-Maximilians-Universität München
Butenandtstraße 5-13, Munich 81377, Germany
hanna.engelke@cup.uni-muenchen.de

The physical properties of the extracellular matrix strongly influence tumor progression and malignancy. This impact can be direct by physically impeding or promoting tumor progression. Furthermore, the ECM properties can also indirectly modulate tumors via interaction with cellular signaling pathways, such as integrin or YAP/TAZ signaling. Conversely, tumors remodel and rebuild the extracellular matrix. This leads to a strong mutual influence of tumor and extracellular matrix including feedback loops and cascades of mutual interaction. Combinations of therapies that treat tumor signaling directly and indirectly via modulation of the tumor's interaction with the extracellular matrix may thus leverage the success of cancer therapy.

Keywords: Extracellular matrix; tumor; invasion; mechanosignalling; matrix remodelling; YAP; integrin.

1. Introduction

Tumor cells are characterized by altered intracellular and extracellular signaling. Beyond that, they also show a perturbed extracellular matrix (ECM). The ECM plays an important role in cancer. It influences a range of processes from tumor progression, onset of invasion and migration of invading cells all the way via intravasation and extravasation to the extent of metastatic outgrowth.[1] This influence is mediated via chemical and physical interactions. Here, I will focus on physical interactions. They are mainly of mechanical nature. Some of them influence tumor processes directly, e.g., by sterical hindrance; others modify intracellular signaling and thereby alter the tumor's behavior rather indirectly. In this chapter, I will describe this interaction of tumor and physical properties of the ECM. Although the general principles hold true for the tumor microenvironment as well as the vasculature and the metastasis environment, I will focus the examples on tumor progression, dissemination, and invading cancer cells.

2. Physical Properties of the ECM that Influence Tumors

The ECM consists of a meshwork of various fibrous proteins and glycoproteins.[2] The way in which these proteins are assembled determines the physical properties of the ECM. The proteins may form a fairly homogenous network of fibers of certain fiber size, density, and mesh size. They may also arrange in strongly aligned fiber bundles or other less homogenous formations.[3] These different topologies combined with the chemical composition lead to distinct rheological properties. The interplay of the different chemical and physical parameters of the fibers and proteins and the resulting physical properties of the matrix are highly complex. Mathematical models combined with experimental results elucidate these dependencies and suggest that the fibrous nature accounts for much of the properties. They further reveal that stress governs the nonlinear rheology and the so-called strain-stiffening, which is an important effect for cell-ECM interactions.[4,5]

Both topology and rheology of the ECM have been shown to influence tumors.[1] For example, the mesh size of the ECM network influences migration of invasive cancer cells as has been shown in *in-vitro* assays,[6,7] *in silico*,[8] but also *in vivo*.[9] The fiber density often correlates with the mesh size and high-fiber densities leading to small mesh sizes can impede cell migration and tumor progression. Furthermore, the dimensionality, i.e., a three-dimensional matrix or a quasi-two-dimensional matrix such as present on any epithelium is a key modulator of migration of invasive cancer cells.[10] The fiber size of a fiber of proteins that offer adhesion sites for cells controls the amount of adhesion sites presented to the cell. This can modulate migratory phenotype as well as migration parameters.[11] The amount of adhesion sites further determines integrin signaling and with that it is an important regulator of tumor growth.

The presence of fibrils and fiber bundles in the ECM has been shown to promote malignancy of tumors.[12] It further directs cancer cell migration and is involved in the onset of invasion.[9,13] The latter is also often associated with inhomogeneities in the ECM.[13,14]

In terms of rheology, the stiffness has a strong impact on tumors. Very soft fibers do not allow for force transmission and thus impede migration, whereas stiffer fibers facilitate migration.[15] The global stiffness of the bulk ECM material, which is probed by the entire cell or cell collective (on the order of $1\,\mathrm{mm}^2$ and larger, as opposed to the local fiber stiffness, which is probed by cell adhesions and on the order of $1\,\mu\mathrm{m}^2$)[16] modulates tumor progression. Tumors on stiffer substrates progress *in vitro* and *in vivo* much faster towards malignancy than controls on soft substrates.[17,18] Also, the nonlinearity of the stress–strain response and the induced strain-stiffening has an important role in tumor progression — it facilitates strong remodeling of the ECM around a tumor, which in turn promotes invasion.[3] An overview of all these physical properties of the ECM, which influence tumor biology is given in Fig. 1.

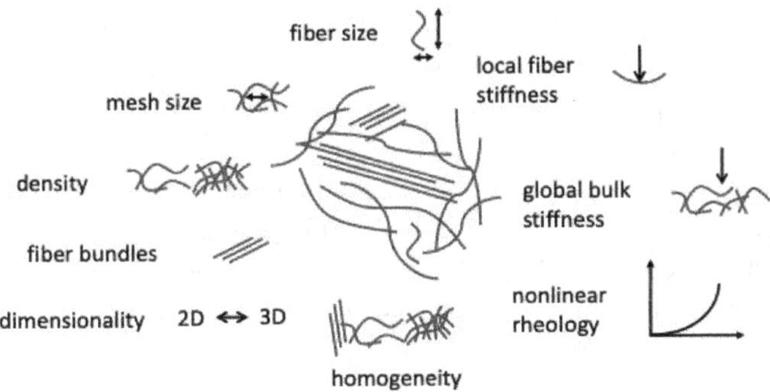

Fig. 1.   Overview of physical properties of the ECM, which influence tumor biology.

2.1. *Direct physical influence*

Some of the above-mentioned properties of the ECM exhibit a direct physical impact on the tumor and its progression to malignancy. One of the direct influences is sterical hindrance due to the geometry of the matrix. This occurs, for example, when single migrating cells encounter a stiff ECM with a mesh size smaller than their nucleus.[6] Due to the limited deformability of the nucleus, such an ECM may block migration in the absence of ECM modulators. Similarly, a dense ECM with small mesh size cannot only impair single cell migration, but also impede tumor growth of the tumor mass in its entirety.

Another direct physical influence is that of a symmetry break. This may be a fiber or any other inhomogeneity in an otherwise homogenous ECM. This is particularly relevant for the onset of invasion of a tumor, which is usually fairly homogenous in shape. Invasion is often observed along such inhomogeneities of the matrix,[13,14,19] which induce the necessary symmetry break to transform the symmetrical shape of the tumor into the branched shape of an invasive tumor. A demonstration of the impact has been shown in a model of mammary acini that were placed from 3D matrigel onto 2D collagen, where they disassembled.[19] They did so even more when they were connected via collagen fibrils that allowed for the necessary symmetry break in topology and of course force transmission. The same study shows that such fibrils may also serve for contact guidance of migrating cells, which travel along the fibers. Such contact guidance has also been observed in other studies *in vitro* and *in vivo*.[9,13,20,21] Theoretical modeling of this process based on minimalistic assumptions suggests that such contact guidance by oriented fibers can strongly promote invasion.[21]

Another important aspect of the physics of the ECM and its direct influence on tumor progression is the mechanics of the ECM. Only a surrounding that provides sufficient force transmission will enable tumor invasion and migration. An extremely soft environment, which does not allow for application of sufficient traction force, impedes any movement—simply for mechanical reasons. Accordingly, migration of invasive cancer cells is blocked in very soft matrices ($\leq 10\,\mathrm{kPa}$ as measured by plate rheometer and atomic force microscopy measurements)[15] and tumor invasion is impaired. Beyond that the nonlinear stress–strain relation of some ECM environments allows for long-range force propagation and facilitates the formation of collagen fibrils as shown by a combination of experiment and theory.[22] It amplifies mechanical interactions of cells with the ECM and thus impacts onset of invasion and migration.

Somewhat connected to force transmission and propagation is the amount of contact sites, which is determined by the size of the protein fibers. The number of contact sites modulates adhesion and with that it controls the ability of cells to apply forces to the ECM. It is thus not strictly physical, but a mixture of chemical and physical parameter. Depending on the migration mode of cells, a strong adhesion may favor motility, since it allows for good force transmission. However, it may lead to trapping of the cells. Low adhesion migratory phenotypes, e.g., amoeboid migration, can also lead to a very efficient migration, which can be explained by adhesion-independent forces that expand rather than contract the substrate in the direction of motion.[11,23] Furthermore, obstructing adhesion of tumors via blocking of the integrin binding sites, impeded tumor growth.[24] In this case, the effect is partially physical, but is also mediated via intracellular integrin signaling inside the tumor. This and similar effects, in which the physics of the ECM modulate cellular signaling, will be described in the next section.

The number of contact sites is also crucial for a phenomenon called haptotaxis, in which cells migrate along a gradient of adhesion sites.[25] A similar phenomenon — durotaxis - describes migration along a gradient of stiffness.[26] Durotaxis is even more efficient and sensitive when performed in cell collectives, since they can sense a gradient over the entire collective as has been shown in a combination of experiment and theoretical modeling.[27] Both haptotaxis and durotaxis are directly mediated via the physics, since better adhesion as well as greater stiffness leads to stronger forces and thus a directional preference. However, both also trigger intracellular signals that amplify the phenomenon.

In many cases of direct physical influences of the ECM on tumor progression, a combination of several physical parameters impacts the tumor. An example is the combination of contact sites and fiber stiffness i.e., ability of force transmission, which controls migration. Only the correct amount of adhesion sites for the given migration phenotype combined with optimal force transmission allows for efficient migration.[28,29] This interplay of stiffness and adhesion can be described by theoretical considerations as a result of force balance.[28,30] The combination of adhesion and rigidity of the ECM also control dissemination of tumors as found in collagen gels[31] and characterized in synthetic polymer hydrogels.[32] Here, a combination of medium range adhesion and medium-sized rigidity promotes tumor dissemination. Also *in silico* studies based on cell–cell, cell-matrix adhesion, and ECM components, predict moderate adhesion and rigidity to promote invasion from tumors.[33] While ECM stiffness and number of adhesion sites can be controlled independently *in silico* and in synthetic hydrogels as used here to dissect their influences on tumors, they are often linked via the fiber size (thick fibers are usually stiffer and at the same time offer more adhesion sites) and density in physiological ECM and protein hydrogels. Thus, the effects of stiffness and adhesion sites are often combined *in vivo*.

3. Indirect Influence Via Cell Signaling

The example on inhibition of tumor growth via integrin blocking that obstructs adhesion already indicated that the physical properties also influence cell signaling beyond their direct physical impact. In this case, blocking integrins and thus adhesion to the matrix, triggered intracellular integrin signaling, which led to a change in phenotype and stopped growth.[24] Integrin signaling also mediates a transformation to a malignant phenotype triggered by an enhanced stiffness of the surrounding matrix. Stiff tissues activate integrin signaling, which in turn activates ERK and Rho signaling that lead to a malignant phenotype.[18] Also, fibronectin as an adhesion site, which is also a crosslinker for collagen enhancing its stiffness, is abundant in tumor tissue and activates a fibronectin-specific integrin signaling leading to invasion.[12]

Next to integrin-signalling, the YAP/TAZ signaling network is an important translator of physical properties of the ECM into cell signaling. YAP/TAZ is nuclear in stiff environments and cytoplasmic in soft environments.[34] In the nucleus, it activates proliferation and mechanical activity, two processes that can drive malignancy. Accordingly, elevated nuclear YAP and in some cases TAZ are correlated with poor prognosis for many kinds of tumors.[35] Moreover, YAP/TAZ knockout or overexpression can prevent or promote tumor growth, respectively. YAP has also been associated with invasion and is nuclear in cells migrating on stiff collagen fibers.[19] MRTF is also a mechanotransduction protein, which translates mechanical signals from the ECM via actin signaling to altered gene expression. MRTF has been associated with cancer, as well. For example, in breast cancer high expression of MRTF was associated with lower survival and *in vitro* studies confirmed this result.[36] Another report shows a similar link between lung metastases stemming from melanoma and MRTF expression.[37]

Another less known mechanosensitive signaling pathway is that of TWIST. Enhanced matrix stiffness drives nuclear translocation of TWIST by releasing it from its cytoplasmic binding partner G3BP2, which otherwise blocks nuclear import of TWIST. Once it is in the nucleus, TWIST triggers epithelial to mesenchymal transition and tumor metastasis.[38] Besides these pathways, a wide range of other signaling pathways can be regulated by the physical properties of the ECM. For example, external mechanical stress exerted on the tumor has been shown to modulate Rho/Rack signaling leading to a malignant phenotype.[39] In mice, external stress has been reported to activate myc and beta-catenin signaling also driving malignancy.[40] Furthermore, external mechanical stress on a tumor can transform the phenotype of cells to a leader cell phenotype and thus induce invasion.[41] Due to the plasticity of tumor cells, such an ECM-induced phenotype transformation can happen based on many other physical stimuli of the ECM. Mesh size, stiffness, and amount of adhesion sites e.g., can change the migratory phenotype of cancer cells.[15,29]

Moreover, the ECM also influences the assembly of invadopodia that are involved in cancer cell invasion and dissemination.[42]

4. Impact of Cells on Matrix

While the matrix has a strong influence on cell behavior as described above, cells in turn constantly remodel and rebuild the ECM turning the matrix into a highly dynamic environment. Specifically, in tumor environments, cells strongly modify the ECM. One such modification is a remodeling of the matrix. It is based on traction forces of cells acting on the matrix. These are amplified by the nonlinear stress–strain behavior (the exact relationship depends on the composition of the matrix; in collagen stress $\sigma \sim \exp(\gamma/\gamma_0)$, with strain γ[43]) of the matrix and the so-called strain-stiffening resulting sometimes in strong fiber bundles that arise. Several computational models exist, which explain and predict mechanical remodeling of the ECM based on the nonlinear properties of these fibrous networks.[44–47] Such mechanical remodeling of the surrounding matrix has been observed *in vitro* resulting from contractions of mammary acini on collagen matrices[19] and of tumors in collagen.[13,48] In pancreatic adenocarcinoma, poor prognosis correlates with stiff matrix fibers around the tumor. This is caused by cellular STAT3 signaling, which increases matricellular fibrosis and tissue tension leading to tumor promotion.[49] Furthermore, tumor spheroids with hyperactivated ERK1/2 and phosphorylation of the actin nucleator WAVE2, exhibit striking collective motility resulting in persistent rotations of the spheroids in the ECM.[14] This rotation remodels the ECM and ultimately leads to invasion. Not only tumors consisting of many cells can remodel the matrix — also single cancer cells have been shown to induce strain stiffening on collagen.[50]

Beyond matrix remodeling by the tumor itself and its disseminated cells, also fibroblasts in the ECM can support matrix modulation. Fibroblasts show even stronger strain stiffening on collagen than cancer cells.[50] They are also able to mechanically deform the ECM in a way that induces collective invasion from squamous carcinoma cell collectives.[51] Also, stromal fibroblasts of breast carcinoma can modulate the fiber organization of the ECM. Specifically, when expressing the cell surface proteoglycan syndecan-1, they could induce formation of fibers, which can promote carcinoma invasion.[52]

Another type of matrix manipulation, which tumor cells employ, is that of fiber cleavage by expression of digesting enzymes, such as collagenases. Matrix cleavage allows cancer cells to migrate in matrices with mesh sizes that would otherwise block their motion.[53,54] Matrix cleavage by a specific metalloproteinase has also been shown to enable invasion of breast carcinoma cells and deleting the proteinase inhibited invasion and metastasis.[55] Next to destruction via cleavage, tumors also produce new matrix proteins to enhance matrix stiffness and to increase matrix crosslinking via fibronectin. Enhanced collagen and fibronectin deposition by tumors have been found to drive the malignant phenotype.[3,12]

5. Mutual Influence of Cells and ECM

The examples above show that physical properties of the ECM, tumor cell signaling and tumor progression mutually influence each other (see also Fig. 2). They are so strongly interconnected that a change in cell behavior or in the ECM triggers many downstream cascading effects that are bouncing between ECM and cell behavior. It also includes feedback loops, which leverage some effects or entire cascades, depending on their position in the cascade. Stiff matrices activate cell signaling that enhances the mechanical activity of cells.[35] Thus, they increase their remodeling of the matrix, which in turn stiffens the matrix. This feedback loop may trigger a set of cascading effects that are strongly interconnected with stiffened matrices. Examples are activation of various signaling pathways, such as integrin signaling, twist and many others, which are associated with onset of invasion and enhanced malignancy.[1] Furthermore, a stiffened ECM and enhanced mechanical cell contractility can stimulate metalloproteinase secretion, which in turn enhances ECM cleavage and thus facilitates tumor dissemination and cancer cell migration.[56] These are just a few of many examples that show how strongly interconnected physical properties of the ECM and tumor cell signaling and behavior are.

Given this strong interconnection between physics of the ECM and biology of the tumor, it is crucial to include both into the picture when searching for new therapies. Only the combination of physics and biology — none alone by itself — will allow for significant improvement with respect to understanding the processes of the disease and finding therapies. For example hijacking of feedback-loops in ECM and tumor interactions can boost the effect of a therapy, since it will automatically self-amplify. However, the modulated system needs to be well understood, since a feedback into the wrong direction will also be amplified. Here, the combination of experiments and

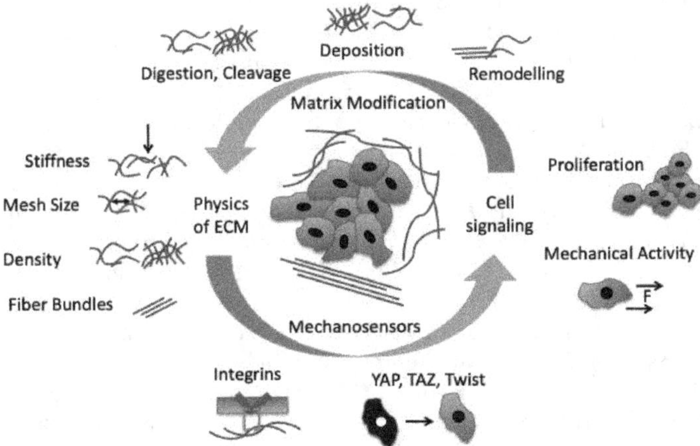

Fig. 2. The physical properties of the ECM and cell signaling influence each other with important consequences for tumor progression to malignancy. The schematic shows examples of physical properties and cell signaling and how they influence each other.

predictions from mathematical models will be essential. Many new mechano-sensitive pathways are currently being discovered, e.g., recently a link between lipid metabolism and ECM mechanics.[57] The discovery and better understanding of these pathways will help to find appropriate targets for tumor therapy. Some first drugs that target tumor-ECM interactions are being tested,[1] but further investigations are needed to achieve successful therapies. A main challenge is tumor plasticity, which allows for adaptation of the cells to new environmental cues. Thus, it is difficult to address exclusively cells or exclusively ECM. The strong connection between physics of ECM and tumor biology rather suggests combinatorial therapies, with one drug acting on the ECM and another acting on the tumor biology. Alternatively, of course, a central regulator could be targeted. An example that is discussed as such a possible key target is YAP/TAZ.[35] Whether it is a target as promising as it is seems now, remains up for future research.

References

1. F. Kai, A. P. Drain and V. M. Weaver, *Dev. Cell* **49**, 332 (2019).
2. J. Mouw, G. Ou and V. Weaver, *Nat. Rev. Mol. Cell Biol.* **15**, 771 (2014).
3. I. Acerbi, L. Cassereau, I. Dean, Q. Shi, A. Au, C. Park, Y. Y. Chen, J. Liphardt, E. S. Hwang and V. M. Weaver, *Integr. Biol. (Camb)* **7**, 1120 (2015).
4. K. A. Jansen, A. J. Licup, A. Sharma, R. Rens, F. C. MacKintosh and G. H. Koenderink, *Biophys. J.* **114**, 2655 (2018).
5. A. J. Licup, S. Münster, A. Sharma, M. Sheinman, L. M. Jawerth, B. Fabry, D. A. Weitz and F. C. MacKintosh, *Proc. Natl. Acad. Sci. USA* **112**, 9573 (2015).
6. K. Wolf, M. Te Lindert, M. Krause, S. Alexander, J. Te Riet, A. L. Willis, R. M. Hoffman, C. G. Figdor, S. J. Weiss and P. Friedl, *J. Cell Biol.* **201**, 1069 (2013).
7. M. Miron-Mendoza, J. Seemann and F. Grinnell, *Biomaterials* **31**, 6425 (2010).
8. J. Zhu and A. Mogilner, *Interface Focus* **6**, 20160040 (2016).
9. J. Condeelis and J. E. Segall, *Nat. Rev. Cancer* **3**, 921 (2003).
10. G. Charras and E. Sahai, *Nat. Rev. Mol. Cell Biol.* **15**, 813 (2014).
11. Y. J. Liu, M. Le Berre, F. Lautenschlaeger, P. Maiuri, A. Callan-Jones, M. Heuzé, T. Takaki, R. Voituriez and M. Piel, *Cell* **160**, 659 (2015).
12. Y. A. Miroshnikova, G. I. Rozenberg, L. Cassereau, M. Pickup, J. K. Mouw, G. Ou, K. L. Templeman, E. I. Hannachi, K. J. Gooch, A. L. Sarang-Sieminski, A. García and V. M. Weaver, *Mol. Biol. Cell* **28**, 2958 (2017).
13. P. Provenzano, K. Eliceiri, D. Inman and P. Keely, *Curr. Protoc. Cell Biol.* **10**, (2010).
14. A. Palamidessi, C. Malinverno, E. Frittoli, S. Corallino, E. Barbieri, S. Sigismund, G. V. Beznoussenko, E. Martini, M. Garre, I. Ferrara, C. Tripodo, F. Ascione, E. A. Cavalcanti-Adam, Q. Li, P. P. Di Fiore, D. Parazzoli, F. Giavazzi, R. Cerbino and G. Scita, *Nat. Mater.* **18**, 1252 (2019).
15. F. Geiger, D. Rüdiger, S. Zahler and H. Engelke, *PLoS One* **14**, e0225215 (2019).
16. A. Doyle, N. Carvajal, A. Jin, K. Matsumoto, K. Yamada, *Nat. Commun.* **6**, 8720 (2015).
17. K. Leventhal, H. Yu, L. Kass, J. Lakins, M. Egeblad, J. Erler, S. Fong, K. Csiszkar, A. Giaccia, W. Weninger, M. Yamauchi, D. Gasser and V. Weaver, *Cell* **139**, 891 (2009).
18. M. J. Pasze *et al.*, *Cancer Cell* **8**(3) 241 (2005).
19. Q. Shi, R. Ghosh, H. Engelke, C. Rycroft, L. Cassereau, J. Sethian, V. Weaver and J. Liphardt, *Proc. Natl. Acad. Sci. USA* **111**, 658 (2014).

20. P. P. Provenzano, D. R. Inman, K. W. Eliceiri, S. M. Trier and P. J. Keely, *Biophys J.* **95**, 5374 (2008).
21. W. Han, S. Chen, W. Yuan, Q. Fan, J. Tian, X. Wang, L. Chen, X. Zhang, W. Wei, R. Liu, J. Qu, Y. Jiao, R. H. Austin and L. Liu, *Proc. Natl. Acad. Sci. USA* **113**, 11208 (2016).
22. Y. L. Han, P. Ronceray, G. Xu, A. Malandrino, R. D. Kamm, M. Lenz, C. P. Broedersz and M. Guo, *Proc. Natl. Acad. Sci. USA* **115**, 4075 (2018).
23. M. Bergert, A. Erzberger, R. A. Desai, I. M. Aspalter, A. C. Oates, G. Charras, G. Salbreux and E. K. Paluch, *Nat. Cell Biol.* **17**, 524 (2015).
24. V. M. Weaver, O. W. Petersen, F. Wang, C. A. Larabell, P. Briand, C. Damsky and M. J. Bissell, *J. Cell Biol.* **137**, 231 (1997).
25. M. J. Oudin, O. Jonas, T. Kosciuk, L. C. Broye, B. C. Guido, J. Wyckoff, D. Riquelme, J. M. Lamar, S. B. Asokan, C. Whittaker, D. Ma, R. Langer, M. J. Cima, K. B. Wisinski, R. O. Hynes, D. A. Lauffenburger, P. J. Keely, J. E. Bear and F. B. Gertler, *Cancer Discov.* **6**, 516 (2016).
26. B. J. DuChez, A. D. Doyle, E. K. Dimitriadis and K. M. Yamada, *Biophys. J.* **116**, 670 (2019).
27. R. Sunyer, V. Conte, J. Escribano, A. Elosegui-Artola, A. Labernadie, L. Valon, D. Navajas, J. M. García-Aznar, J. J. Muñoz, P. Roca-Cusachs and X. Trepat, *Science* **353**, 1157 (2016).
28. M. H. Zaman, L. M. Trapani, A. L. Sieminski, D. Mackellar, H. Gong, R. D. Kamm, A. Wells, D. A. Lauffenburger and P. Matsudaira, *Proc. Natl. Acad. Sci. USA* **103**, 13897 (2006).
29. Y. J. Liu, M. Le Berre, F. Lautenschlaeger, P. Maiuri, A. Callan-Jones, M. Heuzé, T. Takaki, R. Voituriez and M. Piel, *Cell* **160**, 659 (2015).
30. M. H. Zaman, R. D. Kamm, P. Matsudaira and D. A. Lauffenburger, *Biophys. J.* **89**, 1389 (2005).
31. K. V. Nguyen-Ngoc, K. J. Cheung, A. Brenot, E. R. Shamir, R. S. Gray, W. C. Hines, P. Yaswen, Z. Werb and A. J. Ewald, *Proc. Natl. Acad. Sci. USA* **109**, E2595 (2012).
32. J. N. Beck, A. Singh, A. R. Rothenberg, J. H. Elisseeff and A. J. Ewald, *Biomaterials* **34**, 9486 (2013).
33. B. M. Rubenstein and L. Kaufman, *Biophys. J.* **95**, 5661 (2008).
34. S. Dupont, L. Morsut, M. Aragona, E. Enzo, S. Giulitti, M. Cordenonsi, F. Zanconato, J. Digabel, M. Forcato, S. Bicciato, N. Elvassore and S. Piccolo, *Nature* **474**, 179 (2011).
35. F. Zanconato, M. Cordenonsi and S. Piccolo, *Cancer Cell* **29**, 783 (2016).
36. A. Seifert and G. Posern, *Breast Cancer Res.* **19**, 68 (2017).
37. A. J. Haak, K. M. Appleton, E. M. Lisabeth, S. A. Misek, Y. Ji, S. M. Wade, J. L. Bell, C. E. Rockwell, M. Airik, M. A. Krook, S. D. Larsen, M. Verhaegen, E. R. Lawlor and R. R. Neubig, *Mol. Cancer Ther.* **16**, 193 (2017).
38. S. C. Wei, L. Fattet, J. H. Tsai, Y. Guo, V. H. Pai, H. E. Majeski, A. C. Chen, R. L. Sah, S. S. Taylor, A. J. Engler and J. Yang, *Nat. Cell Biol.* **17**, 678 (2015).
39. S. T. Boyle, J. Kular, M. Nobis, A. Ruszkiewicz, P. Timpson and M. S. Samuel, *Small GTPases* 1 (2018).
40. M. E. Fernández-Sánchez, S. Barbier, J. Whitehead, G. Béalle, A. Michel, H. Latorre-Ossa, C. Rey, L. Fouassier, A. Claperon, L. Brullé, E. Girard, N. Servant, T. Rio-Frio, H. Marie, S. Lesieur, C. Housset, J. L. Gennisson, M. Tanter, C. Ménager, S. Fre, S. Robine and E. Farge, *Nature* **523**, 92 (2015). DOI: 10.1080/21541248.2017.1413496.
41. J. M. Tse, G. Cheng, J. A. Tyrrell, S. A. Wilcox-Adelman, Y. Boucher, R. K. Jain and L. L. Munn, *Proc. Natl. Acad. Sci. USA* **109**, 911 (2012).

42. R. J. Eddy, M. D. Weidmann, V. P. Sharma and J. S. Condeelis, *Trends Cell Biol.* **27**, 595 (2017).

43. A. J. Licup, S. Münster, A. Sharma, M. Sheinman, L. M. Jawerth, B. Fabry and D. A. Weitz, *Proc. Natl. Acad. Sci. USA* **112**, 9573 (2015).

44. E. Ban, J. M. Franklin, S. Nam, L. R. Smith, H. Wang, R. G. Wells, O. Chaudhuri, J. T. Liphardt and V. B. Shenoy, *Biophys. J.* **114**, 450 (2018).

45. T. K. Ohsumi, J. E. Flaherty, M. C. Evans and V. H. Barocas, *Biomech. Model. Mechanobiol.* **7**, 53 (2008).

46. J. Feng, H. Levine, X. Mao and L. M. Sander, *Phys. Rev. E Stat. Nonlin. Soft Matter Phys.* **91**, 042710 (2015).

47. D. Vader, A. Kabla, D. Weitz and L. Mahadevan, *PLoS One* **4**, e5902 (2009).

48. L. J. Kaufman, C. P. Brangwynne, K. E. Kasza, E. Filippidi, V. D. Gordon, T. S. Deisboeck and D. A. Weitz, *Biophys. J.* **89**, 635 (2005).

49. H. Laklai, Y. A. Miroshnikova, M. W. Pickup, E. A. Collisson, G. E. Kim, A. S. Barrett, R. C. Hill, J. N. Lakins, D. D. Schlaepfer, J. K. Mouw, V. S. LeBleu, N. Roy, S. V. Novitskiy, J. S. Johansen, V. Poli, R. Kalluri, C. A. Iacobuzio-Donahue, L. D. Wood, M. Hebrok, K. Hansen, H. L. Moses and V. M. Weaver, *Nat. Med.* **22**, 497 (2016).

50. S. van Helvert and P. Friedl, *ACS Appl. Mater. Inter.* **8**, 21946 (2016).

51. C. Gaggioli, S. Hooper, C. Hidalgo-Carcedo, R. Grosse, J. F. Marshall, K. Harrington and E. Sahai, *Nat. Cell Biol.* **9**(12), 1392 (2007), doi: 10.1038/ncb1658.

52. N. Yang, R. Mosher, S. Seo, D. Beebe and A. Friedl, *Am. J. Pathol.* **178**, 325 (2011).

53. M. Dietrich, H. Le Roy, D. B. Brückner, H. Engelke, R. Zantl, J. O. Rädler and C. P. Broedersz, *Soft Matter* **14**, 2816 (2018).

54. R. Ferrari, G. Martin, O. Tagit, A. Guichard, A. Cambi, R. Voituriez, S. Vassilopoulos and P. Chavrier, *Nat. Commun.* **10**, 4886 (2019).

55. T. Y. Feinberg, H. Zheng, R. Liu, M. S. Wicha, S. M. Yu and S. J. Weiss, *Dev. Cell* **47**, 145 (2018).

56. A. Haage and I. C. Schneider, *FASEB J.* **28**, 3589 (2014).

57. P. Romani, I. Brian, G. Santinon, A. Pocaterra, M. Audano, S. Pedretti, S. Mathieu, M. Forcato, S. Bicciato, J. B. Manneville, N. Mitro and S. Dupont, *Nat. Cell Biol.* **21**, 338 (2019).

Chapter 7

Computational Modeling and Experimental Investigations to Enhance the Successful Response of Anti-PD-1 Cancer Immunotherapies

Damijan Valentinuzzi[1,2,*], Katja Uršič[3], Matea Maruna[3], Simon Buček[3], Urban Simončič[1,2], Martina Vrankar[3], Maja Čemažar[3], Gregor Serša[3] and Robert Jeraj[1,2,4,†]

[1]*Jožef Stefan Institute*
Ljubljana, Slovenia

[2]*Faculty of Mathematics and Physics*
University of Ljubljana
Ljubljana, Slovenia

[3]*Institute of Oncology Ljubljana*
Ljubljana, Slovenia

[4]*Department of Medical Physics*
University of Wisconsin
Madison, WI, USA
**damijan.valentinuzzi@ijs.si*
†rjeraj@wisc.edu

Immunotherapies with anti-programmed death-1 antibodies (anti-PD-1) have revolutionized cancer treatment. However, the majority of patients still fail to respond and the reasons remain largely unknown. In order to identify tumor characteristics associated with treatment failure, we developed a computational model capable of simulating tumor response to anti-PD-1 immunotherapy, validated with on-site *in vitro* and *in vivo* experiments.

The experiments were conducted using non-responsive and responsive murine cell lines, 4T1 mammary carcinoma and CT26 colon carcinoma, respectively. Tumors were induced in athymic nude mice, wild-type mice, and wild-type mice receiving anti-PD-1 immunotherapy to: (1) assess Gompertzian parameters describing intrinsic tumor growth in the absence of T cells, (2) assess the immune-related parameters, and (3) validate model's predictive ability, respectively. Flow-cytometric analyses of major histocompatibility complex (MHC) class I and PD-1 ligand (PD-L1) expression were performed to assess initial heterogeneity in tumor cell subpopulations. Model parameter sensitivity studies were performed to establish importance of different tumor characteristics on the observed tumor response to anti-PD-1 immunotherapy.

The model correctly predicted the observed trend of response to anti-PD-1 in 4T1, while the predictions in CT26, where dichotomous responses were observed experimentally, were less accurate. Sensitivity studies of measured model parameters and initial conditions suggested that complete responses to anti-PD-1 immunotherapy might be possible only in the case of homogeneous MHC class I positive tumors. On the other hand, heterogeneous tumors containing MHC class I negative tumor cell subpopulations were associated with resistance to anti-PD-1 therapy.

Keywords: Computational model; mathematical model; tumor modeling; anti-PD-1; immunotherapy; MHC class I; PD-L1; 4T1 mammary carcinoma; CT26 colon carcinoma; tumor biology; T cells.

1. Introduction

Modern immunotherapies with immune checkpoint inhibitors (ICI), such as cytotoxic T lymphocyte-associated antigen-4 (CTLA-4) and programmed-death-1 (PD-1) antibodies, have revolutionized oncology. The ICI-induced blockade of the negative feedback loop of the patient's immune system has proven to be highly efficient for the treatment of various malignancies, even in the deadliest types of cancer, such as metastatic melanoma and non-small-cell lung cancer (NSCLC), which have been considered almost incurable with conventional cytotoxic therapies.[1,2] The responding patients often achieve durable remission of the disease, and even disease-free outcomes are possible. ICI have been proclaimed as the "scientific breakthrough of the year for 2013" by Science journal, and in 2018, James P. Allison and Tasuku Honjo won the Nobel Prize in Physiology or Medicine for their foundational work in immune checkpoints.[3,4]

Despite the undoubted success of the aforementioned drugs, numerous challenges and pitfalls remain to be resolved in order to further improve the benefit for future cancer patients. While immunotherapy is extremely effective in some cancer types, the outcomes in some other types (e.g., pancreatic carcinoma) are still poor.[5] Moreover, even in the cancer types sensitive to immunotherapy, the majority of patients fail to respond.[6] Thus, it would be of utmost importance to (1) identify more reliable predictive biomarkers of response to immunotherapy compared to the existing ones, (2) develop new methods for treatment response assessment, (3) identify optimal combination therapies, and (4) develop new generations of ICI, as well as other immuno-oncology (IO) agents. However, these tasks will hardly be accomplished by relying solely on the costly and time-consuming trial-and-error approach of clinical trials, since the number of currently available IO agents (not to mention all possible combinations or dosing regimens) has increased significantly over the last few years. 2004 immuno-oncology agents against 303 targets, by 864 companies, tested in 3042 active clinical trials were identified in 2017.[7]

Computational models are promising complementary tools for guiding future immuno-oncology research, because they are fast, powerful, and cost-efficient. However, while the first computational models of cancer immunotherapy appeared as early as the late 1980s, their routine use in pre-clinical and clinical settings is still limited.[8] One reason for this is certainly the lack of proper experimental validation of such models. Likewise, most of the existing computational models of cancer immunotherapy with ICI are either purely theoretical or fitted to experimental data, while their predictive ability of independent data in most cases remains unconfirmed.[9–16]

In our previous study, we developed a bottom-up deterministic population model to simulate tumor response to anti-PD-1 immunotherapy.[17] The model was built with validation in mind; therefore, it was minimalistic (to avoid overfitting), yet with a reasonable number of parameters reflecting the essential biology involved in anti-PD-1 immunotherapy. It was tuned and verified on independent experiments from different published studies performed in B16 murine melanoma.[18–20] The model suggested that downregulation of major histocompatibility complex (MHC) class I (transmembrane proteins responsible for presentation of self and foreign antigens) might be one of the main drivers of resistance to anti-PD-1 immunotherapy.[21]

The aim of this study was to further explore the tumor characteristics associated with resistance to anti-PD-1 immunotherapy by using computational modeling. The previously developed model was modified to better account for the heterogeneity of tumor cells in MHC class I expression and PD-1 ligand (PD-L1) expression; however, the number of model parameters has not increased. It was validated with our own tumor growth experiments conducted beyond murine melanoma. Additionally, we performed direct or indirect on-site measurements of some specific model parameters and initial conditions. Finally, we conducted sensitivity analyses to explore the impact of the measured tumor characteristics on the observed tumor response to anti-PD-1 immunotherapy.

2. Methods

2.1. *Computational model*

The interplay between tumor cells (C_{1-4}) and tumor-infiltrating lymphocytes (*TIL*) (consisting only of T cells) is described with the following bottom-up deterministic population model (Eq. (1)–(7)).

$$\frac{dC_1}{dt} = k \cdot C_1 \cdot ln\left(\frac{C_{max}}{\sum_{i=1}^{4} C_i}\right)$$
$$- a \cdot SNV \cdot (1 - PD1 \cdot (1 - PD1_{occup}(D,t))) \cdot TIL \cdot C_1 \tag{1}$$

$$\frac{dC_2}{dt} = k \cdot C_2 \cdot ln\left(\frac{C_{max}}{\sum_{i=1}^{4} C_i}\right) - a \cdot SNV \cdot TIL \cdot C_2 \tag{2}$$

$$\frac{dC_3}{dt} = k \cdot C_3 \cdot ln\left(\frac{C_{max}}{\sum_{i=1}^{4} C_i}\right) \tag{3}$$

$$\frac{dC_4}{dt} = k \cdot C_4 \cdot ln\left(\frac{C_{max}}{\sum_{i=1}^{4} C_i}\right) \tag{4}$$

$$\frac{dTIL}{dt} = b \cdot SNV \cdot \left(\sum_{i=1}^{4} C_i + C_{dead} \right)^{\frac{2}{3}}$$

$$- a \cdot SNV \cdot PD1 \cdot (1 - PD1_{occup}(D,t)) \cdot TIL \cdot (C_1 + C_3) - d \cdot TIL \qquad (5)$$

$$\frac{dC_{dead}}{dt} = a \cdot SNV \cdot (1 - PD1 \cdot (1 - PD1_{occup}(D,t))) \cdot TIL \cdot C_1$$

$$+ a \cdot SNV \cdot TIL \cdot C_2 - l \cdot C_{dead} \qquad (6)$$

$$V(t) = V_c \cdot \left(\sum_{i=1}^{4} C_i(t) + C_{dead}(t) \right) + V_{TIL} \cdot TIL(t)$$

$$= \frac{4}{3}\pi r_c^3 \cdot \left(\sum_{i=1}^{4} C_i(t) + C_{dead}(t) \right) + \frac{4}{3}\pi r_{TIL}^3 \cdot TIL(t) \qquad (7)$$

All variables, model parameters, and initial conditions, obtained either by our own measurements or from the available literature, are summarized in Table 1.

Compared to our previous model,[17] the current one also accounts for heterogeneity in tumor cell population, associated with different levels of resistance to anti-PD-1 immunotherapy (Eqs. (1)–(4)). This increase in model complexity (additional variables) was supported by on-site experiments, where for each studied cell line (4T1 mammary carcinoma, CT26 colon carcinoma), the initial number of cells in each tumor cell subpopulation was measured using flow cytometry (Table 2). A detailed description of the captured biology, as well as in-depth justification of modeling assumptions, together with the corresponding references, can be found in our previous paper,[17] while here we present only a brief summary of the model terms, parameters, and initial conditions:

2.1.1. *Tumor cell subpopulations*

- The first group of tumor cells (C_1, Eq. (1)) is characterized as major histocompatibility complex (MHC) class I positive and PD-1 ligand (PD-L1) positive (MHC-I$^+$ PD-L1$^+$), meaning that these cells express both MHC class I as well as PD-L1 on their membranes. MHC class I molecules are transmembrane proteins responsible for presentation of self and foreign antigens; therefore, their expression on tumor cells is a prerequisite for T cells to recognize the target to attack.[21] PD-L1 is a ligand for PD-1 (expressed on T cells). After a successful ligation of PD-1 and PD-L1, a negative feedback loop of the immune response is activated, resulting in inactivated and exhausted T cells.[22] Therefore, PD-L1 expression on tumor cells serves as a resistance mechanism, but can be disabled with anti-PD-1 immunotherapy. To sum up, C_1 is an MHC-I$^+$ PD-L1$^+$ subpopulation of sensitive tumor cells which can be recognized and eliminated by T cells, but can also have a negative effect on the immune response via the activation of the PD-1/PD-L1 pathway if not being treated with anti-PD-1 immunotherapy.

Table 1. Summary of model variables, parameters, and initial conditions with the corresponding references.

Variable	Description	Initial condition ($t=0$) (4T1)	Initial condition ($t=0$) (CT26)	Reference
C_1	Initial fraction of MHC-I$^+$ PD-L1$^+$ tumor cells	$10.6 \pm 7.5\%$	$17.3 \pm 10.4\%$	Measured
C_2	Initial fraction of MHC-I$^+$ PD-L1$^-$ tumor cells	$89.0 \pm 7.5\%$	$82.6 \pm 10.4\%$	Measured
C_3	Initial fraction of MHC-I$^-$ PD-L1$^+$ tumor cells	$0.2 \pm 0.2\%$	$0.0 \pm 0.0\%$	Measured
C_4	Initial fraction of MHC-I$^-$ PD-L1$^-$ tumor cells	$0.3 \pm 0.5\%$	$0.04 \pm 0.09\%$	Measured
TIL	Number of tumor-infiltrating lymphocytes	0	0	
C_{dead}	Number of dead tumor cells	0	0	
V	Tumor volume	0	0	

Parameter	Description	Value (4T1)	Value (CT26)	Reference
k	Tumor growth rate	$0.11 \pm 0.01\,\mathrm{day}^{-1}$	$0.077 \pm 0.004\,\mathrm{day}^{-1}$	Tuned to experimental data
C_{max}	Maximum number of cells in tumor	$(0.22 \pm 0.01) \times 10^9$	$(3.95 \pm 0.29) \times 10^9$	Tuned to experimental data
a	Interaction rate between tumor cells and tumor-infiltrating lymphocytes	$(1.59 \pm 0.04) \times 10^{-8}\,\mathrm{day}^{-1}$	$(0.45 \pm 0.18) \times 10^{-8}\,\mathrm{day}^{-1}$	Tuned to experimental data
b	Infiltration rate of TIL into tumor	$(2.24 \pm 1.19) \times 10^{-2}\,\mathrm{day}^{-1}$	$(0.46 \pm 0.41) \times 10^{-2}\,\mathrm{day}^{-1}$	Tuned to experimental data
SNV	Single nucleotide variations	293	3023	27
$PD1$	PD-1 expression	76%	79%	28
$PD1_{occup}$	Occupancy of PD-1 by anti-PD-1 antibodies	Dose- and schedule-dependent	Dose- and schedule-dependent	29
l	Dead tumor cells removal rate	$0.094\,\mathrm{day}^{-1}$	$0.094\,\mathrm{day}^{-1}$	26
d	Natural TIL death rate	$0.41\,\mathrm{day}^{-1}$	$0.41\,\mathrm{day}^{-1}$	30
r_c	Tumor cell radius	$9.5 \pm 0.8\,\mu\mathrm{m}$	$9.9 \pm 0.3\,\mu\mathrm{m}$	Measured
r_{TIL}	TIL radius	$5\,\mu\mathrm{m}$	$5\,\mu\mathrm{m}$	31

- The second group of tumor cells (C_2, Eq. (2)) is MHC class I positive, PD-L1 negative (MHC-I$^+$ PD-L1$^-$). This is the most sensitive subpopulation, which can be recognized and eliminated by T cells even if no immunotherapy has been administered. This subpopulation is not able to suppress the immune response via the activation of the PD-1/PD-L1 pathway.
- The third group of tumor cells (C_3, Eq. (3)) is MHC class I negative, PD-L1 positive (MHC-I$^-$ PD-L1$^+$), therefore the most resistant subpopulation which cannot be recognized and eliminated by T cells. Additionally, such cells are also able to suppress the immune response via the PD-1/PD-L1 pathway.
- The fourth subpopulation (C_4, Eq. (4)) is MHC class I negative, PD-L1 negative (MHC-I$^-$ PD-L1$^-$). It is a resistant subpopulation as well. Since the target ligand (PD-L1) is missing, anti-PD-1 immunotherapy does not have any effect on this subpopulation.

2.1.2. *Intrinsic tumor growth, tumor-immune interactions, and tumor volume calculation*

- Intrinsic growth of tumor cells in the absence of T cells is assumed to be Gompertzian with parameters tumor growth rate k and the maximum number of cells in tumor C_{max}, which are assumed to have the same values for all four tumor cell subpopulations (first terms in Eqs. (1)–(4)).[23]
- Tumor cell elimination due to T cell-specific responses is represented with second terms in Eqs. (1) and (2). It is proportional to the number of single nucleotide variations (SNV) of the tumor, which are associated with tumor mutational burden.[24] In Eq. (1), tumor cells (C_1) can be eliminated only by the proportion of T cells which either do not express PD-1, or their PD-1 receptors are occupied by PD-1 antibodies ($PD1_{occup}$). The details about anti-PD-1 pharmacokinetics and pharmacodynamics, associated with the time- and dose-dependent value of parameter $PD1_{occup}$, can be found in our previous study.[17]
- Infiltration of TIL into the tumor is proportional to SNV and the tumor surface area, since it is assumed that tumor vasculature (which is more abundant at the tumor surface) is the main bottleneck for successful infiltration[25] (first term of Eq. (5)). TIL which express PD-1 can be eliminated by PD-L1$^+$ C_1 and C_3 subpopulations via PD-1/PD-L1 pathway (second term of Eq. (5)). All TIL are also being removed with a death rate d due to their natural life span (third term of Eq. (5)).
- Tumor cells which are eliminated by the immune system are not removed from the tumor instantly; therefore, we move them into the compartment of dead tumor cells (C_{dead}, Eq. (6)). C_{dead} are then removed from the tumor with a removal rate l (third term of Eq. (6)).[26]
- Equation (7) is used to calculate macroscopic tumor volume by assuming all tumor cells and TIL are spherical with radius r_C and r_{TIL}, respectively.

2.2. *Model tuning and validation*

To assess Gompertzian parameters k and C_{max}, the immune-related parameters a and b, initial conditions (initial fraction of each tumor cell subpopulation C_{1-4} $(t = 0)$), as well as to validate model predictions, we performed our own *in vitro* and *in vivo* experiments. Two different murine cell lines were used, namely 4T1 mammary carcinoma ("cold tumor" – known to be non-responsive to anti-PD-1 immunotherapy) and CT26 colon carcinoma ("hot tumor" – known to be responsive to anti-PD-1 immunotherapy).[32] The experiments were designed in such a way as to allow for decoupling of biological phenomena described by the aforementioned parameters and initial conditions:

- To assess heterogeneity of tumor cells injected in mice, the initial fraction of each tumor cell subpopulation (C_{1-4}) was measured using flow cytometry.
- In the second set of experiments, intrinsic tumor growth in the absence of T cells was measured *in vivo* by inducing 4T1 and CT26 tumors in athymic nude mice (e.g., mice strain lacking T cells). This data enabled us to assess the Gompertzian parameters k and C_{max} of each tumor cell line. The model was fit to the obtained tumor growth curves by using the nonlinear mixed-effects regression model (*nlmefitsa*, Matlab, R2019a, The Mathworks, Inc., Natick, MA) assuming that all immune-related model parameters (a, b, l, d) were equal to zero.[33]
- In the third set of experiments, 4T1 and CT26 tumors were induced in wild-type mice with fully functional immune system, which enabled us to assess the immune-related parameters a and b. Again, the nonlinear mixed-effects regression model was used to fit our model to the obtained tumor growth curves. All parameters (except the fitted parameters a and b) were kept to the values shown in Table 1, including the previously measured k and C_{max}. Since the induced tumors received no therapy, the parameter $PD1_{occup}$ was set to zero.
- Finally, to validate model predictions, the fourth set of *in vivo* experiments was performed. 4T1 and CT26 tumors were induced in wild-type mice receiving daily anti-PD-1 immunotherapy when the tumor volumes reached 35–40 mm^3 (see methods, Section 2.6). The obtained experimental tumor growth curves were compared to the growth curves predicted by the model. Model predictions were generated with fixed parameters from Table 1. No free parameters or additional fitting was allowed at this stage.

2.3. *Sensitivity analyses*

To explore the impact of the measured model parameters and initial conditions (k, C_{max}, a, b, C_3 $(t = 0)$, C_4 $(t = 0)$) on the predicted tumor growth curves, sensitivity studies were performed by modulating the mean value μ_i of each parameter i by $\pm$ the measured standard deviation σ_i. When modulating initial conditions C_3 $(t = 0)$ or C_4 $(t = 0)$ by $\pm$ their standard deviations σ_i, the same amount of cells were subtracted/added to the compartments C_1 and C_2, respectively, to preserve the same total number of the implanted tumor cells.

2.4. *Tumor cells and animals*

Murine 4T1 mammary carcinoma cells and CT26 colon carcinoma cells (American Type Culture Collection, Manassas, VA, USA) were cultured as monolayers at 37°C in a humidified incubator containing 5% CO_2 using advanced RPMI (Roswell Park Memorial Institute) 1640 Medium, supplemented with 5% fetal bovine serum (FBS) (Gibco, Thermo Fisher Scientific, Waltham, MA, USA), 10 mM L-glutamine (GlutaMAX, Gibco), 100 U/mL penicillin (Grünenthal, Aachen, Germany) and 50 mg/mL gentamicin (Krka, Novo mesto, Slovenia).

Six- to eight-week-old female BALB/c (BALB/cAnNCrl) or athymic nude BALB/c mice (Crl:NU(NCr)-Foxn1^{nu}) (Charles River, Calco, Italy) were maintained in specific pathogen-free conditions with 12 h light/dark cycle at 20–24°C with 55% $\pm$ 10% relative humidity and given food and water *ad libitum*. For nude mice, food, water, bedding, cages and other housing equipment were sterilized by autoclaving. Personnel involved in the care of nude mice wore sterilized gloves and full laboratory equipment, including masks. All procedures were performed in compliance with the guidelines for animal experiments of the EU Directives and the permission of the Ministry of Agriculture, Forestry and Food of the Republic of Slovenia (Permission No. U34401-1/2015/43) based on the approval by the National Ethics Committee for Experiments on Laboratory Animals.

2.5. *Flow-cytometric analysis of MHC class I and PD-L1 expression*

When 4T1 and CT26 cells reached 80% confluency, they were trypsinized. After centrifugation (1400 rpm, 5 min, 4°C), cells were washed with ice-cold 1× phosphate buffer saline (PBS) twice and 1.5×10^6 4T1 or CT26 cells were incubated with anti-MHC-I (MHC Class I (H-2Kd) Monoclonal Antibody (SF1-1.1.1), APC; eBioscience^{TM}, Thermo Fischer Scientific) and anti-PD-L1 (CD274 (PD-L1, BH-H1) Monoclonal antibody (MIH5), PE; eBioscience^{TM}, Thermo Fischer Scientific) antibodies. After 15 min incubation on ice, cells were washed with ice-cold 1× PBS. To exclude dead cells from the analysis, cells were additionally stained with FITC Annexin V Apoptosis Detection Kit with 7-AAD (BioLegend, San Diego, CA, USA) according to manufacturer's instructions. Samples were measured using a FACS-Canto II flow cytometer (BD Biosciences, San Jose, CA, USA) equipped with two lasers (488 nm and 633 nm), and the results were analyzed with the BD FACSDiva 8.0 software (BD Biosciences). At least 100,000 events were measured. Debris and double cells were excluded using the FSC/SSC dot plot. FMOs and an isotype control (Mouse IgG2a kappa Isotype Control (eBM2a), APC, eBioscience^{TM}, Thermo Fisher Scientific) were used to determine the gating regions. For each cell line, the experiment was repeated five times (biological repetitions), and then the mean value and standard deviation were calculated.

2.6. *Tumor induction and therapy*

One day prior to tumor induction, mice were shaved on their right flanks. For tumor induction, suspensions of 0.5×10^6 CT26 or 4T1 cells in $100\,\mu l$ of 0.9% saline were injected subcutaneously into right flanks of mice. Treatment was started when tumors reached 35–$40\,\mathrm{mm}^3$. Tumor volume was calculated using the following formula: $d_1 \times d_2 \times d_3 \times \pi/6$; where d_1, d_2, and d_3 represent three mutually orthogonal tumor diameters. The treatment consisted of a daily intraperitoneal injection of $50\,\mu g$ of anti-PD-1 antibodies in $100\,\mu l$ of 0.9% saline (α PD-1; clone 29F.1A12; BioXCell, West Lebanon, NH, USA) until tumors reached 300–$350\,\mathrm{mm}^3$ or disappeared (i.e., complete response). Tumor size was measured daily with a caliper from day 0 (start of the therapy). Additionally, animal weights were measured and their behavior was observed.

3. Results

3.1. *Flow-cytometric analysis of MHC class I and PD-L1 expression*

The results of flow-cytometric analysis of 4T1 and CT26 cell lines are presented in Table 2. In both cell lines, the most sensitive MHC-I$^+$ PD-L1$^-$ subpopulation dominated with $89.0\% \pm 7.5\%$ and $82.6\% \pm 10.4\%$ of the total number of 4T1 and CT26 tumor cells, respectively. The MHC-I$^+$ PD-L1$^+$ subpopulation was represented with $10.6\% \pm 7.5\%$ and $17.3\% \pm 10.4\%$ of the total number of 4T1 and CT26 tumor cells, respectively. On the other hand, interestingly, we did not detect the resistant MHC-I$^-$ PD-L1$^+$ subpopulation in any of the five measurements in CT26, while in

Table 2. Flow-cytometric analysis of MHC class I and PD-L1 expression.

Measurement	MHC-I$^+$ PD-L1$^+$ [%]	MHC-I$^+$ PD-L1$^-$ [%]	MHC-I$^-$ PD-L1$^+$ [%]	MHC-I$^-$ PD-L1$^-$ [%]
4T1				
1	11.5	86.9	0.4	1.2
2	3.4	96.5	0.1	0.0
3	6.6	93.4	0.0	0.0
4	22.9	77.1	0.0	0.0
5	8.5	90.9	0.3	0.3
mean $\pm$ standard deviation	10.6 ± 7.5	89.0 ± 7.5	0.2 ± 0.2	0.3 ± 0.5
CT26				
1	29.3	70.5	0.0	0.2
2	15.4	84.6	0.0	0.0
3	5.6	94.4	0.0	0.0
4	9.7	90.3	0.0	0.0
5	26.7	73.3	0.0	0.0
mean $\pm$ standard deviation	17.3 ± 10.4	82.6 ± 10.4	0.0 ± 0.0	0.04 ± 0.09

4T1, this subpopulation was present in three of five measurements with an average of $0.2\% \pm 0.2\%$. Similarly, the MHC-I$^-$ PD-L1$^-$ tumor cell subpopulation, which is also not expected to be recognized by T cells, was present only in one of five measurements in CT26 with an average of $0.04\% \pm 0.09\%$. In 4T1, this subpopulation was found in two of five measurements with an average of $0.3\% \pm 0.5\%$.

3.2. *Untreated tumor growth in athymic nude mice*

The results of *in vivo* experiments with 4T1 and CT26 tumors induced in nude mice (athymic mice strain lacking T cells) are presented in Fig. 1. This series of experiments was performed to assess intrinsic tumor growth in the absence of T cells, because it is well-known that the immune system has an important impact on tumor growth even if the absence of treatment. When the T cells are excluded from our model ($a = b = l = d = 0$), the model reduces to Gompertz terms with parameters k and C_{max} only (first terms of Eqs. (1)–(4)). The results obtained by the nonlinear mixed-effects regression model for tumor growth rate k were $0.11 \pm 0.01 \, \mathrm{day}^{-1}$ and $0.077 \pm 0.004 \, \mathrm{day}^{-1}$ for 4T1 and CT26, respectively. The maximum numbers of tumor cells in 4T1 and CT26 tumors C_{max} were assessed to be $(0.22 \pm 0.01) \times 10^9$ and $(3.95 \pm 0.29) \times 10^9$, respectively.

3.3. *Untreated tumor growth in wild-type mice*

To assess the impact of T cells on the growth of untreated tumors, 4T1 and CT26 tumors were induced in wild-type (BALB/c) mice with fully functional immune

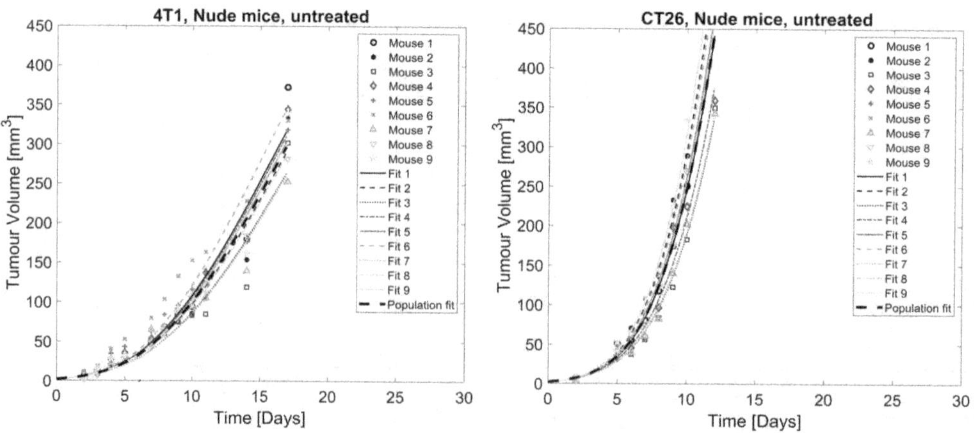

Fig. 1.   Untreated tumor growth in nude mice (mice strain lacking T cells). Left: 4T1 mammary carcinoma, right: CT26 colon carcinoma. Markers represent tumor growth data of individual mice. The dashed thick line represents population fit to the data, while the thin lines represent individual fits. Fitting was performed using the nonlinear mixed-effects regression model, where k and C_{max} were the parameters fit, while all immune-related parameters (a, b, l, d) were set to 0.

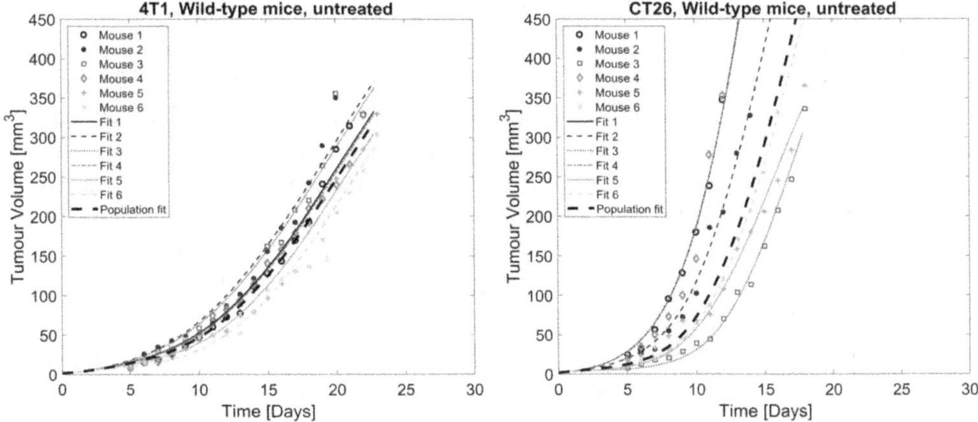

Fig. 2. Untreated tumor growth in wild-type mice (mice strain with fully functional immune system). Left: 4T1 mammary carcinoma, right: CT26 colon carcinoma. Markers represent tumor growth data of individual mice. The dashed thick line represents population fit to the data, while the thin lines represent individual fits. Fitting was performed using the nonlinear mixed-effects regression model, where a and b were the parameters fit, while all other parameters were kept to the values from Table 1, except the parameter $PD1_{occup}$ (describing the effect of anti-PD-1 immunotherapy), which was set to 0.

systems (Fig. 2). The model with free parameters a, representing interaction rate between tumor cells and T cells, and b, representing infiltration rate of T cells into the tumor, was fitted to the obtained tumor growth curves. All other parameters were kept fixed to the values from Table 1, including the measured values of k and C_{max}. The exception was the parameter $PD1_{occup}$ describing the effect of anti-PD-1 immunotherapy, which was set to zero. The obtained results were $a = (1.59 \pm 0.04) \times 10^{-8}\,\mathrm{day}^{-1}$, and $a = (0.45 \pm 0.18) \times 10^{-8}\,\mathrm{day}^{-1}$, for 4T1 and CT26, respectively. The obtained infiltration rates were $b = (2.24 \pm 1.19) \times 10^{-2}\,\mathrm{day}^{-1}$, and $b = (0.46 \pm 0.41) \times 10^{-2}\,\mathrm{day}^{-1}$.

Comparison of tumor growth curves in Fig. 1 (nude mice) and Fig. 2 (wild-type mice) clearly shows that the effect of the immune system could be important even if tumor is not being treated. All nude mice bearing 4T1 mammary carcinoma were sacrificed on day 17 (tumor volume reached 300–$350\,\mathrm{mm}^3$) (Fig. 1, left), while while-type mice bearing the same tumor type were sacrificed between days 20–23 (Fig. 2, left). Similarly, nude mice bearing CT26 colon carcinoma were sacrificed between days 10-12 (Fig. 1, right), while wild-type mice bearing CT26 tumors were sacrificed between days 12–18 (Fig. 2, right).

3.4. *Growth of tumors receiving anti-PD-1 immunotherapy and sensitivity analyses*

The last series of *in vivo* experiments were performed to validate the model's ability to predict responses of 4T1 and CT26 tumors to anti-PD-1 immunotherapy.

Ultimately, to study which of the measured tumor characteristics has the major impact on tumor response or resistance to anti-PD-1 immunotherapy, we performed sensitivity studies of the parameters and initial conditions obtained by previous experiments (k, C_{max}, a, b, and the initial number of tumor cells in the resistant subpopulations C_3 ($t = 0$), and C_4 ($t = 0$)). 4T1 and CT26 tumors were induced in wild-type (BALB/c) mice which received daily anti-PD-1 immunotherapy (50 μg intraperitoneally) once the tumor volume reached 35–40 mm^3. Model predictions were generated with fixed parameters summarized in Table 1 (none of the parameters was fitted to this set of experimental data). The results of 4T1 and CT26 tumors are presented in Figs. 3 and 4, respectively.

The experiments revealed that 4T1 tumors were mostly insensitive to anti-PD-1 immunotherapy. Although the growth of tumors treated with anti-PD-1 immuno-therapy (Fig. 3) was delayed compared to the untreated tumors (Fig. 2, left), none of the 4T1 tumors demonstrated complete response. Also the predictions of our model suggested that anti-PD-1 immunotherapy should not have been effective in 4T1 tumors. Finally, the sensitivity analyses revealed that modulation of the aforemen-tioned parameters and initial conditions by $\pm\sigma_i$ resulted only in accelerated or delayed tumor growth, while the general trend (insensitivity to anti-PD-1 therapy) was preserved.

On the other hand, the observed response to anti-PD-1 immunotherapy of CT26 tumors was dichotomous (Fig. 4). Two of six tumors were insensitive to anti-PD-1 immunotherapy, while four tumors responded well to therapy (three tumors completely disappeared, and one tumor responded with long term stable disease with tumor volume below 50 mm^3). The predictions of our model suggested an initial delay of tumor growth with a subsequent regrowth, which was not observed experimentally (Fig. 4). However, sensitivity studies demonstrated that the out-come of the therapy might be highly dependent particularly on two parameters. First, the model was very sensitive to parameter b,describing infiltration rate of T cells into tumor (Fig. 4, middle right). If the parameter b was decreased by one standard deviation ($b = \mu_b - \sigma_b$), the predicted delay in tumor growth disappeared, which explained the observed trend of two completely non-responsive tumors. Secondly, the modulation of the initial condition C_4 ($t = 0$), describing initial fraction of the resistant MHC-I$^-$ PD-L1$^-$ tumor cell subpopulation, was the only way to explain the observed favorable responses to anti-PD-1 immunotherapy (Fig. 4, bottom right). Such responses were possible only if C_4 ($t = 0$) was equal to 0, describing the scenario with no resistant MHC-I$^-$tumor cells in the injected tumor cell population (C_3 was always equal to 0 in CT26). This could be a plau-sible scenario in reality, because in our flow-cytometric measurements (Table 2) of CT26, we detected the MHC-I$^-$ PD-L1$^-$ tumor cell subpopulation only in one of five cases (Table 2, CT26, Measurement 1), and MHC-I$^-$ PD-L1$^+$ in zero of five cases.

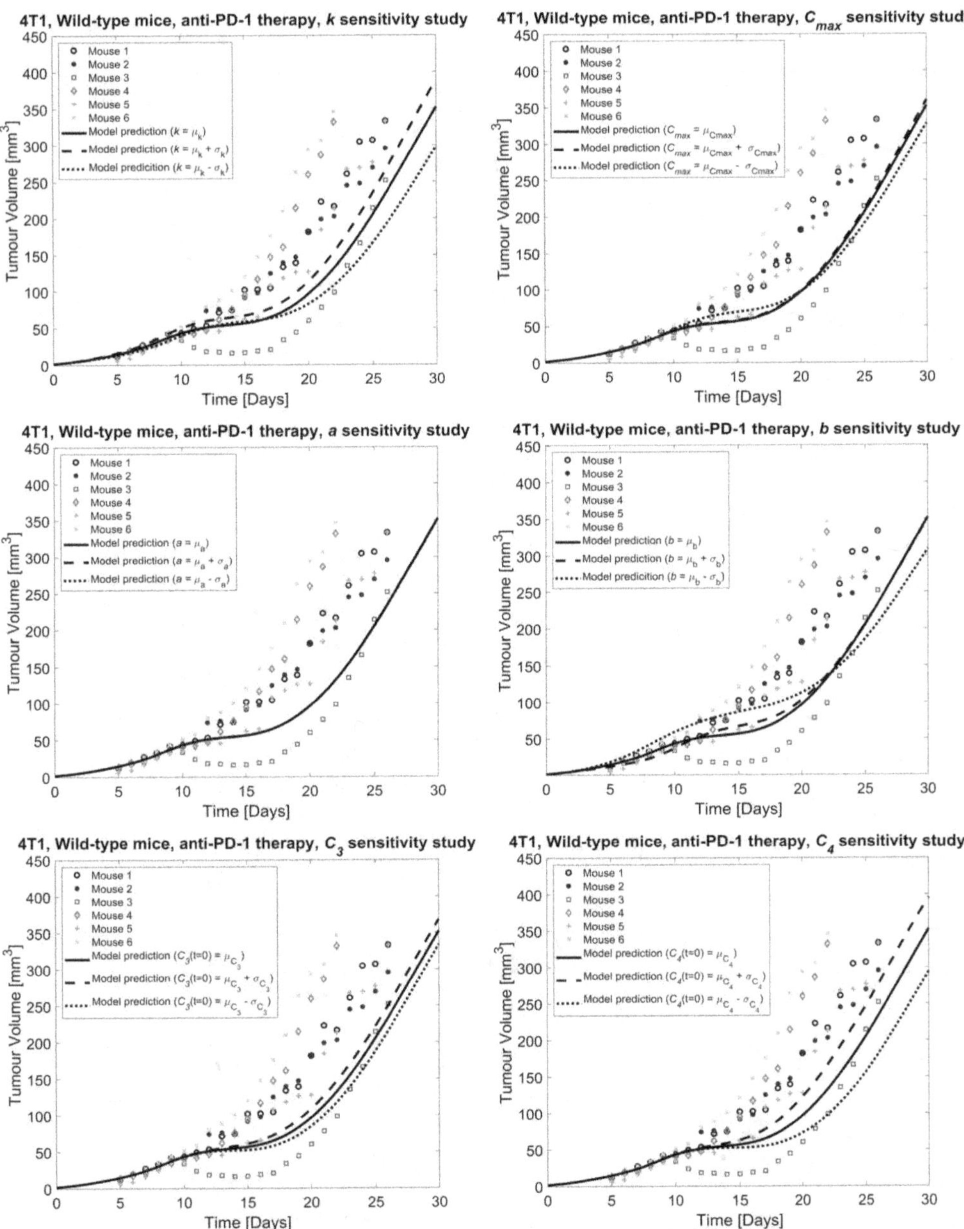

Fig. 3. Growth of 4T1 tumors receiving anti-PD-1 immunotherapy and sensitivity analyses. Markers represent tumor growth data of individual mice. Solid line represents model prediction using mean parameter values μ_i from Table 1. Dashed lines represent sensitivity studies where the mean value μ_i of i-th parameter was increased for one standard deviation σ_i. Dotted lines represent sensitivity studies where the mean value μ_i of i-th parameter was decreased for one standard deviation σ_i. Top left: k sensitivity study, top right: C_{max} sensitivity study, middle left: a sensitivity study, middle right: b sensitivity study, bottom left: C_3 $(t = 0)$ sensitivity study, bottom right: $C_4(t = 0)$ sensitivity study.

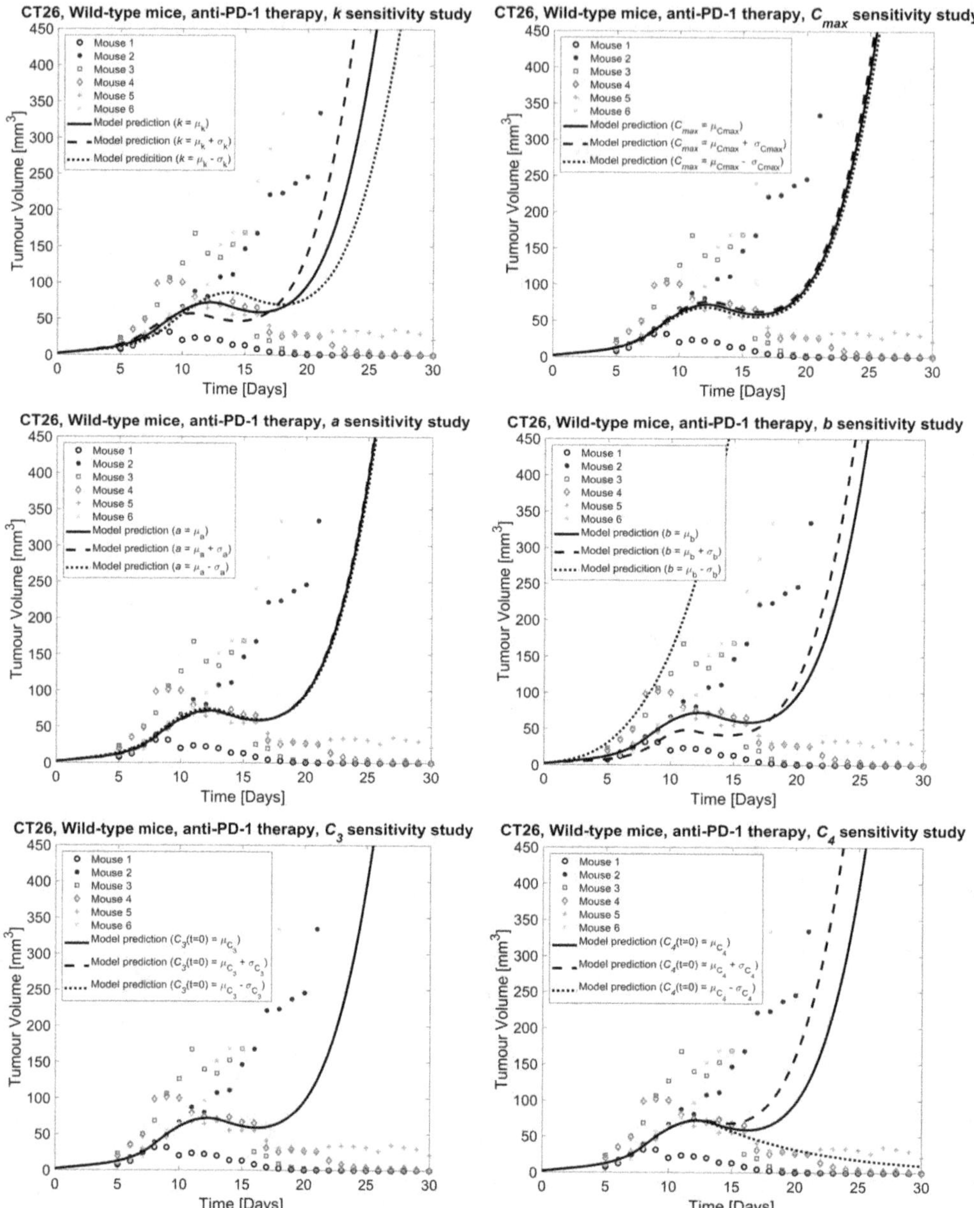

Fig. 4. Growth of CT26 tumors receiving anti-PD-1 immunotherapy and sensitivity analyses. Markers represent tumor growth data of individual mice. Solid line represents model prediction using mean parameter values μ_i from Table 1. Dashed lines represent sensitivity studies where the mean value μ_i of i-th parameter was increased for one standard deviation σ_i. Dotted lines represent sensitivity studies where the mean value μ_i of i-th parameter was decreased for one standard deviation σ_i. Top left: k sensitivity study, top right: C_{max} sensitivity study, middle left: a sensitivity study, middle right: b sensitivity study, bottom left: C_3 ($t = 0$) sensitivity study, bottom right: C_4 ($t = 0$) sensitivity study.

4. Discussion

This study was designed to explore the question why the majority of patients still fail to respond to anti-PD-1 immunotherapy. In order to identify tumor characteristics associated with treatment failure, we developed a computational model capable of simulating tumor response to anti-PD-1 immunotherapy, which was validated with on-site *in vitro* and *in vivo* experiments.

The computational model used in this work is an extension of our previous computational modeling of anti-PD-1 immunotherapy, where we identified major histocompatibility complex (MHC) class I expression and, conditionally, pro-grammed-death-1 ligand (PD-L1) expression on tumor cells as possible biomarkers of response to anti-PD-1 immunotherapy.[17] This motivated us to measure the expression of these proteins on tumor cell membranes experimentally using flow cytometry. The ability of flow cytometry to measure the expression of both proteins (MHC class I, and PD-L1) on each individual tumor cell simultaneously, led us to modify the previous model to better describe the physical reality. In the previous model we assumed only one group of tumor cells, and their elimination due to interaction with the immune system (as well as vice versa) was, among other, proportional to the fraction of tumor cells expressing MHC class I, and proportional to the fraction of tumor cells (not) expressing PD-L1. Thus, MHC class I and PD-L1 expression were model parameters regulating strength of the immune response. Unlike the previous model, in the current study we assumed (consistently with our flow-cytometric measurements) four different groups of tumor cells (MHC-I$^+$ PD-L1$^+$, MHC-I$^+$ PD-L1$^-$, MHC-I$^-$ PD-L1$^+$, and MHC-I$^-$ PD-L1$^-$). MHC class I and PD-L1 expression were incorporated in the model as initial conditions, quantifying the number of cells in each tumor cell subpopulation at t=0. Although it may seem like a minor modification of the model, it reflects an important conceptual difference. Namely, the two tumor cell subpopulations with negative MHC class I expression (MHC-I$^-$ PD-L1$^+$, and MHC-I$^-$ PD-L1$^-$) are now assumed to be intrinsically resistant, because MHC-I$^-$ tumor cells cannot be recognized by T cells.[21] Therefore, they are allowed to proliferate unrestricted by the T cell response. Such type of resistance was not modeled in the previous study, where, according to the model equations, all tumor cells could be eventually eliminated if the T cell response was strong enough.

The existence of such resistant subpopulations of tumor cells (confirmed by flow cytometry) and their important impact on the results of simulations (as shown in sensitivity analyses), are the major findings of this study. Although the measured fractions of MHC-I$^-$ subpopulations at $t = 0$ were very low (order of magnitude 0.01–0.1%) (Table 2), the results of simulations showed that even such small amount of resistant tumor cells was sufficient to ultimately prevail in the latter stages of tumor growth due to the immune selection pressure.[34] Flow-cytometric analysis of both cell lines (4T1 mammary carcinoma, and CT26 colon carcinoma) revealed that MHC-I$^-$ tumor subpopulations are more likely to be present in the initial population of 4T1 compared to the CT26, in which we detected MHC-I$^-$ PD-L1$^-$ subpopulation only in one of five measurements, and MHC-I$^-$ PD-L1$^+$ subpopulation in zero of five measurements (Table 2). This might explain why complete responses to anti-PD-1 immunotherapy were observed only in CT26 cell line (Fig. 4). Also the results of sensitivity analyses suggested a hypothesis that complete response to anti-PD-1 immunotherapy might be possible only in the case of homogeneous MHC-I$^+$ tumors, where all tumor cells present their antigens via MHC class I ($C_3 = C_4 = 0$). An interesting future study would be to surgically excise the responding CT26 tumors slightly before they disappear, and flow-cytometrically analyze MHC class I expression of the individual tumor cells.

In reality this hypothesis might be less stringent, because MHC-I$^-$ tumor cells can be recognized and eliminated by natural killer (NK) cells, which function in a non-MHC-restricted fashion and are not included in our model.[35,36] NK cells could have been included in the model similar to T cells, except that they would have interacted only with MHC-I$^-$ tumor cell subpopulations, but at the cost of additional free parameters. However, NK cells are also known for their poor ability to infiltrate solid tumors; therefore, their exclusion from our current model was reasonable.[37] Additionally, tumor cells have developed several strategies to avoid NK immune surveillance. For example, tumor cells can inhibit NK cells via the PD-1/PD-L1 pathway (similar to the interaction with T cells), secrete different cytokines (e.g., transforming growth factor-beta (TGF-β), and kynurenine) which downregulate NKG2D – the main activating receptor on NK cells, or alternatively, shed MHC class I polypeptide-related sequences A and B (MICA, MICB) from tumor cell surface, which normally act as ligands for NKG2D.[36] Nevertheless, since our model highlights the importance of the resistant MHC-I$^-$ tumor cell subpopulations, even if initially present in very small relative amount, one possible strategy to enhance the benefit of anti-PD-1 immunotherapy could be to combine anti-PD-1 immunotherapy with NK cell-based immunotherapies (e.g., agonist antibodies, blockade of inhibitory receptors, and NK adoptive therapy).[38] For example, one such combination therapy targeting both the innate (NK cells) and adaptive immune system (T cells), has shown promising results for treatment of relapsed follicular lymphoma.[39] Unfortunately, NK cell-based immunotherapies are currently in the shadow of T cell-based immunotherapies, with the majority of them still being in the research phase.

Another possible therapeutic approach, which has already been extensively studied and proved to be beneficial in some pre-clinical and clinical studies, is the combination of anti-PD-1 immunotherapy (or other immune checkpoint inhibitors) and radiotherapy.[40] Besides the direct cytotoxic effect, radiotherapy also promotes the release of tumor neoantigens via the special type of cell death termed "immunogenic cell death (ICD)".[41] Therefore, radiotherapy acts as an *in situ* cancer vaccine, which, in combination with anti-PD-1 immunotherapy, additionally enhances the immune response. However, based on the results of our simulations, we could speculate that an equally important beneficial effect of radiotherapy is the time- and dose-dependent increase of MHC class I expression, which was observed in the study by Reits and colleagues.[42] An enriched number of tumor neoantigens alone would not be expected to be a sufficient prerequisite for a successful combination therapy, if tumor cells would not express functional MHC class I. In such case, the resistant MHC-I$^-$ subpopulations could ultimately prevail, similarly as they prevail in our simulations of anti-PD-1 monotherapy.

Interesting as well are the results of sensitivity studies of parameter b (infiltration rate of T cells into tumor) in the case of CT26 cell line (Fig. 4, middle right). High sensitivity of the parameter b can explain the behavior of two tumors, which did not respond to anti-PD-1 therapy at all (Fig. 4). According to our model, in these two cases infiltration of T cells into tumor was too weak to observe at least an initial delay in tumor growth, which was predicted by the simulations performed with average parameter values. In general, high sensitivity of parameter b in CT26 is a consequence of relatively high standard deviation of parameter b (σ_b), obtained by fitting the model to the growth curves of untreated tumors in wild-type mice (Fig. 2, right). Growth of these tumors was surprisingly variable among the six examined mice. From these results we could speculate that intrinsic growth of CT26 tumors, as well as the response to anti-PD-1 therapy, also depends on some random biological effects associated with infiltration of T cells, possibly on randomness in vasculature structure development, or spatial accumulation of immune suppressive cells in the tumor microenvironment (e.g., regulatory T cells (T_{reg}),[43] myeloid-derived suppressor cells (MDSC),[44] etc.).

Tumor responses to anti-PD-1 immunotherapy most likely depend on many other (known or unknown) biological phenomena not included in our (intentionally) simplistic computational model, which remains the main limitation of our work. However, our future aim is to build more complex models hand in hand with new *in vitro* or *in vivo* experiments, in order to avoid additional free parameters. A high number of free parameters usually makes the model prone to overfitting, and consequently reduces its predictive power. Therefore, in our view, this study represents an exemplary case of an in-depth collaboration between the researchers in immuno-oncology developing theoretical computational/mathematical models, and the researchers working in pre-clinical and clinical settings. Such collaborations are absolutely necessary in an effort to translate computational models into routine pre-clinical and clinical practice.

5. Conclusion

After successful integration of the data obtained by on-site *in vitro* and *in vivo* experiments, our model showed satisfactory predictive power. The results of sensitivity analyses highlighted the impact of resistant (MHC class I negative) tumor cell subpopulations, which could be, even if initially present in very small amounts, responsible for treatment failure. This hypothesis could be used as a rationale for combining anti-PD-1 immunotherapy with a therapy aimed to stimulate natural killer (NK) cell activity. Such combined therapy could be studied with future extensions of our model; however, additional model parameters describing NK cell infiltration into tumor, NK-mediated tumor cell killing, and the effect of NK-stimulating immunotherapeutic, should be measured experimentally. Similarly, any other increase in model complexity should be supported by experiments.

Acknowledgments

The authors acknowledge the financial support from the Slovenian Research Agency (research core funding P1-0389, and P3-0003) and University of Wisconsin Carbone Cancer Center core grant (P30 CA014520).

References

1. A. Ribas, I. Puzanov, R. Dummer, D. Schadendorf, O. Hamid, C. Robert *et al.*, Pembrolizumab versus investigator-choice chemotherapy for ipilimumab-refractory melanoma (KEYNOTE-002): A randomised, controlled, phase 2 trial, *Lancet Oncol.* **16**, 908–918 (2015).
2. E. B. Garon, N. A. Rizvi, R. Hui, N. Leighl, A. S. Balmanoukian, J. P. Eder *et al.*, Pembrolizumab for the Treatment of Non–Small-Cell Lung Cancer, *N. Engl. J. Med.* **372**, 2018–2028 (2015).
3. J. Couzin-Frankel, Breakthrough of the year 2013. Cancer immunotherapy, *Science* **342**, 1432–1433 (2013).
4. J. Wolchok, Putting the immunologic brakes on cancer, *Cell* **175**, 1452–1454 (2018).
5. D. Kabacaoglu, K. J. Ciecielski, D. A. Ruess and H. Algül, Immune checkpoint inhibition for pancreatic ductal adenocarcinoma: Current limitations and future options, *Front Immunol.* **9**, 1878 (2018).
6. R. S. Herbst, P. Baas, D.-W. Kim, E. Felip, J. L. Pérez-Gracia, J.-Y. Han *et al.*, Pembrolizumab versus docetaxel for previously treated, PD-L1-positive, advanced non-small-cell lung cancer (KEYNOTE-010): A randomised controlled trial, *Lancet.* **387**, 1540–1550 (2016).
7. J. Tang, A. Shalabi and V. M. Hubbard-Lucey, Comprehensive analysis of the clinical immuno-oncology landscape, *Ann. Oncol.* **29**, 84–91 (2018).
8. R. Dillman and J. Koziol, A mathematical model of monoclonal antibody therapy in leukemia, *Math. Model.* **9**, 29–35 (1987).
9. C. Gong, O. Milberg, B. Wang, P. Vicini, R. Narwal, L. Roskos *et al.*, A computational multiscale agent-based model for simulating spatio-temporal tumour immune response to PD1 and PDL1 inhibition, *J. R. Soc. Interface* **14**, 20170320 (2017).
10. X. Lai and A. Friedman, Combination therapy for melanoma with BRAF/MEK inhibitor and immune checkpoint inhibitor: A mathematical model, *Lai Friedman BMC Syst. Biol.* **11** (2017).
11. X. Lai and A. Friedman, Combination therapy of cancer with cancer vaccine and immune checkpoint inhibitors: A mathematical model, Haass NK, editor, *PLoS One* **12**, e0178479 (2017).
12. R. Serre, S. Benzekry, L. Padovani, C. Meille, N. André, J. Ciccolini *et al.*, Mathematical modeling of cancer immunotherapy and its synergy with radiotherapy, *Cancer Res.* **76**, 4931–4940 (2016).
13. Y. Kosinsky, S. J. Dovedi, K. Peskov, V. Voronova, L. Chu, H. Tomkinson *et al.*, Radiation and PD-(L)1 treatment combinations: Immune response and dose optimization via a predictive systems model, *J. Immunother Cancer* **6**, 17 (2018).

14. H. Wang, O. Milberg, I. H. Bartelink, P. Vicini, B. Wang, R. Narwal *et al.*, *In silico* simulation of a clinical trial with anti-CTLA-4 and anti-PD-L1 immunotherapies in metastatic breast cancer using a systems pharmacology model, *R. Soc. Open Sci.* **6**, 190366 (2019).

15. J.-L. Yu and S. R.-J. Jang, A mathematical model of tumor-immune interactions with an immune checkpoint inhibitor, *Appl. Math. Comput.* **362**, 124523 (2019).

16. M. Jafarnejad, C. Gong, E. Gabrielson, I. H. Bartelink, P. Vicini, B. Wang *et al.*, A computational model of neoadjuvant PD-1 inhibition in non-small cell lung cancer, *AAPS J.* **21**, 79 (2019).

17. D. Valentinuzzi, U. Simončič, K. Uršič, M. Vrankar, M. Turk and R. Jeraj, Predicting tumour response to anti-PD-1 immunotherapy with computational modelling, *Phys. Med. Biol.* **64**, 025017 (2019).

18. S. Kleffel, C. Posch, S. R. Barthel, H. Mueller, C. Schlapbach, E. Guenova *et al.*, Melanoma cell-intrinsic PD-1 receptor functions promote tumor growth, *Cell* **162**, 1242–1256 (2015).

19. A. R. Sanchez-Paulete, F. J. Cueto, M. Martinez-Lopez, S. Labiano, A. Morales-Kastresana, M. E. Rodriguez-Ruiz *et al.*, Cancer immunotherapy with immunomodulatory Anti-CD137 and Anti-PD-1 monoclonal antibodies requires BATF3-dependent dendritic cells, *Cancer Discov.* **6**, 71–79 (2016).

20. N. E. Scharping, A. V. Menk, R. D. Whetstone, X. Zeng and G. M. Delgoffe, Efficacy of PD-1 blockade is potentiated by metformin-induced reduction of tumor hypoxia, *Cancer Immunol. Res.* **5**, 9–16 (2017).

21. R. N. Germain, MHC-dependent antigen processing and peptide presentation: Providing ligands for T lymphocyte activation, *Cell* **76**, 287–299 (1994).

22. H. Dong, S. E. Strome, D. R. Salomao, H. Tamura, F. Hirano, D. B. Flies *et al.*, Tumor-associated B7-H1 promotes T-cell apoptosis: A potential mechanism of immune evasion, *Nat. Med.* **8**, 793–800 (2002).

23. S. Benzekry, C. Lamont, A. Beheshti, A. Tracz, J. M. L. Ebos, L. Hlatky *et al.*, Classical mathematical models for description and prediction of experimental tumor growth, Mac Gabhann F, editor, *PLoS Comput. Biol.* **10**, e1003800 (2014).

24. N. A. Rizvi, M. D. Hellmann, A. Snyder, P. Kvistborg, V. Makarov, J. J. Havel *et al.*, Mutational landscape determines sensitivity to PD-1 blockade in non-small cell lung cancer, *Science* (80-). **348**, 124–128 (2015).

25. M. W. L. Teng, S. F. Ngiow, A. Ribas and M. J. Smyth, Classifying cancers based on T-cell infiltration and PD-L1, *Cancer Res.* **75**, 2139–2145 (2015).

26. A. V. Chvetsov, Tumor response parameters for head and neck cancer derived from tumor-volume variation during radiation therapy, *Med. Phys.* **40**, 034101 (2013).

27. J. C. Castle, M. Loewer, S. Boegel, A. D. Tadmor, V. Boisguerin, J. de Graaf *et al.*, Mutated tumor alleles are expressed according to their DNA frequency, *Sci. Rep.* **4**, 4743 (2015).

28. K. Sakuishi, L. Apetoh, J. M. Sullivan, B. R. Blazar, V. K. Kuchroo and A. C. Anderson, Targeting Tim-3 and PD-1 pathways to reverse T cell exhaustion and restore anti-tumor immunity, *J. Exp. Med.* **207**, 2187–2194 (2010).

29. J. R. Brahmer, C. G. Drake, I. Wollner, J. D. Powderly, J. Picus, W. H. Sharfman *et al.*, Phase I study of single-agent anti–programmed death-1 (MDX-1106) in refractory solid tumors: Safety, clinical activity, pharmacodynamics, and immunologic correlates, *J. Clin. Oncol.* **28**, 3167–3175 (2010).

30. R. J. De Boer, D. Homann and A. S. Perelson, Different dynamics of CD4+ and CD8+ T cell responses during and after acute lymphocytic choriomeningitis virus infection, *J. Immunol.* **171**, 3928–3935 (2003).

31. D. M. Gakamsky, I. F. Luescher, A. Pramanik, R. B. Kopito, F. Lemonnier, H. Vogel *et al.*, CD8 kinetically promotes ligand binding to the T-cell antigen receptor, *Biophys. J.* **89**, 2121–2133 (2005).

32. S. I. S. Mosely, J. E. Prime, R. C. A. Sainson, J.-O. Koopmann, D. Y. Q. Wang, D. M. Greenawalt *et al.*, Rational selection of syngeneic preclinical tumor models for immunotherapeutic drug discovery, *Cancer Immunol. Res.* **5**, 29–41 (2017).

33. M. J. Lindstrom and D. M. Bates, Nonlinear mixed effects models for repeated measures data, *Biometrics* **46**, 673–687 (1990).

34. F. Garrido, F. Ruiz-Cabello and N. Aptsiauri, Rejection versus escape: The tumor MHC dilemma, *Cancer Immunol. Immunother.* **66**, 259–271 (2017).

35. G. Trinchieri, Biology of natural killer cells, *Adv. Immunol.* 187–376 (1989).

36. K. Choucair, J. R. Duff, C. S. Cassidy, M. T. Albrethsen, J. D. Kelso, A. Lenhard *et al.*, Natural killer cells: A review of biology, therapeutic potential and challenges in treatment of solid tumors, *Future Oncol.* **15**, 3053–3069 (2019).

37. G. Sconocchia, R. Arriga, L. Tornillo, L. Terracciano, S. Ferrone and G. C. Spagnoli, Melanoma cells inhibit NK cell functions–letter, *Cancer Res.* **72**, 5428–5429 (2012).

38. C. Guillerey, N. D. Huntington and M. J. Smyth, Targeting natural killer cells in cancer immunotherapy, *Nat. Immunol.* **17**, 1025–1036 (2016).

39. J. R. Westin, F. Chu, M. Zhang, L. E. Fayad, L. W. Kwak, N. Fowler *et al.*, Safety and activity of PD1 blockade by pidilizumab in combination with rituximab in patients with relapsed follicular lymphoma: A single group, open-label, phase 2 trial, *Lancet. Oncol.* **15**, 69–77 (2014).

40. S. C. Formenti and S. Demaria, Combining radiotherapy and cancer immunotherapy: A paradigm shift, *J. Natl. Cancer Inst.* 256–265 (2013).

41. G. Kroemer, L. Galluzzi, O. Kepp and L. Zitvogel, Immunogenic Cell Death in Cancer Therapy, *Annu. Rev. Immunol.* **31**, 51–72 (2013).

42. E. A. Reits, J. W. Hodge, C. A. Herberts, T. A. Groothuis, M. Chakraborty, E. K. Wansley *et al.*, Radiation modulates the peptide repertoire, enhances MHC class I expression, and induces successful antitumor immunotherapy, *J. Exp. Med.* **203**, 1259–1271 (2006).

43. S. Sakaguchi, T. Yamaguchi, T. Nomura and M. Ono, Regulatory T cells and immune tolerance, *Cell* **133**, 775–787 (2008).

44. D. I. Gabrilovich and S. Nagaraj, Myeloid-derived suppressor cells as regulators of the immune system, *Nat. Rev. Immunol.* **9**, 162–174 (2009).

Chapter 8

The Physical Microenvironment of Tumors: Characterization and Clinical Impact

Matthew R. Zanotelli[1,2,*], Neil C. Chada[2,†], C. Andrew Johnson[2,‡]
and Cynthia A. Reinhart-King[2,§]

[1]*Nancy E. and Peter C. Meinig School of Biomedical Engineering
Cornell University, Weill Hall
Ithaca, NY 14583, USA*

[2]*Department of Biomedical Engineering
Vanderbilt University
2414 Highland Avenue
Nashville, TN 37235, USA*
**mrz36@cornell.edu*
†neil.chada@vanderbilt.edu
‡andrew.johnson@vanderbilt.edu
§cynthia.reinhart-king@vanderbilt.edu

The tumor microenvironment plays a critical role in tumorigenesis and metastasis. As tightly controlled extracellular matrix homeostasis is lost during tumor progression, a dysregulated extracellular matrix can significantly alter cellular phenotype and drive malignancy. Altered physical properties of the tumor microenvironment alter cancer cell behavior, limit delivery and efficacy of therapies, and correlate with tumorigenesis and patient prognosis. The physical features of the extracellular matrix during tumor progression have been characterized; however, a wide range of methods have been used between studies and cancer types resulting in a large range of reported values. Here, we discuss the significant mechanical and structural properties of the tumor microenvironment, summarizing their reported values and clinical impact across cancer type and grade. We attempt to integrate the values in the literature to identify sources of reported differences and commonalities to better understand how aberrant extracellular matrix dynamics contribute to cancer progression. An intimate understanding of altered matrix properties during malignant transformation will be crucial in effectively detecting, monitoring, and treating cancer.

Keywords: Tumor microenvironment; extracellular matrix; mechanobiology; prognostic markers.

1. Introduction

The tumor microenvironment is mechanically and biologically dynamic and is progressively remodeled by tumor cells during tumor progression to further drive malignancy.[1–5] Cancer cells actively remodel the tumor microenvironment resulting in altered biophysical cues that change cell behavior and often lead to further remodeling. These bi-directional mechanical interactions between cancer cells and the surrounding microenvironment create a reciprocal exchange and positive tumorigenic feedback loop.[2] Solid tumors are composed of a heterogeneous collection of fibroblasts, immune cells, and cancer cells confined within a stroma containing an evolving extracellular matrix (ECM), vasculature, and lymphatic vessels.[6–8] A critical component of the tumor microenvironment is the ECM.[1–3,9] The ECM, which is a complex assembly of macromolecules with distinct biophysical properties, is largely composed of interconnected fibrillar and nonfibrillar collagens, elastic fibers, and glycosaminoglycan (GAG)-containing noncollagenous glycoproteins; however, its specific composition is dependent on the tissue.[2,10–13] Notably, this collection of proteins not only serves a structural and biochemical role, but it can also provide contextual information responsible for cell behavior such as proliferation, differentiation, survival, and migration.[1,2,9,14,15] Together, the physical properties of the matrix play a crucial role in regulating tissue development as well as maintaining tissue homeostasis, and dysregulation of the ECM can lead to many pathological states including cancer.[1–3,15,16]

Due to the significant impact of altered matrix mechanics on tumorigenesis, characterizing the physical tumor microenvironment is critically important to better understand the dynamic mechanoreciprocity in cancer. Several notable mechanical and structural properties of tumor tissue have been correlated with tumor malignancy.[1,2,17–20] Many studies have attempted to characterize the physical microenvironment of tumors and recent developments in microscale interrogation of tissue mechanics have aimed to more accurately measure and quantify critically important ECM remodeling that occurs during tumor progression.[21–24] While it is clear that physical properties differ between normal and tumor tissue, there is a large range of values reported in the literature. These differences in reported values for tumor-associated ECM physical properties can be attributed to the wide variety of available modalities used to collect mechanical data, differences in settings used during sample measurement, sample-to-sample variation, and heterogeneity within samples.[25–27] Together, these variations in physical characterization of the tumor microenvironment obscure specific and numerical interpretations of the influence of the tumor-associated ECM on malignancy. Accurately characterizing the physical changes to the tumor microenvironment during tumorigenesis is an important step in developing a quantitative understanding of the dynamic ECM alterations that drive malignancy.

Here, we describe ECM changes during solid tumor progression across tumor types as well as grades and highlight quantitative efforts to characterize the mechanical and structural properties of the tumor microenvironment. Other excellent reviews have outlined the values of various mechanical properties within the tumor microenvironment across cancer types.[4,24,28–31] In this review, we focus on the characterization of important clinical markers of tumor progression and aggressiveness, matrix stiffness, anisotropy, and porosity, summarizing the relevant ranges reported in the literature and describing their clinical impact. We discuss the methodologies used to obtain physical measurements of tumor samples and discuss commonalities as well as differences in reported values to better identify ECM alterations during tumor progression. Additionally, we attempt to identify the current limitations of existing methods and how they can be improved upon to better our knowledge of the physical properties of the tumor microenvironment. An improved understanding of the significant biophysical characteristics of the tumor-associated ECM will be critical in not only the detection of tumor progression but also a response to a therapeutic treatment. Finally, we examine attempts to therapeutically target the physical microenvironment of tumors and cellular response to aberrant matrix mechanics for cancer treatment.

2. Altered Physical Properties of the Tumor Microenvironment

In the last few decades, there has been increasing interest in how physical cues in the tumor microenvironment can influence cancer cells and promote tumorigenesis. The ECM commonly becomes dysregulated and disorganized during tumor progression, creating tissue with markedly distinct mechanical and architectural properties compared to normal tissue.[1,2,4,18,32–34] Now, it is well established that altered ECM can drive malignancy (Fig. 1).[1] Tissue mechanics are sensed through cell-generated force loading on integrins adhered to the matrix, linking the actomyosin cytoskeleton with the matrix.[1] Altered biophysical cues from the matrix significantly influence cytoskeletal regulation, contractility, and epithelial-to-mesenchymal transition to drive a more aggressive phenotype.[2,5,14] Together, the biophysical cues from the tumor-associated ECM can promote the hallmarks of cancer and, as such, an intimate understanding of ECM dynamics and interactions during tumor progression is vital to developing successful therapy programs.[15]

3. Mechanical Properties of the Tumor ECM

Altered tissue mechanics are observed in many pathologies, including cancer, and ECM stiffness is now appreciated as a prominent mechanical cue that not only precedes disease states but also drives disease progression by altering cellular behavior.[1,4,35] The biomechanical properties of the ECM are tightly regulated to maintain tissue homeostasis,[36,37] and sustained dysregulation of ECM dynamics is critically important in promoting tumorigenesis.[2,14,38] Increases in matrix stiffness have been found to be critical in disrupting tensional homeostasis and driving a malignant phenotype.[39] Furthermore, changes in matrix stiffness can not only alter the behavior of cancer cells but also significantly influence surrounding stromal cells to further promote tumor progression.[40–42]

The mechanical properties of the tissue are derived, in part, from the combined components of the basement membrane and interstitial ECM.[1] The basement

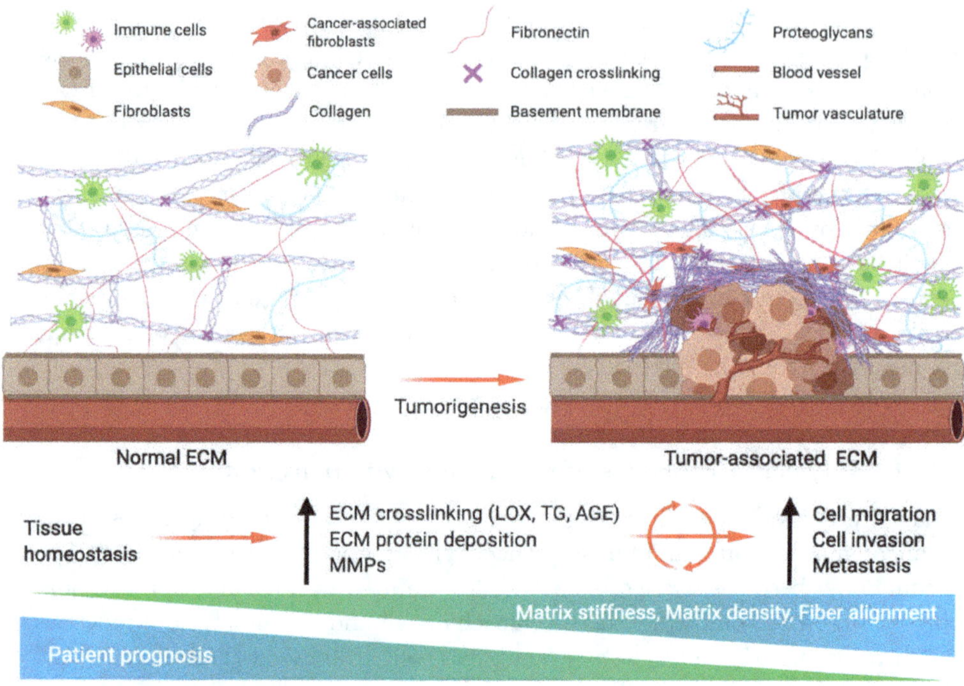

Fig. 1.   The altered physical microenvironment of tumors. Alterations to the mechanical and structural properties of the tumor microenvironment drive malignancy, impact therapeutic efficacy, and are correlated with poor patient prognosis.

membrane is a specialized meshwork of proteins largely composed of collagen type IV and VII, laminin, fibronectin, and linker proteins that provide anchorage and separate cells from the stroma.[1,43] While basement membrane stiffening occurs with age, the increase in stiffness can be exacerbated in disease states like cancer.[44,45] Clinically, there have been attempts to use structural proteins of the basement membrane-like collagen type IV, laminin, and fibronectin as prognostic markers; however, the results have been somewhat inconclusive, as some studies have shown increased protein deposition correlated with lower survival, while others have shown the opposite.[46] The interstitial ECM is generally composed of fibrous collagen type I, proteoglycans, and glycoproteins and provides much of the tensile strength of tissue.[1] ECM homeostasis is maintained through a complex balance of matrix modification, degradation, and synthesis. During tumor progression, the interstitial ECM is remodeled and increasingly stiffens through two primary mechanisms: (1) ECM modification via crosslinking and degradation and (2) ECM synthesis and deposition.[2,47] While both mechanisms act to stiffen tumor tissue, they have profoundly different effects on cell function.

3.1. *ECM modification and degradation*

Collagen crosslinking within the matrix can occur through more regulated, enzymatic methods as well as nonregulated, nonenzymatic methods.[2] Together, these processes of matrix crosslinking act to dynamically stiffen the matrix, a significant ECM modification observed during cancer progression. Additionally, the ECM is actively degraded by proteases to produce significant ECM remodeling during tumorigenesis.[1,37,48]

3.1.1. *Enzymatic crosslinking*

Regulated crosslinking is predominately driven by lysyl oxidase (LOX) and the LOX family of secreted amine oxidases.[49,50] The LOX family consists of LOX and lysyl oxidase-like (LOXL) 1–4. LOX is overexpressed in a wide range of cancers,[51] and its overexpression is correlated with poor prognosis in breast,[51,52] colon,[53] gastric,[54] lung cancer,[55] as well as head and neck cancer.[56,57] LOX is produced and modified in the cytosol of cells, before being secreted into the surrounding ECM where it crosslinks both collagen and elastin to increase matrix stiffness.[58] Mechanistically, LOX acts by oxidizing collagen lysine residues and creating reactive aldehydes that then interact with lysine and hydroxylysine residues producing crosslinks.[50] Secreted LOX can also be transferred back into the cytosol to drive intracellular signaling and invasion through increased focal adhesion activity and altered matrix adhesion.[52] In addition to crosslinking and stiffening the primary tumor, LOX is proposed to play an important role in remodeling the pre-metastatic niche and recruiting additional cancer cells.[36] LOX has also been shown to be essential in driving hypoxia-induced metastasis.[52,59,60]

Tissue transglutaminase 2 (TG2) overexpression has also been implicated in cancer progression and linked to poor patient outcomes. Analysis of the Cancer Genome Atlas (TGCA) of pancreatic tumors demonstrated that TG2 overexpression is associated with reduced patient survival.[61] Furthermore, *in vivo* data of a xenograft pancreatic cancer model has shown that increased TG2 secretion by pancreatic cancer cells stiffens the surrounding matrix and promotes fibroblast proliferation, together which stimulate cancer cell proliferation.[61] Similarly, TG2 activated fibroblasts have been observed to increase secretion of laminin 1 which aids in protecting pancreatic cancer cells from gemcitabine treatment.[62] In renal cell carcinoma (RCC), TG2 overexpression has also been associated with disease progression mediated through increased actin stress fiber formation and increased $\beta1$ integrin binding.[63] *In vitro* work with human osteoblasts and human foreskin dermal fibroblasts shows the treatment of collagen type I matrices with TG2 improved spreading and proliferation and well as reduced protease degradation of the matrix.[64] More recent work has also demonstrated TG2 is an important regulator of tumor cell tensional homeostasis.[65]

3.1.2. *Nonenzymatic crosslinking*

Nonenzymatic collagen crosslinking typically occurs via glycation,[66] and the family of advanced glycation end products (AGEs) are responsible for much of the nonenzymatic crosslinking found in solid tumors.[67] Signaling through AGEs and the receptor for advanced glycation end products (RAGEs) has been shown to be pro-inflammatory and is implicated in a variety of diseases, including cancer.[68] While these matrix crosslinks act to stiffen the matrix, they act on a longer timescale compared to enzymatic crosslinking, as structural ECM proteins demonstrate significant longevity *in vivo*.[69,70] AGEs accumulate normally with age and can have a particularly significant effect on ECM components due to their low turnover rate.[70,71] Hyperglycemia, such as seen in diabetes, also contributes to AGE formation. In collagen, there are several AGE crosslinks that can occur. Glucosepane is the most commonly observed AGE collagen crosslink found in diabetic individuals.[72] Recent work has shown that LOX-mediated and sugar mediated collagen crosslinking occur via the same lysine sites in type I collagen.[73]

The body has natural reductase enzymes to remove AGE precursors, and several inhibitors have been developed to target AGEs before the formation of irreversible AGE adducts. The inhibitor MnmC can reverse carboxyethyl-lysine and carboxymethyl-lysine AGEs.[74] Alagebrium (ATL-711) disrupts the dicarbonyl crosslink intermediate, and amadoriases and DJ-1 enzymes remove glycated moieties before mature AGEs can form.[75–77] Many of these inhibitors have been labeled AGE-breakers; however, all of these inhibitors target AGE intermediates and no inhibitor has been reported yet that can break, or reverse mature AGEs once formed. As such, it is not yet known whether reversing AGE-based crosslinks would affect tumor progression or slow metastasis.

3.1.3. *Enzymatic degradation*

During tumor progression, tissue homeostasis is disrupted, and the healthy balance between ECM synthesis and enzymatic degradation is lost due to excessive matrix synthesis and increased protease activity.[78,79] Matrix metalloproteinase (MMP) activity is upregulated in almost every type of cancer and MMP expression is often correlated with poor patient prognosis. MMPs are proteolytic enzymes responsible for basement membrane and interstitial ECM degradation as well as processing bioactive mediators.[79] These enzymes are secreted as proenzymes and subsequently activated by oxidation or proteolytic cleavage in the ECM. The abnormal biochemical features of the tumor microenvironment, such as elevated hypoxia and the high level of reactive oxygen species, also contribute to increased MMP expression.[80,81] The adamalysin family of proteins includes ADAMs (a disintegrin and metalloproteinase) and ADAMTs (ADAMs with a thrombospondin motif) and are membrane-associated metalloproteinases that have also been implicated in tumorigenesis and tumor progression.[37,82] Tissue inhibitors of metalloproteinases (TIMPs) regulate MMP activity and the ratio of MMPs/TIMPs in tissue controls the rate of

ECM degradation.[83] In normal tissue, this ratio is low, but in many pathological conditions, including cancer, the MMP/TIMP ratio is increased.[83] The change in MMP/TIMP ratio in tumors is due to increased secretion of MMPs by cancer cells and stromal cells including cancer-associated fibroblasts and macrophages.[83]

In addition to ECM remodeling, proteases play a significant role in cell migration. MMP-2 and MMP-9 are key contributors to cancer cell proteolytic activity. These MMPs are released by cancer cell invadopodia to degrade a path through the matrix and enhance invasion.[84] MMP-2 and MMP-9 are both gelatinases also called type IV collagenases and preferentially degrade type IV collagen and laminin, which are the primary components of the basement membrane.[85] Degradation of the basement membrane is thought to be a critical step for metastasis; however, some evidence indicates that the basement membrane is softened and pushed apart during extravasation rather than being completely degraded.[86] During the degradation of the matrix, MMPs can also unmask hidden binding sites and release sequestered growth factors stored in the ECM, which further promotes invasion and reduces the anti-tumor immune response.[87]

3.2. *ECM synthesis and deposition*

Tumor tissue is markedly stiffer than healthy tissue,[21,31,88,89] and increased ECM deposition accounts for a majority of the increased stiffness in tumor and surrounding tissue.[19] Desmoplasia, the growth of dense stroma, of tumor tissue results in significant deposition of the ECM proteins collagen type I and fibronectin,[19] altered organization,[15] and enhanced post-translational modifications of ECM proteins.[1] While desmoplasia is observed in other pathological states including fibrosis and is associated with cancer,[90] this fibrotic state is rarely seen in benign tumor and is typically only observed in malignant tumors.[32] Such enhanced ECM deposition is often coupled with the sustained presence of MMPs to progressively destroy the normal ECM and establish a pro-tumorigenic tumor-associated ECM.[2]

ECM density has been identified as both a risk factor in developing cancer as well as a prognostic indicator for diagnosed patients.[91,92] Notably, increased matrix density has been associated with more advanced disease or poor prognostic in breast cancer,[93,94] gastric cancer,[95] RCC,[92] and ovarian cancer.[96] *In vitro* studies have demonstrated that increasing collagen concentration can promote cancer cell proliferation,[97] as well as increase tumor formation and metastasis in a mouse model of mammary tumors.[19] In addition to increased tissue stiffness, increased matrix density can also have a profound effect on integrin binding. Integrin expression controls many cellular functions that are important for the initiation, progression, and metastasis of solid tumors.[98,99] Integrins are prominent cellular adhesion receptors and play key roles as signaling molecules, mechanotransducers, and regulators of cell migration machinery.[100] Altered integrin expression is commonly observed in tumors.[100] Elevated tumor tissue stiffness has been shown to increase integrin clustering, producing heightened mechanotransduction and cell migration.[39] Together, these combined effects of altered ECM mechanics act to drive malignancy and create an altered tumor microenvironment that further promotes tumorigenesis.

3.3. *Measurement and quantification of tumor mechanical properties*

While the cancer biology field now widely accepts that tissue stiffening affects tumor progression, the exact range of stiffness that occurs physiologically and the range at which cells in a particular tumor type can sense changes in stiffness are less clear. To sufficiently understand the effects of dynamic ECM remodeling observed during tumor progression, quantitative assessments of ECM composition and mechanical properties are collected using a variety of techniques (Table 1).

Staining of tissue sections for biochemical signatures via immunohistochemistry or immunofluorescence is the most common and widely used method to visualize ECM composition.[2,112] More specialized techniques including second harmonic generation (SHG), confocal reflectance, and multi-photon imaging can be used to better visualize collagen structure and organization within tissue.[2,58,113] Clinically, methods such as optical coherence tomography (OCT), ultrasound, magnetic resonance imaging (MRI), and sonoelastography have been used to noninvasively assess mechanical properties of tissue.[2,114,115] More sensitive techniques such as atomic force microscopy (AFM), optical and magnetic tweezers, and deformable microdroplet can provide high-resolution images of ECM structure from biopsies,[2,24,116] and force-mapping AFM can provide stiffness measurements at the micron scale.[25] The viscoelastic properties of tissue, which is dictated by its ECM and individual cellular components, also present the opportunity to perform noninvasive imaging and quantitatively assess matrix stiffness. Shear rheology, mechanical compression, and nanoindentation allow for measurements of tissue stiffness and elasticity.[2] However, with all of these techniques, careful consideration must be taken when reviewing the quantitative data on mechanical properties, as differing experimental set-up, collection parameters, and scale can greatly influence results. Below we detail changes in ECM stiffness that occur with tumorigenesis for major solid tumor types and the different modalities used to characterize their mechanical properties (Table 2).

3.3.1. *Breast cancer*

The ECM directs various developmental stages of the breast while perturbation of ECM composition and architecture significantly influences breast tumor progression.[14] During breast tumor progression, the mammary gland becomes increasingly stiff and more collagen-rich through desmoplasia.[39] Increased mammographic density, driven by elevated collagen deposition, is also associated with increased risk of developing breast cancer.[133] Moreover, increased matrix stiffness and ECM remodeling have been shown to contribute to malignant transformation of breast tissue, with the stroma progressively increasing in stiffness from 150 Pa, $\sim$350 Pa, and $\sim$400 Pa for normal, pre-malignant, and tumor tissue, respectively. Some reports

Table 1. Methods used to characterize the mechanical and structural properties of the physical tumor microenvironment (AFM = atomic force microscopy, US = ultrasound, SWE = shear wave elastography, MRE = magnetic resonance elastography, OCT = optical coherence tomography, SHG = second harmonic generation).

	Technique	Input	Output	Comments	Reference
Rheology	Indentation	Stress–strain	Young's modulus	Bulk stiffness measurement, moderate spatial resolution	89
	Confined compression	Stress–strain	Young's modulus	Bulk stiffness measurement	101
	Unconfined compression	Stress–strain	Young's modulus	Bulk stiffness measurement	101
Active microrheology	AFM	Cantilever tip deflection	Local cellular/tissue viscoelasticity	High-resolution, continuous mapping limited to surface measurements	102
Shear imaging	Compression elastography	Probe pressure shear wave generation	Shear modulus	Low-resolution, limited depth penetration	103, 104
	Transient elastography	US measured shear wave velocity	Shear modulus	Low-resolution, limited depth penetration, commonly used clinically	103–105
	SWE	Acoustic radiation force impulse shear wave generation	Shear modulus	Improved spatial resolution, limited penetration depth	103, 104
	MRE	Pneumatic shear wave generation	Shear modulus	High resolution, high penetration depth	106, 107
Imaging	OCT	Low-coherence light images	Matrix structure	High tissue penetration Lower resolution than confocal microscopy	108
	SHG	Second-harmonic light generation	Collagen fiber organization	Noninvasive, label free	109
	Confocal reflectance microscopy	Transmitted light	Collagen fiber organization	High resolution, lower tissue penetration than OCT, noninvasive, label free	110
	Polarized microscopy	Optical phase retardance	Relative stress distribution	Resolution same as optical diffraction limit, only relative stress output, noninvasive, label free	111

Table 2. Summary of changes in tumor tissue stiffness across cancer type and grade, and the methods used for measurement (DCIS = ductal carcinoma *in situ*, IDC = invasive ductal carcinoma, MRE = magnetic resonance elastography, SWE = shear wave elastography, AFM = atomic force microscopy).

Tissue	Sub-type	Measurement method	Young's modulus (kPa)	Reference
Mammary	Adipose Tissue	Indentation	1.9	117
			3.25 ± 0.91	89
	Fibroglandular Tissue		1.8	117
			3.24 ± 0.61	89
	Fibroadenoma		6.41 ± 2.86	89
	DCIS		16.38 ± 1.55	
	Low-Grade IDC		10.40 ± 2.6	
	Intermediate-Grade IDC		19.99 ± 4.2	
	High-Grade IDC		42.52 ± 12.47	
			12	117
Prostate	Normal	Dynamic compression (strain)	15.9	118
		Unconfined compression (stress)	3.8 (0.1 Hz)	119
			16 (150 Hz)	
	Tumor	Dynamic compression (strain)	40.4	118
			110	120
		Unconfined compression (stress)	7.8 (0.1 Hz)	119
			40.6 (150 Hz)	
Pancreas	Normal	Indentation	1.06 ± 0.25	121
		MRE 3T at 40 Hz	2.47 ± 0.11	122
		MRE 1.5T at 40 Hz	1.15 ± 0.17	123
		MRE 1.5T at 60 Hz	2.09 ± 0.33	123
	Pancreatitis	Indentation	2.15 ± 0.41	121
		MRE 1.5T at 60 Hz	2.24–3.56	124
	Tumor	Indentation	5.46 ± 3.18	121
		MRE 1.5T at 60 Hz	3.22–5.11	124
		MRE 1.5T at 60 Hz	6.06 ± 0.49	122
Brain	Normal	SWE	7.3 ± 2.1	125
	Meningioma	SWE	33.1 ± 5.9	125
		Indentation	3.97 ± 3.66	126
	Low-Grade Glioma	SWE	23.7 ± 4.9	125
		Indentation	2.75 ± 1.4	126
		AFM	0.5–1.4	127
	Gliotic Tissue	AFM	0.01–0.18	127
	High-Grade Glioma	SWE	11.4 ± 3.6	125
	Glioblastoma	AFM	0.07–13.5	127
	Metastatic Lesion	SWE	16.7 ± 2.5	125
Colorectal	Normal	Indentation (Venustron)	0.936	128
		AFM	0.8 ± 0.4	129
	Tumor	Indentation (Venustron)	7.51	128
		AFM	2.4 ± 1.83	129
Liver	Normal	MRE	2.2 ± 0.31	130
		MRE 1.5T	2.3 ± 0.3	131
	Fibrosis	MRE	12.1 ± 1.2	130
		MRE 1.5T	5.9 ± 2.5	131
	Tumor	MRE 1.5T	10.1 ± 3.6	131

Table 2. (*Continued*)

Tissue	Sub-type	Measurement method	Young's modulus (kPa)	Reference
	Hepatocellular	MRE 1.5T	10.3 ± 2	131
	Carcinoma	Transient elastography (Fibroscan)	55	132
	Hemangioma	MRE 1.5T	2.7 ± 0.5	131
	Benign Tumor	MRE 1.5T	2.7 ± 0.4	
	Cholangiocarcinoma	MRE 1.5T	16.2 ± 3.4	131
		Transient elastography (Fibroscan)	73.9	132
	Metastatic Lesion	MRE 1.5T	7.6 ± 1.7	131
		Transient elastography (Fibroscan)	66.5	132

have identified tumor stroma to be even as high as 10-fold stiffer than healthy breast tissue.[32,58,134]

Due to the significant increase in tissue stiffness observed with breast cancer progression, breast tissue stiffness has used a measure of breast tumorigenesis and is regularly evaluated via physical palpitation, OCT, sonoelastography, MRI, elastography, and tissue Doppler imaging (TDI) analysis.[25,135,136] The wide variety of methods used to evaluate breast tissue stiffness can produce variability in measurements and highlights the importance of selecting not only method but also input parameters and modeling techniques. Young's modulus for high-grade invasion ductal carcinoma (IDC) has been reported to range from 12 kPa.[137] to 42.52 kPa.[117] The lower range of values was obtained from small pieces of tumor removed from a larger resected tumor; however, much higher values were collected when testing an intact resected tumor. Structural components, such as the margins of a tumor, contribute to stiffness and should be considered when measuring tissue stiffness. Several other factors can impact these types of measurements including normal variations in tissue stiffness, thickness of the tissue slice, age, and menstrual cycle stage of the patient.[138] During the evaluation, processing considerations impact measurement including pre-compression strain, frequency, and correctly selecting "normal" or "tumor" regions in a tissue section to analyze.[89] A more recent study using OCT imaging and axial compressive loading with resected breast cancer samples corresponded more closely with the lower range measurements for IDC at 16.45 ± 1.103 kPa.[135]

3.3.2. *Prostate cancer*

Data on the mechanical properties of human prostate cancer is limited; however, some characterization of clinical samples has been performed. Early experiments using compressive loading to evaluate the mechanical behavior of prostate tissue samples showed a measurable elevation in Young's modulus when comparing cancerous prostate tissue to normal tissue.[139] More recently, mechanical testing has been used with material modeling to better characterize the mechanical properties of human tissue.[140] Characterizing the frequency-dependent complex Young's modulus through stress relaxation testing in combination with viscoelastic tissue modeling of samples from normal and cancerous prostate tissue demonstrated an average complex Young's moduli of 15.9 ± 5.9 kPa for normal tissue and 40.4 ± 15.7 kPa for prostate cancer tissue at a 150 Hz indentation rate.[118] In this study, a Kelvin–Voigt Fractional Derivative (KVFD) model was used to model the tissue, which has been commonly applied for mechanical testing of viscoelastic materials.[140] These values closely align with uniaxial unconfined compression tests measuring stress relaxation data fitted to the KVFD model that found average Young's moduli of 16.0 ± 5.7 kPa for normal tissue and 40.6 ± 15.9 kPa for cancerous tissue.[119] However, prostate tissue has a frequency-dependent modulus that can change with indentation frequency. When measuring these samples using a lower indentation rate of 0.1 Hz, a lower elastic constant was observed with Young's moduli of 3.8 ± 1.8 kPa and 7.8 ± 3.3 kPa for normal and tumor prostate tissue, respectively.[119] Changes in prostate tissue stiffness, determined by dynamic indentation, have additionally been correlated with tissue composition including prostate stroma.[120] A piezoelectric resonance sensor has also been used to evaluate the stiffness of fresh prostate tissue.[141] The measured stiffness, modeled by the weighted tissue proportion, was increased in prostate cancer tissue compared to healthy glandular tissue and verified with histological analysis.[141] When detecting variations in tissue stiffness from the resonance frequency, parameters such as impression depth and impression speed can impact measurements.[142] Notably, using the lowest possible impression speed can allow for the largest amount of measurements and provides higher resolution, but the time required for pathological routine work must also be considered.[141] While changes in testing parameters such as indentation rate during mechanical testing, indenter size, and indenter geometry can affect stiffness values collected from tissue samples, mechanical testing produces relatively similar values across many studies.[143] However, the development of common testing parameters for biopsy samples will significantly aid clinicians in monitoring and treating cancer progression.

Mechanical testing of biopsies provides the most accurate measurement of the stiffness; however, it requires destructive testing on *ex vivo* samples.[21,144] As such, recent research efforts have focused on noninvasive and nondestructive imaging of the elastic properties of biological tissue to evaluate the mechanical properties of prostate tumor tissue. Crawling wave sonoelastography, which excites mechanical sources at offset frequencies to produce shear wave interference patterns that propagate through tissue,[145,146] can be applied to quantitatively assess the elasticity in the prostate.[119] Together, these findings suggest that quantitative sonoelastography can be implemented as a simple and effective real-time method to characterize prostate tissue properties; however, further work is needed to implement this technique *in vivo*.

3.3.3. *Pancreatic cancer*

Pancreatic cancer is extremely lethal, with a 5-year relative survival of only 5%.[147] Like other solid tumors, clinical data indicate that pancreatic tumors stiffen during tumorigenesis,[148] and pancreatic ductal adenocarcinoma (PDAC), which comprises most cases of primary pancreatic tumors, is defined by highly dense areas of fibrosis surrounding cancer cells.[149] As such, its early detection is critical. Currently, standard imaging techniques for detection include ultrasound in combination with contrast-enhanced computed tomography (CT), contrast-enhanced MRI, as well as endoscopic ultrasound.[122] Recently, more direct and quantitative measurements have been made. Resected patient samples measured via mesoscale indentation force–displacement fitted to standard linear solid model showed steady-state modulus of $1.06 \pm 0.25\,kPa$, $2.15 \pm 0.41\,kPa$, and $5.46 \pm 3.18\,kPa$ for normal tissue, inflamed chronic pancreatitis regions, and malignant PDAC tumors, respectively.[121] While this indentation method requires resection of the tissue, the incorporation of both elastic and viscous parameters greatly helps in differentiating between normal and diseased tissue. Shear wave elastography has been used to noninvasively measure pancreatic tumor stiffness; however, it has reported much larger values compared to direct measurements made via indentation.[150] To more accurately measure tissue mechanical properties, elastography algorithms that utilized time-dependent, relaxation parameters are currently being developed.[151–153] Magnetic resonance elastography (MRE) has also been utilized to accurately detect changes in stiffness between normal and pancreatic tumor tissue. Using MRE, normal tissue and cancerous pancreatic tissue demonstrated stiffnesses of $2.47 \pm 0.11\,kPa$ and $6.06 \pm 0.49\,kPa$, respectively.[122] Other groups have also used MRE and found an average stiffness of $2.67\,kPa$ for PDAC tissue and $1.24\,kPa$ for normal tissue and correlated mechanical changes with histological features.[124] Stiffness values for tumor are higher; however, MRE is sensitive to measurement settings that can produce large variations in values and offers lower resolution than standard MR imaging pulse sequences.

3.3.4. *Brain cancer*

Similar to other human tissue, human brain tumors of sufficient size for mechanical characterization can be challenging to acquire. Portions of obtained brain tumors are used for clinical and diagnostic testing and access to the patient can be limited.[154] However, the characterization of human brain tumors that has been done indicates that tissue mechanical properties may play a large role in the progression of brain tumors, as gliomas develop a dense and mechanically challenging ECM.[155] Due to the difficulties in obtaining brain samples, noninvasive methods have been used to estimate the mechanical properties of brain tumors.[155–157] For example, shear wave elastography has been used to differentiate between brain tumors. Using shear wave elastography, normal brain tissue has been measured to have a stiffness of $7.3 \pm 2.1\,kPa$, while brain tumors have been measured to have $33.1 \pm 5.9\,kPa$, $23.7 \pm 4.9\,kPa$, $11.4\,kPa$, and $16.7\,kPa$ for meningiomas, low-grade gliomas, high-grade gliomas, and metastasis, respectively.[155] The relatively high sensitivity and specificity of shear wave elastography make it a good candidate for differentiating benign and malignant lesions as well as quantitatively classifying low-grade and high-grade gliomas *in vivo*. Classifying glioma grade is extremely clinically important, as tumor grade is tied to prognosis and treatment strategies.[92,158] Such increased differentiation of tissue mechanical properties with shear wave elastography would improve upon the use of MR imaging and intraoperative features currently used to differentiation gliomas based on progression and grade.[155]

Recently, tissue-level mechanical testing of freshly isolated human brain tumors has also been conducted.[144,159] Microscale indentation using AFM of fresh-frozen human brain biopsies demonstrated increasing ECM stiffness with increasing grade and aggressiveness of brain cancer.[144] ECM stiffness progressively increased from 10–180 Pa, 50–1,400 Pa, and 70–13,500 Pa for gliotic tissue, diffuse lower-grade gliomas (LGGs), and glioblastomas (GBMs), respectively. Force-displacement data has also been collected using a custom-built indenter to quantify the effective modulus for various pathologies and identified a greater than two-fold increase in elastic modulus of brain tumor samples compared to healthy brain tissue.[159] Such increases in brain tissue stiffness with tumor progression are in agreement with both elastography[155] and AFM measurements;[144] however, force-displacement testing produced values over a much wider range and may not be highly accurate. Human brain tumors were found to range from 0.17–16.06 kPa for various pathologies, with human meningiomas presenting an average elastic modulus of 3.97 ± 3.66 kPa.

3.3.5. *Colorectal cancer*

Colon carcinomas have also been demonstrated to be stiffer than normal tissue. Normal colorectal tissue is 0.936 kPa, while colorectal cancer tissue has been reported to be 7.51 kPa when calculated by a tactile sensor of the Venustron system.[128] To collect Young's modulus values, the resected intestinal tract was opened longitudinally, the mucosal side was placed upward so that the probe could be above the surface, and measurements were taken every 0.005 mm of indentation depth using a resonance frequency of 50 Hz.[128] Notably, changes in stiffness are correlated with the stage of the Union of International Cancer Control (UICC) TNM classification of malignant tumors. Here, T describes the size of the primary tumor and whether it has invaded surrounding tissue, N describes local lymph nodes that are involved, and M describes distant metastasis. Similarly, biomechanical measurement collected using AFM nanoindentation showed an increase from 0.8 ± 0.4 kPa to 2.40 ± 1.83 kPa in control and tumor tissue, respectively.[21]

Furthermore, mechanical properties of colorectal cancer samples have been identified to significantly impact cancer progression and aggressiveness. Using 16.6 kPa as a cut-off value to differentiate a hard tumor group (HTG) from a soft tumor group (STG), identified that patients with HTG had shorter disease-free survival and elevated 2-year recurrence rates.[128] Categorizing colorectal cancer tissue based on such stiffness groups may be helpful in identifying more aggressive tumors. The elastic modulus of colorectal cancer tissue correlates with many clinicopathological factors including the size of the tumor, venous invasion, perineural invasion, poorly differentiated clusters, and elastic laminal invasion as well as clinical findings such as an obstructive tumor.[160] Changes in stiffness have also been identified to particularly profound in areas of high collagen deposition in colorectal cancer, where stiffness ranges can be up to four-fold higher than normal tissue.[21] Patients with high expression of collagen type I have significantly lower survival rates compared to patients with low expression.[161] Together, this work shows the potential value of evaluating matrix stiffness to approximate tumor stage and predict clinical behavior in colorectal cancer.

3.3.6. *Liver cancer*

Normal liver tissue has an elastic modulus of ~ 300–600 Pa, and stiffness values can increase to 20 kPa or higher during the progression of pathological conditions such as fibrosis and cirrhosis.[162] Similar increases in stiffness are also observed during the progression of liver cancer. To characterize the mechanical properties of liver tissue, noninvasive imaging methodologies that quantitatively assess the elasticity of the tissue have been used. MRE has been used to noninvasively characterize solid liver tumors and identified significantly higher mean shear stiffness among tumor samples compared to normal tissue.[88] Stiffness ranges of 2.3 ± 0.3 kPa, 2.7 ± 0.4 kPa, and 10.1 ± 3.6 kPa were observed for normal liver samples, benign tumors, and, malignant tumors respectively. Notably, malignant tumors could be accurately differentiated from benign tumors and normal liver parenchyma using a cut-off value of 5 kPa tissue.[88] Using these values could aid in identifying tumors in patients and assessing tumor response to therapeutic treatment. However, fibrotic tissue demonstrated overlap in stiffness values for both benign and malignant tumors, possibly leading to difficulties in distinguishing between the two disease states. Additionally, MRE offers some limitations as planar wave imaging is performed with two-dimensional (2D) wave inversion that can yield incorrectly low stiffness values for smaller structures. To overcome this, higher frequency and smaller wavelength acoustic waves are necessary, which rapidly get attenuated in the liver.[163] Future work will need to develop 3D acquisition to overcome these limitation.[164]

Other measurements of liver tissue using transient elastography identified a much higher median stiffness of 55 kPa in hepatocellular carcinoma (HCC), 75 kPa in cholangiocellular carcinoma (CCC), 66.5 kPa in metastatic tumors, and 16.9 kPa in malignant lymphoma.[132] While both studies use elastography, these differences highlight the variances in stiffness ranges acquired with different imaging modalities. Interestingly, these values were collected using a commercially available device designed to measure liver stiffness and could be readily adapted.[132] However, one drawback of using this device is that stiffness can only be measured in very large liver tumors that occupy a large portion of the right lobe, limiting sensitivity and use for the detection of smaller tumors.

3.4. *Clinical impact of altered tumor ECM mechanics*

The altered mechanical properties of tumor tissue significantly impact therapeutic intervention and malignancy. In breast cancer, increased matrix stiffness enhances ERK activation and increases ROCK-generated contractility to drive a malignant phenotype.[39] Elevated ECM deposition and changes in ECM concentration also affect integrin signaling. Integrin expression has been correlated with tumor aggression.[165] and provides proliferation, differentiation, and migration signals to tumor cells.[100] Integrin signaling has also been implicated in the chemoresistance of tumors.[166] PDAC-associated fibrosis increases matrix stiffness and activates integrin-dependent signaling and YAP to drive tumor progression and malignancy.[148] Additionally, increased deposition of ECM proteins, including collagen, GAGs, and decorin, can create a protective barrier within and around tumors that significantly limits the delivery of chemotherapeutic agents.[167,168]

4. Structural Properties of the Tumor ECM

During tumor progression, the continuous and dynamic remodeling of the ECM results in not only macroscopic changes in matrix composition and mechanical features but also microscopic, cell-scale changes in matrix topology (Table 3).[2,113] These changes in matrix topology architecture can provide strong pro-invasive and pro-migratory cues to cancer cells.[169–171]

4.1. *Fiber alignment*

Elevated fiber alignment has been observed across many cancer types including breast,[173] gastric,[95] ovarian,[175] and pancreatic[177] and is correlated with increasing aggressiveness within individual cancer types including kidney,[92] prostate,[112] and colon.[174] As collagen type I is a major component of the interstitial matrix that, together with fibronectin, provides mechanical strength to tissues,[36] collagen fiber reorientation has been identified to be critically important during tumor-associated ECM remodeling and reorganization.[178–180] Tumor-associated collagen signatures (TACS), which describe collagen reorganization at the tumor–stromal interface, have been well characterized in breast cancer.[173] Originally described in mouse models, TACS provide standard hallmarks to locate and characterize tumors.[18,181] TACS-1 is the presence of dense collagen around the tumor identifying small tumor regions. TACS-2 is the presence of tense and straightened collagen fibers parallel to the tumor indicating growth. TACS-3 is defined as thick, radially aligned collagen fibers, which facilitate invasion and indicate invasion and metastasis.[18] Notably, TACS-3 morphology has been correlated with poorer patient outcomes.[173,176] and a higher probability of recurrence.[182] These differences in matrix alignment are observed

Table 3. Summary of changes in matrix structure during tumor progression across cancer type and grade (SHG = second harmonic generation, TACS = tumor-associated collagen signatures).

Tissue	Sub-type	Measurement method	Observation	Reference
Brain	Glioblastoma	SHG microscopy, CT-FIRE software	Increased collagen fiber length and alignment as patient mortality increases	172
Breast	Carcinoma	SHG microscopy	Increased TACS-3 as patient mortality increases	173
Colorectal	Adenocarcinoma	Scanning Electron Microscopy Software	Increased roughness and porosity as well as loss of crypt organization as adenocarcinoma differentiation decreases	174
	Secondary colorectal carcinoma	Immunohistochemistry	Increased type IV collagen density in secondary tumors compared to primary tumors	91
Gastric	Adenocarcinoma	Picrosirius Red staining, SHG microscopy, immunohistochemistry	Increased collagen deposition, fiber alignment, density, width, length and straightness in tumor tissue compared to normal tissue	95
Ovarian	Tumor	SHG microscopy	Increased collagen density in late-stage (III and IV) ovarian cancer	175
	Tumor	Picrosirius Red staining, SHG microscopy	Increased TACS-3 signal and fiber nisotropy in malignant tumors	176
Pancreatic	Ductal adenocarcinoma	SHG microscopy	Increased collagen alignment around tumorous ducts	177
Prostate	Adenocarcinoma	Immunohistochemistry	Increased focal adhesion kinase as adenocarcinoma different decreases	112
Renal	Carcinoma	SHG microscopy, CT-FIRE Software	Increased collagen fiber density and alignment in late (IV) stage cancer	92

across cancer type and grade, making it an important physical marker for patient prognostics. For example, grade 4 RCC has a higher collagen density and alignment compared to grade 1 RCC.[92] While it has not yet been identified in all solid cancers, fiber alignment provides a useful biomarker to predict patient prognosis in several cancers and should be further investigated and quantified.

At the cellular scale, fiber alignment provides significant pro-invasive cues to surrounding cells and can provide "highways" in the matrix for rapid dissemination from the primary tumor during metastasis.[183–185] Fiber alignment allows cancer cells to migrate in a one-dimensional manner along fibers as opposed to much more challenging movement in three-dimensional spaces.[186] This process results in more directed and persistent migration. Fiber alignment can also increase the fraction of cells within a population that is capable of migrating.[187] Additionally, matrix alignment can influence surrounding stromal cells in a cell-specific manner. Furthermore, stem cell differentiation, endothelial cell infiltration, and immune cell migration are all influenced by matrix architecture.[185,188,189] These findings indicate that fiber alignment can drive migration and regulate cancer and stromal cell movement into and out of the tumor, impacting the cellular composition of the tumor microenvironment.

4.2. *Porosity*

Similar to fiber orientation, matrix porosity is another architectural feature of the tumor microenvironment that is modified during tumorigenesis and impacts cell function and treatment. The active remodeling and replacement of normal ECM with tumor-associated ECM during tumor progression, from dysregulation of ECM synthesis and secretion as well as matrix-remodeling enzymes, can significantly alter ECM composition and porosity. Pores within the tumor microenvironment are created naturally as interstitial spaces, spaces adjacent to the vasculature and nerves, and spaces created by cells themselves.[190] These pores can be highly heterogeneous, contributing to the spatial complexity of the tumor-associated ECM and making it difficult to assign a single value of pore size for a given tumor. Notably, such variable and tight pores, interstitial spaces, cross-sectional areas, and channel-like tracks in the matrix serve as passageways and provide physical guides for migrating cells.[191,192] In some cases, cells can generate their own tunnels through the matrix using MMPs and thus contribute to the complex topographies within tissues.[193–195]

The physical properties of the ECM are key mediators of cell migration and function.[190] Heterogeneity in the porosity of the tumor-associated ECM can contribute to heterogeneity in cell populations within the tumor microenvironment as well as altered individual cell behavior. Pore sizes within the matrix present physical barriers and dictate whether or not a cell can successfully transverse the matrix.[196,197] Consequently, changes to the porosity of the ECM can alter the populations of cells within the tumor microenvironment. There is evidence in melanoma that suggests in younger skin, the ECM is denser, and therefore, the size of pores is smaller than in older skin where the ECM is less dense and pore sizes are larger.[185] This results in cancer cells being unable to move in younger skin while only allowing smaller immune cells to infiltrate the tumor. In older skin, the cancer cells are able to move more easily through the tumor microenvironment and can consequently metastasize to secondary sites in the body. Moreover, matrix porosity can serve not only as a physical barrier during migration but can also affect cells at the transcriptional level to alter their migration behavior. There is evidence that suggests decreased pore sizes and increased confinement of cancer cells can lead to genetic instability, as cells experience forces from multiple directions in the process of migrating through a pore.[190,198] Confined migration demonstrates unique mechanisms compared to 2D and 3D migration, and the effects of confinement on cell migration have been well studied.[190] Confinement can alter cellular cytoskeleton architecture,[199] contractility,[200] forces,[201] and speed,[202] and a more restrictive matrix can modulate ameboid–mesenchymal migration.[197,203,204] During migration, a balance between confinement, cell and matrix stiffness, and cellular energetics has also been shown to play a role in determining migration behavior and decision-making.[205–207]

4.3. *Measurement and quantification of tumor structural properties*

Given the importance of matrix organization on cell function, cancer progression, and patient prognosis, great efforts have been made to quantify matrix fiber alignment in and around tumor tissue. Due to the complexity in studying and quantifying *in vivo* samples, many *in vitro* models, primarily fibrillar collagen matrices, have been developed to study the effects of matrix alignment.[208–210] However, several methods to quantify matrix architecture *in vivo* do exist, each with its own advantages and limitations, producing a wide range of values that attempt to summarize the same characteristics.

The most commonly employed methods to measure *ex vivo* collagen fiber alignment include SHG, quantitative polarized microscopy, spatial light interference microscopy, and two-photon microscopy.[211,212] Dyes such as Masson's Trichrome and Picrosirius Red are also commonly employed to help visualize collagen fiber linearization and orientation.[58,95] To quantify the alignment of individual collagen fibrils, researchers have developed software programs such as CurveAlign and CT-FIRE that utilize MATLAB and ImageJ to produce numerical descriptions of fibrillar alignment.[211] While these methods are extremely helpful in identifying matrix organization changes during tumor progression, they provide static images and are unable to capture the complex temporal dynamics of ECM remodeling.

Many of these methods are often used simultaneously to quantify matrix alignment of tumor tissue. For example, histology is often combined with imaging modalities such as SHG and analysis software. To measure collagen alignment of gastric cancer tumors, Picrosirius Red staining and immunohistochemistry was performed on tissue samples from patients who underwent gastrectomy, images of the matrix architecture were obtained using SHG microscopy, and CT-FIRE was then applied to quantify fibril alignment on a scale from 0 to 1.[95] Notably, gastric cancer samples had increased collagen alignment compared to normal tissue samples. Similar measurements have been made in breast cancer, where SHG imaging has been performed on samples obtained from breast cancer patients and the images were then processed using CurveAlign.[182] Prior to image processing, specific boundaries were drawn on each image to avoid collagen from blood vessels and to focus only on the tumor-associated ECM. Results from this study indicated that collagen alignment was more perpendicular in ductal regions of the tumor which often leads to poorer patient prognosis. Differences in collagen alignment were also observed within a single cancer type, specifically between cancer grades. A study examining RCC used SHG microscopy and analyzed images stained with Hematoxylin and Eosin in CT-FIRE to show that grade 4 RCC is more aligned than grade 1 RCC.[92] The development and implementation of these methods have allowed researchers to numerically quantify alignment as opposed to making qualitative judgments.

Although its importance has been established, matrix porosity in tumors is still not well characterized due to the complex environment of the tumor; however, some efforts have been made to study the porosity of the tumor-associated ECM. Because of the heterogeneities that exist with regard to pore sizes and the spaces within the tumor microenvironment, *in vivo* quantification has proven challenging. Current methods do not have the spatial resolution to allow for pore measurement at the micron scale and also lack the penetration depth to study a tumor sample *in vivo*.[213,214] To overcome these limitations, researchers often use *in vitro* models to study the effects of confinement and its influence on motility and metastasis.[215–217] Physical characterization of tumor tissue has been utilized to design models that recapitulate the tumor-associated ECM and mimic porosity.[226,227] During *in vitro* modeling of the tumor microenvironment, changes in porosity arise as a by-product of either stiffness or the method of matrix fabrication.[218] The ability to achieve fine control over matrix properties such as pore size with *in vitro* models has identified unique migration mechanisms during confined migration and the physical limits of MMP-independent migration through interstitial spaces in the matrix.[34,196] While *in vitro* models have provided insight into the effects of porosity on cell function, these models often examine multiple matrix properties at once as porosity can impact other physical such as stiffness and density.[218,219] Together, these studies have been critical in understanding the impact of a spatially complex microenvironment on cancer cell migration.

4.4. *Clinical impact of altered tumor ECM structure*

The anisotropy of the ECM increases with increasing ECM concentration during tumor progression influencing cell behavior and treatment. Alignment can reduce the local network interconnectivity of fibrillar collagen matrices and decrease stiffness as well as promote crosslinking of collagen fibers to stiffen the matrix.[208,220] The dynamic and active crosslinking, degradation, and synthesis of ECM additionally creates a spatially heterogeneous matrix with areas of high ECM density, small porosity, and limited perfusion.[2,14,221] Such architectural changes in matrix architecture play a significant role in drug delivery as they influence the diffusion of chemotherapeutic agents through and around the tumor.[6,222,223] Many nano- and micro-particle therapeutics are specifically designed to be capable of diffusing through the tissue to the tumor. Additionally, pore size, density, and distribution affect cell adhesion and migration within a tissue and the tumor microenvironment.[224,225] Therefore, matrix porosity does not only inhibit or facilitate migration in the tumor microenvironment, but it is also a factor to consider in cancer drug design.

5. Targeting the Physical Tumor Microenvironment

To improve the delivery and efficacy of cancer treatment, attempts have been made to reengineer the tumor microenvironment by normalizing tumor stroma. Due to their importance in forming the tumor microenvironment, the targeting ECM remodeling enzymes has received significant interest as a possible therapeutic approach to reduce pathogenic ECM remodeling and restore ECM homeostasis.[226] Targeting LOX has gained much attention due to its substantial role in matrix crosslinking in cancer. LOX activity increases the invasiveness of many cancer cell types.[36,52,227] and inhibition of LOX decreases primary tumor growth and mechanosensitivity in the mammary epithelium.[58] However, cancer signaling can develop network compensation and resistance when targeting a single molecule, and LOX inhibition has been ineffective clinically.[2] Treatment with the combination of inhibitors against LOX and LOXL2 demonstrated reduced primary tumor growth and metastasis in a breast cancer mouse model.[228] Similarly, the LOXL2 inhibitor, Simtuzumab, has been used for the treatment of cirrhosis, another disease characterized by increased ECM deposition and tissue hardening, but this drug ultimately failed in phase 2 trials.[229] MMPs are also a tempting therapeutic target because of the crucial role they play in matrix remodeling and metastasis, but clinical trials targeting MMPs have experienced poor clinical outcomes.[230] New highly specific MMP inhibitors may hold promise,[231] but targeting ECM remodeling enzymes while limiting side effects and balancing toxicity is challenging. To therapeutically restore ECM homeostasis, a better understanding of the cell–cell and cell–ECM interactions within tumors and how the microenvironment can influence these interactions and drug effectiveness is necessary.

Other strategies to mitigate the effects of aberrant mechanical cues during tumor progression have sought to target ECM components to alleviate stress and improve perfusion. Matrix-depleting agents including collagenase, relaxin, and MMPs have been used to lower the concentration of collagen and hyaluronan (HA) in tumor tissue.[217] Drug penetration has been improved when collagen-rich tumors have been treated with collagenases.[168] Similarly, reducing HA content in tumor using hyaluronidase has also been shown to reduce tissue stiffness and slow progression.[232,233]

However, the ECM can act as a double agent in cancer, as it is a suppressor at the early stages of cancer progression and a promoter at later stages.[221] As such, treatments targeting the ECM may need to change at different stages of cancer.

More recently, mechanomedicine has aimed to therapeutically target abnormal ECM mechanics by disrupting the cellular response to mechanics.[35] Mechanosignaling through focal adhesion complexes is elevated in tumor tissue due to the altered ECM mechanics and integrin-mediated focal adhesion kinase (FAK) phosphorylation.[35,100,234] Thus, a prominent method to target cellular response to the tumor stroma is inhibition of FAK and downstream signaling events. A FAK1 inhibitor, Defactinib, is currently in pre-clinical models for breast cancer and a FAK1–FAK2 inhibitor has shown promise in inhibiting mechanosignaling and decreasing fibrosis in pancreatic tumors.[235,236] In a mouse model, combination therapy priming tumor tissue via ROCK inhibition with Fasudil followed by chemotherapy treatment showed improved effectiveness of standard-of-care chemotherapy.[237] Combining contractility inhibitors to limit stromal remodeling and sensitize cancer cells to treatment with chemotherapeutics may improve the current efficacy of many treatments. However, these therapies currently have failed to demonstrate therapeutic effects clinically and a successful treatment program targeting the physical tumor microenvironment has yet to be established. Off-target interactions remain a major obstacle for developing inhibitors targeting cell mechanosensing and improved specificity will be needed.

6. Conclusions

Recent developments in the physical sciences, engineering, and oncology have identified tumors not just as aberrant cells, but rather as complex, functionally discrete organs gone rogue that can recruit and corrupt surrounding cells.[3,8,238] The emerging importance of the physical microenvironment of tumors in tumorigenesis creates a critical need to better characterize and understand spatial and temporal dynamics of the ECM during cancer progression. There have been increasing attempts to quantify these mechanical and structural properties leading to the employment of different techniques with a wide range of experimental designs and resolution. While this characterization has contributed to the discovery of various characteristics of the tumor microenvironment as well as how cancer cell behavior can be affected, it has also led to a large range of values claiming to signify the same properties. These differences in values not only exist across cancer types but also persist within the same type of cancer making it difficult for researchers to work collectively on the same goal. To better understand the complex effects of ECM remodeling in cancer progression across many organs, researchers need to more closely investigate the spatial and temporal dynamics of the ECM during tumorigenesis. Improved knowledge of the biophysical character of the ECM including matrix stiffness, density, crosslinking, fiber alignment, and porosity will aid in the detection of tumor development, metastatic spread, and response to therapeutic treatment.

Acknowledgments

This work was supported by NIH grant R01GM131178 and R01HL127499 and the Vanderbilt-Ingram Cancer Center SPORE in Breast Cancer (P50CA098131) to C. A. R.-K. as well as an NSF Graduate Research Fellowship under Grant No. DGE-1650441 to M. R. Z. The authors would like to thank Jacob A. Vanderburgh for his thoughtful review of the manuscript. M. R. Z., N. C. C., and C. A. J. contributed equally to the work.

REFERENCES

1. P. Lu, V. M. Weaver and Z. Werb, *J. Cell Biol.* **196**, 395–406 (2012).
2. T. R. Cox and J. T. Erler, *Dis. Model. Mech.* **4**, 165–178 (2011).
3. F. R. Balkwill, M. Capasso and T. Hagemann, *J. Cell Sci.* **125**, 5591–5596 (2012).
4. H. Yu, J. K. Mouw and V. M. Weaver, *Trends Cell Biol.* **21**, 47–56 (2011).
5. M. R. Zanotelli, F. Bordeleau and C. A. Reinhart-King, *Curr. Opin. Biomed. Eng.* **1**, 8–14 (2017).
6. R. K. Jain, J. D. Martin and T. Stylianopoulos, *Annu. Rev. Biomed. Eng.* **16**, 321–346 (2014).
7. L. M. Coussens and Z. Werb, *Nature* **420**, 860–867 (2002).
8. M. J. Bissell and D. Radisky, *Nat. Rev. Cancer* **1**, 46–54 (2001).
9. R. O. Hynes, *Science* **326**, 1216–1219 (2009).
10. J. K. Mouw, G. Ou and V. M. Weaver, *Nat. Rev. Mol. Cell Bio.* **15**, 771–785 (2014).
11. R. O. Hynes, *Nat. Rev. Mol. Cell Bio.* **15**, 761–763 (2014).
12. F. Rosso, A. Giordano, M. Barbarisi and A. Barbarisi, *J. Cell Physiol.* **199**, 174–180 (2004).
13. G. Huang and D. S. Greenspan, *Trends. Endocrinol. Metab.* **23**, 16–22 (2012).
14. S. Kaushik, M. W. Pickup and V. M. Weaver, *Cancer Metast. Rev.* **35**, 655–667 (2016).
15. M. W. Pickup, J. K. Mouw and V. M. Weaver, *EMBO Rep.* **15**, 1243–1253 (2014).
16. T. H. Barker, *Biomaterials* **32**, 4211–4214 (2011).
17. D. Wirtz, K. Konstantopoulos and P. C. Searson, *Nat. Rev. Cancer* **11**, 512–522 (2011).
18. P. P. Provenzano, K. W. Eliceiri, J. M. Campbell, D. R. Inman, J. G. White and P. J. Keely, *BMC Med.* **4**, 38 (2006).
19. P. P. Provenzano, D. R. Inman, K. W. Eliceiri, J. G. Knittel, L. Yan, C. T. Rueden, J. G. White and P. J. Keely, *BMC Med.* **6**, 11 (2008).
20. C. Rianna and M. Radmacher, *Nanoscale* **9**, 11222–11230 (2017).
21. E. Brauchle, J. Kasper, R. Daum, N. Schierbaum, C. Falch, A. Kirschniak, T. E. Schäffer and K. Schenke-Layland, *Matrix Biol.* **68–69**, 180–193 (2018).
22. C. Rianna, P. Kumar and M. Radmacher, *Semin. Cell Dev. Biol.* **73**, 107–114 (2018).
23. H. T. Nia, L. L. Munn and R. K. Jain, *Clin. Cancer Res.* **25**, 2024–2026 (2019).
24. J. Zhang, N. C. Chada and C. A. Reinhart-King, *Front. Bioeng. Biotech.* **7**, 412 (2019).
25. J. I. Lopez, I. Kang, W.-K. You, D. M. McDonald and V. M. Weaver, *Integr. Biol.* **3**, 910–921 (2011).
26. R. Akhtar, M. J. Sherratt, J. K. Cruickshank and B. Derby, *Mater. Today* **14**, 96–105 (2011).
27. D. Sicard, L. E. Fredenburgh and D. J. Tschumperlin, *J. Mech. Behav. Biomed.* **74**, 118–127 (2017).
28. W. J. Polacheck and C. S. Chen, *Nat. Methods* **13**, 415–423 (2016).
29. P. Roca-Cusachs, V. Conte and X. Trepat, *Nat. Cell Biol.* **19**, 742–751 (2017).

30. K. Sugimura, P.-F. Lenne and F. Graner, *Development* **143**, 186–196 (2016).
31. M. R. Zanotelli and C. A. Reinhart-King, Mechanical forces in tumor angiogenesis, in *Biomechanics in Oncology* (Springer International Publishing, 2018), pp. 91–112.
32. D. T. Butcher, T. Alliston and V. M. Weaver, *Nat. Rev. Cancer* **9**, 108–122 (2009).
33. C. T. Mierke, *Rep. Prog. Phys.* **82**, 064602 (2019).
34. F. Spill, D. S. Reynolds, R. D. Kamm and M. H. Zaman, *Curr. Opin. Biotechnol.* **40**, 41–48 (2016).
35. M. C. Lampi and C. A. Reinhart-King, *Sci. Transl. Med.* **10**, eaao0475 (2018).
36. J. T. Erler and V. M. Weaver, *Clin. Exp. Metastas.* **26**, 35–49 (2009).
37. C. Bonnans, J. Chou and Z. Werb, *Nat. Rev. Mol. Cell Bio.* **15**, 786–801 (2014).
38. S. D. Soysal, A. Tzankov and S. E. Muenst, *Pathobiology* **82**, 142–152 (2015).
39. M. J. Paszek, N. Zahir, K. R. Johnson, J. N. Lakins, G. I. Rozenberg, A. Gefen, C. A. Reinhart-King, S. S. Margulies, M. Dembo, D. Boettiger, D. A. Hammer and V. M. Weaver, *Cancer Cell* **8**, 241–254 (2005).
40. K. A. Myers, K. T. Applegate, G. Danuser, R. S. Fischer and C. M. Waterman, *J. Cell Biol.* **192**, 321–334 (2011).
41. L. Sorokin, *Nat. Rev. Immunol.* **10**, 712–723 (2010).
42. F. Bordeleau, B. N. Mason, E. M. Lollis, M. Mazzola, M. R. Zanotelli, S. Somasegar, J. P. Califano, C. Montague, D. J. Lavalley, J. Huynh, N. Mencia-Trinchant, Y. L. Negrón Abril, D. C. Hassane, L. J. Bonassar, J. T. Butcher, R. S. Weiss and C. A. Reinhart-King, *Proc. Natl. Acad. Sci.* **114**, 492–497 (2017).
43. R. Timpl and J. C. Brown, *Bioessays* **18**, 123–132 (1996).
44. W. Halfter, J. Candiello, H. Hu, P. Zhang, E. Schreiber and M. Balasubramani, *Cell Adh. Migr.* **7**, 64–71 (2013).
45. J. Candiello, G. J. Cole and W. Halfter, *Matrix Biol.* **29**, 402–410 (2010).
46. E. Ioachim, A. Charchanti, E. Briasoulis, V. Karavasilis, H. Tsanou, D. L. Arvanitis, N. J. Agnantis and N. Pavlidis, *Eur. J. Cancer* **38**, 2362–2370 (2002).
47. R. Kalluri and M. Zeisberg, *Nat. Rev. Cancer* **6**, 392–401 (2006).
48. A. Page-Mccaw, A. J. Ewald and Z. Werb, *Nat. Rev. Mol. Cell Bio.* **8**, 221–233 (2007).
49. H. M. Kagan and W. Li, *J. Cell. Biochem.* **88**, 660–672 (2003).
50. K. Csiszar, Lysyl oxidases: A novel multifunctional amine oxidase family, in *Progress in Nucleic Acid Research and Molecular Biology* (Elsevier, 2001), pp. 1–32.
51. B. M. Baker and C. S. Chen, *J. Cell Sci.* **125**, 3015–3024 (2012).
52. J. T. Erler, K. L. Bennewith, M. Nicolau, N. Dornhöfer, C. Kong, Q.-T. Le, J.-T. A. Chi, S. S. Jeffrey and A. J. Giaccia, *Nature* **440**, 1222–1226 (2006).
53. C. Reynaud, L. Ferreras, P. Di Mauro, C. Kan, M. Croset, E. Bonnelye, F. Pez, C. Thomas, G. Aimond, A. E. Karnoub, M. Brevet and P. Clézardin, *Cancer Res.* **77**, 268–278 (2017).
54. L. Peng, Y.-L. Ran, H. Hu, L. Yu, Q. Liu, Z. Zhou, Y.-M. Sun, L.-C. Sun, J. Pan, L.-X. Sun, P. Zhao and Z.-H. Yang, *Carcinogenesis* **30**, 1660–1669 (2009).
55. M.-L. Wilgus, A. C. Borczuk, M. Stoopler, M. Ginsburg, L. Gorenstein, J. R. Sonett and C. A. Powell, *Cancer* **117**, 2186–2191 (2011).
56. Q.-T. Le, J. Harris, A. M. Magliocco, C. S. Kong, R. Diaz, B. Shin, H. Cao, A. Trotti, J. T. Erler, C. H. Chung, A. Dicker, T. F. Pajak, A. J. Giaccia and K. K. Ang, *J. Clin. Oncol.* **27**, 4281–4286 (2009).
57. Q.-T. Le, *Int. J. Radiat. Oncol. Biol. Phys.* **69**, S56–S58 (2007).
58. K. R. Leventall, H. Yu, L. Kass, J. N. Lakins, M. Egeblad, J. T. Erler, S. F. T. Fong, K. Csiszar, A. Giaccia, W. Weninger, M. Yamauchi, D. L. Gasser and V. M. Weaver, *Cell* **139**, 891–906 (2009).
59. B. Muz, P. de la Puente, F. Azab and A. K. Azab, *Hypoxia (Auckl.)* **3**, 83–92 (2015).

60. C. C. L. Wong, D. M. Gilkes, H. Zhang, J. Chen, H. Wei, P. Chaturvedi, S. I. Fraley, C. M. Wong, U. S. Khoo, I. O. L. Ng, D. Wirtz and G. L. Semenza, *Proc. Natl. Acad. Sci.* **108**, 16369–16374 (2011).

61. J. Lee, S. Condello, B. Yakubov, R. Emerson, A. Caperell-Grant, K. Hitomi, J. Xie and D. Matei, *Clin. Cancer Res.* **21**, 4482–4493 (2015).

62. J. Lee, B. Yakubov, C. Ivan, D. R. Jones, A. Caperell-Grant, M. Fishel, H. Cardenas and D. Matei, *Neoplasia* **18**, 689–698 (2016).

63. Y. Bagatur, A. Z. Ilter Akulke, A. Bihorac, M. Erdem and D. Telci, *Cell Adh. Migr.* **12**, 138–151 (2018).

64. D. Y. S. Chau, R. J. Collighan, E. A. M. Verderio, V. L. Addy and M. Griffin, *Biomaterials* **26**, 6518–6529 (2005).

65. F. Bordeleau, W. Wang, A. Simmons, M. A. Antonyak, R. A. Cerione and C. A. Reinhart-King, *J. Cell Sci.* **133**, 231134 (2020).

66. N. C. Avery and A. J. Bailey, *Pathol. Biol.* **54**, 387–395 (2006).

67. N. A. Ansari and Z. Rasheed, *Biochem. (Mosc.) Suppl., Ser. B. Biomed. Chem.* **3**, 335–342 (2009).

68. J. Chaudhuri, Y. Bains, S. Guha, A. Kahn, D. Hall, N. Bose, A. Gugliucci and P. Kapahi, *Cell Metab.* **28**, 337–352 (2018).

69. S.-S. Sivan, E. Wachtel, E. Tsitron, N. Sakkee, F. Van Der Ham, J. Degroot, S. Roberts and A. Maroudas, *J. Biol. Chem.* **283**, 8796–8801 (2008).

70. N. Verzijl, J. Degroot, S. R. Thorpe, R. A. Bank, J. N. Shaw, T. J. Lyons, J. W. J. Bijlsma, F. P. J. G. Lafeber, J. W. Baynes and J. M. Tekoppele, *J. Biol. Chem.* **275**, 39027–39031 (2000).

71. A. Goldin, J. A. Beckman, A. M. Schmidt and M. A. Creager, *Circulation* **114**, 597–605 (2006).

72. D. R. Sell, K. M. Biemel, O. Reihl, M. O. Lederer, C. M. Strauch and V. M. Monnier, *J. Biol. Chem.* **280**, 12310–12315 (2005).

73. D. M. Hudson, M. Archer, K. B. King and D. R. Eyre, *J. Biol. Chem.* **293**, 15620–15627 (2018).

74. N. Y. Kim, T. N. Goddard, S. Sohn, D. A. Spiegel and J. M. Crawford, *Chembiochem.* **20**, 2402–2410 (2019).

75. M. Asif, J. Egan, S. Vasan, G. N. Jyothirmayi, M. R. Masurekar, S. Lopez, C. Williams, R. L. Torres, D. Wagle, P. Ulrich, A. Cerami, M. Brines and T. J. Regan, *Proc. Natl. Acad. Sci. U.S.A.* **97**, 2809–2813 (2000).

76. S. Ferri, S. Kim, W. Tsugawa and K. Sode, *J. Diabetes Sci. Technol.* **3**, 585–592 (2009).

77. G. Richarme, C. Liu, M. Mihoub, J. Abdallah, T. Leger, N. Joly, J.-C. Liebart, U. V. Jurkunas, M. Nadal, P. Bouloc, J. Dairou and A. Lamouri, *Science* **357**, 208–211 (2017).

78. M. Egeblad and Z. Werb, *Nat. Rev. Cancer* **2**, 161–174 (2002).

79. C. M. Overall and C. López-Otín, *Nat. Rev. Cancer* **2**, 657–672 (2002).

80. V. Sosa, T. Moliné, R. Somoza, R. Paciucci, H. Kondoh and M. E. Lleonart, *Ageing Res. Rev.* **12**, 376–390 (2013).

81. D. M. Gilkes, G. L. Semenza and D. Wirtz, *Nat. Rev. Cancer* **14**, 430–439 (2014).

82. G. Murphy, *Nat. Rev. Cancer* **8**, 932–941 (2008).

83. D. Bourboulia and W. G. Stetler-Stevenson, *Semin. Cancer Biol.* **20**, 161–168 (2010).

84. D. E. Kleiner and W. G. Stetler-Stevenson, *Cancer Chemoth. and Pharm.* **43**, S42–S51 (1999).

85. S. Hernandez-Barrantes, M. Bernardo, M. Toth and R. Fridman, *Semin. Cancer Biol.* **12**, 131–138 (2002).

86. L. C. Kelley, L. L. Lohmer, E. J. Hagedorn and D. R. Sherwood, *J. Cell. Biol.* **204**, 291–302 (2014).

87. L. Yang, Y. Pang and H. L. Moses, *Trends Immunol* **31**, 220–227 (2010).

88. S. K. Venkatesh, M. Yin, J. F. Glockner, N. Takahashi, P. A. Araoz, J. A. Talwalkar and R. L. Ehman, *AJR Am. J. Roentgenol.* **190**, 1534–1540 (2008).

89. A. Samani, J. Zubovits and D. Plewes, *Phys. Med. Biol.* **52**, 1565–1576 (2007).

90. C. Chandler, T. Liu, R. Buckanovich and L. G. Coffman, *Transl. Res.* **209**, 55–67 (2019).

91. J. V. Burnier, N. Wang, R. P. Michel, M. Hassanain, S. Li, Y. Lu, P. Metrakos, E. Antecka, M. N. Burnier, A. Ponton, S. Gallinger and P. Brodt, *Oncogene* **30**, 3766–3783 (2011).

92. S. L. Best, Y. Liu, A. Keikhosravi, C. R. Drifka, K. M. Woo, G. S. Mehta, M. Altwegg, T. N. Thimm, M. Houlihan, J. S. Bredfeldt, E. J. Abel, W. Huang and K. W. Eliceiri, *BMC Cancer* **19**, 490 (2019).

93. A. Rizwan, C. Bulte, A. Kalaichelvan, M. Cheng, B. Krishnamachary, Z. M. Bhujwalla, L. Jiang and K. Glunde, *Sci. Rep.* **5**, 10002 (2015).

94. I. Acerbi, L. Cassereau, I. Dean, Q. Shi, A. Au, C. Park, Y. Y. Chen, J. Liphardt, E. S. Hwang and V. M. Weaver, *Integr. Biol. (Camb.)* **7**, 1120–1134 (2015).

95. Z. H. Zhou, C. D. Ji, H. L. Xiao, H. B. Zhao, Y. H. Cui and X. W. Bian, *J. Cancer* **8**, 1466–1476 (2017).

96. Y. Attieh, A. G. Clark, C. Grass, S. Richon, M. Pocard, P. Mariani, N. Elkhatib, T. Betz, B. Gurchenkov and D. M. Vignjevic, *J. Cell Biol.* **216**, 3509–3520 (2017).

97. M. Jang, I. Koh, J. E. Lee, J. Y. Lim, J.-H. Cheong and P. Kim, *Biomater. Sci.* **6**, 2704–2713 (2018).

98. J. S. Desgrosellier and D. A. Cheresh, *Nat. Rev. Cancer* **10**, 9–22 (2010).

99. E. L. Baker and M. H. Zaman, *J. Biomech.* **43**, 38–44 (2010).

100. H. Hamidi and J. Ivaska, *Nat. Rev. Cancer* **18**, 533–548 (2018).

101. J. M. Patel, B. C. Wise, E. D. Bonnevie and R. L. Mauck, *Tissue Eng. Part C Methods* **25**, 593–608 (2019).

102. D. P. Allison, N. P. Mortensen, C. J. Sullivan and M. J. Doktycz, *WIREs Nanomed. Nanobi.* **2**, 618–634 (2010).

103. R. M. S. Sigrist, J. Liau, A. E. Kaffas, M. C. Chammas and J. K. Willmann, *Theranostics* **7**, 1303–1329 (2017).

104. A. S. Klauser, H. Miyamoto, R. Bellmann-Weiler, G. M. Feuchtner, M. C. Wick and W. R. Jaschke, *Radiology* **272**, 622–633 (2014).

105. C. Cassinotto, B. Lapuyade, A. Aït-Ali, J. Vergniol, D. Gaye, J. Foucher, C. Bailacq-Auder, F. Chermak, B. Le Bail and V. de Lédinghen, *Radiology* **269**, 283–292 (2013).

106. Y. K. Mariappan, K. J. Glaser and R. L. Ehman, *Clin. Anat.* **23**, 497–511 (2010).

107. A. Bunevicius, K. Schregel, R. Sinkus, A. Golby and S. Patz, *NeuroImage: Clin.* **25**, 102109 (2020).

108. R. A. Katkar, S. A. Tadinada, B. T. Amaechi and D. Fried, *Dent. Clin. North Am.* **62**, 421–434 (2018).

109. B. E. Cohen, *Nature* **467**, 407–408 (2010).

110. W. B. Amos and J. G. White, *Biol. Cell.* **95**, 335–342 (2003).

111. N. M. Kalwani, C. A. Ong, A. C. Lysaght, S. J. Haward, G. H. McKinley and K. M. Stankovic, *J. Biomed. Opt.* **18**, 26021–26021 (2013).

112. J. D. Rovin, H. F. Frierson, W. Ledinh, J. T. Parsons and R. B. Adams, *Prostate* **53**, 124–132 (2002).

113. L. A. Hapach, J. A. Vanderburgh, J. P. Miller and C. A. Reinhart-King, *Phys. Biol.* **12**, 061002 (2015).

114. J. Ophir, I. Céspedes, H. Ponnekanti, Y. Yazdi and X. Li, *Ultrason. Imaging* **13**, 111–134 (1991).

115. R. Muthupillai, D. J. Lomas, P. J. Rossman, J. F. Greenleaf, A. Manduca and R. L. Ehman, *Science* **269**, 1854 (1995).

116. H. K. Graham, N. W. Hodson, J. A. Hoyland, S. J. Millward–Sadler, D. Garrod, A. Scothern, C. E. M. Griffiths, R. E. B. Watson, T. R. Cox, J. T. Erler, A. W. Trafford and M. J. Sherratt, *Matrix Biol.* **29**, 254–260 (2010).

117. A. Samani, J. Bishop, C. Luginbuhl and D. B. Plewes, *Phys. Med. Biol.* **48**, 2183–2198 (2003).

118. M. Zhang, P. Nigwekar, B. Castaneda, K. Hoyt, J. V. Joseph, A. di Sant'Agnese, E. M. Messing, J. G. Strang, D. J. Rubens and K. J. Parker, *Ultrasound Med. Biol.* **34**, 1033–1042 (2008).

119. K. Hoyt, B. Castaneda, M. Zhang, P. Nigwekar, P. A. di Sant'agnese, J. V. Joseph, J. Strang, D. J. Rubens and K. J. Parker, *Cancer Biomark.* **4**, 213–225 (2008).

120. S. Phipps, T. H. J. Yang, F. K. Habib, R. L. Reuben and S. A. McNeill, *Urology* **66**, 447–450 (2005).

121. A. Rubiano, D. Delitto, S. Han, M. Gerber, C. Galitz, J. Trevino, R. M. Thomas, S. J. Hughes and C. S. Simmons, *Acta. Biomater.* **67**, 331–340 (2018).

122. Y. Itoh, Y. Takehara, T. Kawase, K. Terashima, Y. Ohkawa, Y. Hirose, A. Koda, N. Hyodo, T. Ushio, Y. Hirai, N. Yoshizawa, S. Yamashita, H. Nasu, N. Ohishi and H. Sakahara, *J. Magn. Resonance Imaging* **43**, 384–390 (2016).

123. Y. Shi, K. J. Glaser, S. K. Venkatesh, E. I. Ben-Abraham and R. L. Ehman, *J Magn Reson. Imaging* **41**, 369–375 (2015).

124. Y. Shi, L. Cang, X. Zhang, X. Cai, X. Wang, R. Ji, M. Wang and Y. Hong, *Eur. J. Radiol.* **108**, 13–20 (2018).

125. D. Chauvet, M. Imbault, L. Capelle, C. Demene, M. Mossad, C. Karachi, A. L. Boch, J. L. Gennisson and M. Tanter, *Ultraschall Med.* **37**, 584–590 (2016).

126. D. C. Stewart, A. Rubiano, K. Dyson and C. S. Simmons, *PLoS ONE* **12**, e0177561–e0177561 (2017).

127. Y. A. Miroshnikova, J. K. Mouw, J. M. Barnes, M. W. Pickup, J. N. Lakins, Y. Kim, K. Lobo, A. I. Persson, G. F. Reis, T. R. McKnight, E. C. Holland, J. J. Phillips and V. M. Weaver, *Nat. Cell. Biol.* **18**, 1336–1345 (2016).

128. S. Kawano, M. Kojima, Y. Higuchi, M. Sugimoto, K. Ikeda, N. Sakuyama, S. Takahashi, R. Hayashi, A. Ochiai and N. Saito, *Cancer Sci.* **106**, 1232–1239 (2015).

129. E. Brauchle, J. Kasper, R. Daum, N. Schierbaum, C. Falch, A. Kirschniak, T. E. Schäffer and K. Schenke-Layland, *Matrix Biol.* **68–69**, 180–193 (2018).

130. M. Yin, J. A. Talwalkar, K. J. Glaser, A. Manduca, R. C. Grimm, P. J. Rossman, J. L. Fidler and R. L. Ehman, *Clin. Gastroenterol. Hepatol.* **5**, 1207–1213.e1202 (2007).

131. S. K. Venkatesh, M. Yin, J. F. Glockner, N. Takahashi, P. A. Araoz, J. A. Talwalkar and R. L. Ehman, *AJR Am. J. Roentgenol.* **190**, 1534–1540 (2008).

132. R. Masuzaki, R. Tateishi, H. Yoshida, T. Sato, T. Ohki, T. Goto, H. Yoshida, S. Sato, Y. Sugioka, H. Ikeda, S. Shiina, T. Kawabe and M. Omata, *Hepatol. Int.* **1**, 394–397 (2007).

133. N. F. Boyd, H. Guo, L. J. Martin, L. Sun, J. Stone, E. Fishell, R. A. Jong, G. Hislop, A. Chiarelli, S. Minkin and M. J. Yaffe, *N. Engl. J. Med.* **356**, 227–236 (2007).

134. L. Kass, J. T. Erler, M. Dembo and V. M. Weaver, *Int. J. Biochem. Cell Biol.* **39**, 1987–1994 (2007).

135. A. Srivastava, Y. Verma, K. D. Rao and P. K. Gupta, *Strain* **47**, 75–87 (2011).

136. T. Lehtimäki, M. Lundin, N. Linder, H. Sihto, K. Holli, T. Turpeenniemi-Hujanen, V. Kataja, J. Isola, H. Joensuu and J. Lundin, *Breast Cancer Res.* **13**, R134 (2011).

137. J. R. Staunton, W. Vieira, K. L. Fung, R. Lake, A. Devine and K. Tanner, *Cell. Mol. Bioeng.* **9**, 398–417 (2016).

138. J. Lorenzen, R. Sinkus, M. Biesterfeldt and G. Adam, *Invest. Radiol.* **38**, 236–240 (2003).
139. T. A. Krouskop, T. M. Wheeler, F. Kallel, B. S. Garra and T. Hall, *Ultrason. Imaging* **20**, 260–274 (1998).
140. H. Zhang, Q. Z. Zhang, L. Ruan, J. Duan, M. Wan and M. F. Insana, *Meas. Sci. Technol.* **29**, 035701 (2018).
141. V. Jalkanen, B. M. Andersson, A. Bergh, B. Ljungberg and O. A. Lindahl, *Med. Biol. Eng. Comput.* **44**, 593–603 (2006).
142. J. A. Motherway, P. Verschueren, G. Van Der Perre, J. Vander Sloten and M. D. Gilchrist, *J. Biomech.* **42**, 2129–2135 (2009).
143. D. W. Good, A. Khan, S. Hammer, P. Scanlan, W. Shu, S. Phipps, S. H. Parson, G. D. Stewart, R. Reuben and S. A. McNeill, *PLoS ONE* **9**, e112872 (2014).
144. Y. A. Miroshnikova, J. K. Mouw, J. M. Barnes, M. W. Pickup, Johnathan, Y. Kim, K. Lobo, A. I. Persson, G. F. Reis, T. R. McKnight, Eric, J. J. Phillips and V. M. Weaver, *Nat. Cell Biol.* **18**, 1336–1345 (2016).
145. A. Partin, Z. Hah, C. T. Barry, D. J. Rubens and K. J. Parker, *Ultrasound Med. Biol.* **40**, 685–694 (2014).
146. Z. Wu, L. S. Taylor, D. J. Rubens and K. J. Parker, *Phys. Med. Biol.* **49**, 911–922 (2004).
147. R. L. Siegel, K. D. Miller and A. Jemal, *CA Cancer J. Clin.* **67**, 7–30 (2017).
148. H. Laklai, Y. A. Miroshnikova, M. W. Pickup, E. A. Collisson, G. E. Kim, A. S. Barrett, R. C. Hill, J. N. Lakins, D. D. Schlaepfer, J. K. Mouw, V. S. LeBleu, N. Roy, S. V. Novitskiy, J. S. Johansen, V. Poli, R. Kalluri, C. A. Iacobuzio-Donahue, L. D. Wood, M. Hebrok, K. Hansen, H. L. Moses and V. M. Weaver, *Nat. Med.* **22**, 497–505 (2016).
149. A. Bhaw-Luximon and D. Jhurry, *Cancer Lett.* **369**, 266–273 (2015).
150. T. Kuwahara, Y. Hirooka, H. Kawashima, E. Ohno, H. Sugimoto, D. Hayashi, T. Morishima, M. Kawai, H. Suhara, T. Takeyama, T. Yamamura, K. Funasaka, M. Nakamura, R. Miyahara, O. Watanabe, M. Ishigami, Y. Shimoyama, S. Nakamura, S. Hashimoto and H. Goto, *Pancreatology* **16**, 1063–1068 (2016).
151. S. Y. Kim, J. H. Cho, Y. J. Kim, E. J. Kim, J. Y. Park, T. J. Jeon and Y. S. Kim, *J. Gastroenterol. Hepatol.* **32**, 1115–1122 (2017).
152. S. Kazemirad, S. Bernard, S. Hybois, A. Tang and G. Cloutier, *IEEE Trans. Ultrason. Ferroelectr. Freq. Control* **63**, 1399–1408 (2016).
153. A. Ouared, E. Montagnon, S. Kazemirad, L. Gaboury, A. Robidoux and G. Cloutier, *IEEE Trans. Ultrason. Ferroelectr. Freq. Control* **62**, 1453–1466 (2015).
154. R. Grützmann and C. Pilarsky, *Cancer Gene Profiling: Methods and Protocols* (Springer, 2010).
155. D. Chauvet, M. Imbault, L. Capelle, C. Demene, M. Mossad, C. Karachi, A. L. Boch, J. L. Gennisson and M. Tanter, *Ultraschall Med.* **37**, 584–590 (2015).
156. S. A. Kruse, G. H. Rose, K. J. Glaser, A. Manduca, J. P. Felmlee, C. R. Jack and R. L. Ehman, *NeuroImage* **39**, 231–237 (2008).
157. U. Hamhaber, I. Sack, S. Papazoglou, J. Rump, D. Klatt and J. Braun, *Acta. Biomater.* **3**, 127–137 (2007).
158. J. T. Grayhack and D. G. Assimos, *Prostate* **4**, 13–31 (1983).
159. D. C. Stewart, A. Rubiano, K. Dyson and C. S. Simmons, *PLoS ONE* **12**, e0177561 (2017).
160. M. Kawano, S. Mabuchi, Y. Matsumoto, T. Sasano, R. Takahashi, H. Kuroda, K. Kozasa, K. Hashimoto, A. Isobe, K. Sawada, T. Hamasaki, E. Morii and T. Kimura, *Sci. Rep.* **5**, 18217 (2015).
161. B. Wei, X. Zhou, C. Liang, X. Zheng, P. Lei, J. Fang, X. Han, L. Wang, C. Qi and H. Wei, *Int. J. Biol. Sci.* **13**, 1450–1457 (2017).
162. R. G. Wells, *Hepatology* **47**, 1394–1400 (2008).

163. S. Hoodeshenas, M. Yin and S. K. Venkatesh, *Top. Magn. Reson. Imaging* **27**, 319–333 (2018).
164. Y. K. Mariappan, K. J. Glaser and R. L. Ehman, *Clin. Anat.* **23**, 497–511 (2010).
165. M. A. Chrenek, P. Wong and V. M. Weaver, *Breast Cancer Res.* **3**, 224 (2001).
166. A. W. Holle, J. L. Young and J. P. Spatz, *Adv. Drug Deliv. Rev.* **97**, 270–279 (2016).
167. M. Magzoub, S. Jin and A. S. Verkman, *FASEB J.* **22**, 276–284 (2008).
168. P. A. Netti, D. A. Berk, M. A. Swartz, A. J. Grodzinsky and R. K. Jain, *Cancer Res.* **60**, 2497–2503 (2000).
169. P. G. Gritsenko, O. Ilina and P. Friedl, *J. Pathol.* **226**, 185–199 (2012).
170. F. Calvo and E. Sahai, *Curr. Opin. Cell Biol.* **23**, 621–629 (2011).
171. S. I. Fraley, P.-H. Wu, L. He, Y. Feng, R. Krisnamurthy, G. D. Longmore and D. Wirtz, *Sci. Rep.* **5**, 14580 (2015).
172. K. B. Pointer, P. A. Clark, A. B. Schroeder, M. S. Salamat, K. W. Eliceiri and J. S. Kuo, *J. Neurosurg.* **126**, 1–10 (2016).
173. M. W. Conklin, J. C. Eickhoff, K. M. Riching, C. A. Pehlke, K. W. Eliceiri, P. P. Provenzano, A. Friedl and P. J. Keely, *Am. J. Pathol.* **178**, 1221–1232 (2011).
174. R. Anderson, E. Anderson, L. Shakir and S. Glover, *Microsc. Anal.* **114**, 5 (2006).
175. O. Nadiarnykh, R. B. Lacomb, M. A. Brewer and P. J. Campagnola, *BMC Cancer* **10**, 94 (2010).
176. J. Adur, V. B. Pelegati, A. A. De Thomaz, M. O. Baratti, L. A. L. A. Andrade, H. F. Carvalho, F. Bottcher-Luiz and C. L. Cesar, *J. Biophotonics* **7**, 37–48 (2014).
177. C. R. Drifka, A. G. Loeffler, K. Mathewson, A. Keikhosravi, J. C. Eickhoff, Y. Liu, S. M. Weber, W. John Kao and K. W. Eliceiri, *Oncotarget* **7**, 76197–76213 (2016).
178. T. Hompland, A. Erikson, M. Lindgren, T. Lindmo and C. De Lange Davies, *J. Biomed. Opt.* **13**, 054050 (2008).
179. H. F. Dvorak, V. M. Weaver, T. D. Tlsty and G. Bergers, *J. Surg. Oncol.* **103**, 468–474 (2011).
180. R. Malik, P. I. Lelkes and E. Cukierman, *Trends Biotechnol.* **33**, 230–236 (2015).
181. J. Locker and J. E. Segall, *Am. J. Pathol.* **178**, 966–968 (2011).
182. M. W. Conklin, R. E. Gangnon, B. L. Sprague, L. Van Gemert, J. M. Hampton, K. W. Eliceiri, J. S. Bredfeldt, Y. Liu, N. Surachaicharn, P. A. Newcomb, A. Friedl, P. J. Keely and A. Trentham-Dietz, *Cancer Epidemiol. Biomark. Prev.* **27**, 138–145 (2018).
183. A. S. Piotrowski-Daspit, B. A. Nerger, A. E. Wolf, S. Sundaresan and C. M. Nelson, *Biophys. J.* **113**, 702–713 (2017).
184. K. M. Riching, B. L. Cox, M. R. Salick, C. Pehlke, A. S. Riching, S. M. Ponik, B. R. Bass, W. C. Crone, Y. Jiang, A. M. Weaver, K. W. Eliceiri and P. J. Keely, *Biophys. J.* **107**, 2546–2558 (2014).
185. A. Kaur, B. L. Ecker, S. M. Douglass, C. H. Kugel, M. R. Webster, F. V. Almeida, R. Somasundaram, J. Hayden, E. Ban, H. Ahmadzadeh, J. Franco-Barraza, N. Shah, I. A. Mellis, F. Keeney, A. Kossenkov, H.-Y. Tang, X. Yin, Q. Liu, X. Xu, M. Fane, P. Brafford, M. Herlyn, D. W. Speicher, J. A. Wargo, M. T. Tetzlaff, L. E. Haydu, A. Raj, V. Shenoy, E. Cukierman and A. T. Weeraratna, *Cancer. Discov.* **9**, 64–81 (2019).
186. A. Ray, Z. M. Slama, R. K. Morford, S. A. Madden and P. P. Provenzano, *Biophys. J.* **112**, 1023–1036 (2017).
187. S. P. Carey, Z. E. Goldblatt, K. E. Martin, B. Romero, R. M. Williams and C. A. Reinhart-King, *Integr. Biol.* **8**, 821–835 (2016).
188. K. T. Kurpinski, J. T. Stephenson, R. R. R. Janairo, H. Lee and S. Li, *Biomaterials* **31**, 3536–3542 (2010).
189. S. D. Subramony, B. R. Dargis, M. Castillo, E. U. Azeloglu, M. S. Tracey, A. Su and H. H. Lu, *Biomaterials* **34**, 1942–1953 (2013).

190. C. D. Paul, P. Mistriotis and K. Konstantopoulos, *Nat. Rev. Cancer* **17**, 131–140 (2017).

191. M. Lintz, A. Muñoz and C. A. Reinhart-King, *J. Biomech. Eng.* **139**, 021005 (2017).

192. M. Mak, F. Spill, R. D. Kamm and M. H. Zaman, *J. Biomech. Eng.* **138**, 021004–021004 (2016).

193. C.-H. Hsin, Y.-E. Chou, S.-F. Yang, S.-C. Su, Y.-T. Chuang, S.-H. Lin and C.-W. Lin, *Oncotarget* **8**, 32783–32793 (2017).

194. R. Li, J. D. Hebert, T. A. Lee, H. Xing, A. Boussommier-Calleja, R. O. Hynes, D. A. Lauffenburger and R. D. Kamm, *Cancer Res.* **77**, 279–290 (2017).

195. V. Salvatore, G. Teti, S. Focaroli, M. C. Mazzotti, A. Mazzotti and M. Falconi, *Oncotarget* **8**, 9608–9616 (2017).

196. K. Wolf, M. Te Lindert, M. Krause, S. Alexander, J. Te Riet, A. L. Willis, R. M. Hoffman, C. G. Figdor, S. J. Weiss and P. Friedl, *J. Cell Biol.* **201**, 1069–1084 (2013).

197. S. Van Helvert, C. Storm and P. Friedl, *Nat. Cell Biol.* **20**, 8–20 (2018).

198. T. A. Potapova, J. Zhu and R. Li, *Cancer Metast. Rev.* **32**, 377–389 (2013).

199. C. G. Rolli, T. Seufferlein, R. Kemkemer and J. P. Spatz, *PLoS ONE* **5**, e8726 (2010).

200. E. M. Balzer, Z. Tong, C. D. Paul, W. C. Hung, K. M. Stroka, A. E. Boggs, S. S. Martin and K. Konstantopoulos, *FASEB J.* **26**, 4045–4056 (2012).

201. J. A. Mosier, A. Rahman-Zaman, M. R. Zanotelli, J. A. Vanderburgh, F. Bordeleau, B. D. Hoffman and C. A. Reinhart-King, *Biophys. J.* **117**, 1692–1701 (2019).

202. S. P. Carey, A. Rahman, C. M. Kraning-Rush, B. Romero, S. Somasegar, O. M. Torre, R. M. Williams and C. A. Reinhart-King, *Am. J. Physiol. Cell Physiol.* **308**, C436–C447 (2015).

203. K. Talkenberger, E. A. Cavalcanti-Adam, A. Voss-Böhme and A. Deutsch, *Sci. Rep.* **7**, 9237–9237 (2017).

204. V. te Boekhorst, L. Preziosi and P. Friedl, *Annu. Rev. Cell Dev. Biol.* **32**, 491–526 (2016).

205. M. R. Zanotelli, A. Rahman-Zaman, J. A. Vanderburgh, P. V. Taufalele, A. Jain, D. Erickson, F. Bordeleau and C. A. Reinhart-King, *Nat. Comm.* **10**, 1–12 (2019).

206. M. R. Zanotelli, Z. E. Goldblatt, J. P. Miller, F. Bordeleau, J. Li, J. A. VanderBurgh, M. C. Lampi, M. R. King and C. A. Reinhart-King, *Mol. Biol. Cell* **29**, 1–9 (2018).

207. J. Zhang, K. F. Goliwas, W. Wang, P. V. Taufalele, F. Bordeleau and C. A. Reinhart-King, *Proc. Natl. Acad. Sci.* **116**, 7867–7872 (2019).

208. P. V. Taufalele, J. A. VanderBurgh, A. Muñoz, M. R. Zanotelli and C. A. Reinhart-King, *PLoS ONE* **14**, e0216537 (2019).

209. A. Mukherjee, A. Jana, B. Koons and A. Nain, *Adv. Exp. Med. Biol.* **1092**, 289–318 (2018).

210. W. Y. Wang, A. T. Pearson, M. L. Kutys, C. K. Choi, M. A. Wozniak, B. M. Baker and C. S. Chen, *APL Bioeng.* **2**, 046107 (2018).

211. Y. Liu, A. Keikhosravi, G. S. Mehta, C. R. Drifka and K. W. Eliceiri, Methods for quantifying fibrillar collagen alignment, in *Fibrosis — Methods and Protocols* (Springer, New York, 2017), pp. 429–451.

212. H. Majeed, C. Okoro, A. Kajdacsy-Balla, K. Toussaint and G. Popescu, *J. Biomed. Opt.* **22**, 046004 (2017).

213. W. Goth, J. Lesicko, M. S. Sacks and J. W. Tunnell, *Annu. Rev. Biomed. Eng.* **18**, 357–385 (2016).

214. J. K. Williams, D. Entenberg, Y. Wang, A. Avivar-Valderas, M. Padgen, A. Clark, J. A. Aguirre-Ghiso, J. Castracane and J. S. Condeelis, *Intravital* **5**, e1182271 (2016).

215. T. Fischer, A. Hayn and C. T. Mierke, *Sci. Rep.* **9**, 8352–8352 (2019).

216. A. Malandrino, R. D. Kamm and E. Moeendarbary, *ACS Biomater. Sci. Eng.* **4**, 294–301 (2018).

217. T. Stylianopoulos, L. L. Munn and R. K. Jain, *Trends Cancer* **4**, 292–319 (2018).

218. S. J. Hollister, *Nat. Mater.* **4**, 518–524 (2005).

219. C. Y. Lin, N. Kikuchi and S. J. Hollister, *J. Biomech.* **37**, 623–636 (2004).

220. L. I. Smith-Mungo and H. M. Kagan, *Matrix Biol.* **16**, 387–398 (1998).

221. B. Emon, J. Bauer, Y. Jain, B. Jung and T. Saif, *Comput. Struct. Biotec.* **16**, 279–287 (2018).

222. M. Kai, A. Ziemys, Y. T. Liu, M. Kojic, M. Ferrari and K. Yokoi, *Transl. Oncol.* **12**, 1196–1205 (2019).

223. V. P. Chauhan, T. Stylianopoulos, Y. Boucher and R. K. Jain, *Annu. Rev. Chem. Biomol. Eng.* **2**, 281–298 (2011).

224. S.-W. Choi, Y. Zhang and Y. Xia, *Langmuir* **26**, 19001–19006 (2010).

225. Q. L. Loh and C. Choong, *Tissue Eng. Part B Rev.* **19**, 485–502 (2013).

226. H. E. Barker, T. R. Cox and J. T. Erler, *Nat. Rev. Cancer* **12**, 540–552 (2012).

227. D. A. Kirschmann, E. A. Seftor, S. F. T. Fong, D. R. C. Nieva, C. M. Sullivan, E. M. Edwards, P. Sommer, K. Csiszar and M. J. C. Hendrix, *Cancer Res.* **62**, 4478–4483 (2002).

228. J. Chang, M. C. Lucas, L. E. Leonte, M. Garcia-Montolio, L. Babloo Singh, A. D. Findlay, M. Deodhar, J. S. Foot, W. Jarolimek, P. Timpson, J. T. Erler and T. R. Cox, *Oncotarget* **8**, 26066–26078 (2017).

229. S. A. Harrison, M. F. Abdelmalek, S. Caldwell, M. L. Shiffman, A. M. Diehl, R. Ghalib, E. J. Lawitz, D. C. Rockey, R. A. Schall, C. Jia, B. J. McColgan, J. G. McHutchison, G. M. Subramanian, R. P. Myers, Z. Younossi, V. Ratziu, A. J. Muir, N. H. Afdhal, Z. Goodman, J. Bosch and A. J. Sanyal, *Gastroenterology* **155**, 1140–1153 (2018).

230. B. Fingleton, *Semin. Cell Dev. Biol*, **19**, 61–68 (2008).

231. G. S. Mack and A. Marshall, *Nat. Biotechnol.* **28**, 214–229 (2010).

232. K. Beckenlehner, S. Bannke, T. Spruß, G. Bernhardt, H. Schönenberger and W. Schiess, *J. Cancer Res. Clin. Oncol.* **118**, 591–596 (1992).

233. S. Shuster, G. I. Frost, A. B. Csoka, B. Formby and R. Stern, *Int. J. Cancer* **102**, 192–197 (2002).

234. S. K. Mitra and D. D. Schlaepfer, *Curr. Opin. Cell Biol.* **18**, 516–523 (2006).

235. V. N. Kolev, W. F. Tam, Q. G. Wright, S. P. McDermott, C. M. Vidal, I. M. Shapiro, Q. Xu, M. S. Wicha, J. A. Pachter and D. T. Weaver, *Oncotarget* **8**, 51733–51747 (2017).

236. H. Jiang, S. Hegde, B. L. Knolhoff, Y. Zhu, J. M. Herndon, M. A. Meyer, T. M. Nywening, W. G. Hawkins, I. M. Shapiro, D. T. Weaver, J. A. Pachter, A. Wang-Gillam and D. G. Denardo, *Nat. Med.* **22**, 851–860 (2016).

237. C. Vennin, V. T. Chin, S. C. Warren, M. C. Lucas, D. Herrmann, A. Magenau, P. Melenec, S. N. Walters, G. Del Monte-Nieto, J. R. W. Conway, M. Nobis, A. H. Allam, R. A. McCloy, N. Currey, M. Pinese, A. Boulghourjian, A. Zaratzian, A. A. S. Adam, C. Heu, A. M. Nagrial, A. Chou, A. Steinmann, A. Drury, D. Froio, M. Giry-Laterriere, N. L. E. Harris, T. Phan, R. Jain, W. Weninger, E. J. McGhee, R. Whan, A. L. Johns, J. S. Samra, L. Chantrill, A. J. Gill, M. Kohonen-Corish, R. P. Harvey, A. V. Biankin, T. R. J. Evans, K. I. Anderson, S. T. Grey, C. J. Ormandy, D. Gallego-Ortega, Y. Wang, M. S. Samuel, O. J. Sansom, A. Burgess, T. R. Cox, J. P. Morton, M. Pajic and P. Timpson, *Sci. Transl. Med.* **9**, eaai8504 (2017).

238. M. J. Bissell and W. C. Hines, *Nat. Med.* **17**, 320–329 (2011).

Chapter 9

The Architecture of Co-Culture Spheroids Regulates Tumor Invasion Within a 3D Extracellular Matrix

Yu Ling Huang[1], Carina Shiau[1], Cindy Wu[1],
Jeffrey E. Segall[2,*] and Mingming Wu[1,†]

[1]*Department of Biological and Environmental Engineering*
306 Riley-Robb Hall, Cornell University
Ithaca, NY 14853, USA

[2]*Anatomy and Structural Biology*
Albert Einstein College of Medicine
1300 Morris Park Avenue, Bronx, NY 10461, USA
[*]jeffrey.segall@einsteinmed.org
[†]mw272@cornell.edu

Tumor invasion, the process by which tumor cells break away from their primary tumor and gain access to vascular systems, is an important step in cancer metastasis. Most current 3D tumor invasion assays consisted of a single tumor cell embedded within an extracellular matrix (ECM). These assays taught us much of what we know today on how key biophysical (e.g., ECM stiffness) and biochemical (e.g., cytokine gradients) parameters within the tumor microenvironment guided and regulated tumor invasion. One limitation of the single tumor cell invasion assays was that it did not account for cell–cell adhesion within the tumor. In this chapter, we developed a micrometer scale 3D co-culture spheroid invasion assay that recapitulated physiologically realistic tumor microenvironment and was compatible with microscopic imaging. Micrometer scale co-culture spheroids (1:1 ratio of metastatic breast cancer MDA-MB-231 and non-tumorigenic epithelial MCF-10A cells) were made using an array of microwells, and then were embedded within a collagen matrix in a microfluidic platform. Real time imaging of tumor spheroid invasion revealed that the spatial distribution of the two cell types within the tumor spheroid critically regulated tumor invasion. This work linked tumor architecture with tumor invasion and highlighted the importance of the biophysical cues within the bulk of the tumor in tumor invasion.

Keywords: Tumor invasion; tumor spheroids; cancer; microfluidics; collagen.

1. Introduction

Visual inspection of cell/nucleus shape along with spatial distribution of different cell types has always been a critical component of cancer diagnosis. An important method of identifying tumor progression stage is to inspect a 2D slice of tumor tissue under a microscope, with infiltration of immune cells and large nucleus size indicating a poor prognosis.[1] For breast cancer, the cell–cell spatial arrangement, or the architecture of the tumor, has been used for diagnosing tumor invasiveness. An early form of breast cancer is ductal carcinoma *in situ* (DCIS) where the presence of abnormal cells are inside the milk duct.[2,3] Viewing from a cross-section, DCIS can be described as a hollow circular epithelial cell shell with abnormal cells at the center of the shell. DCIS is typically non-invasive, however, it can be transformed into an invasive form when the abnormal cells break away from the normal epithelial cell layer and invade into the surrounding tissue (see Fig. S1).[4,5]

Tumor invasion is an important step of cancer metastasis because tumor cells need to break away from the primary tumor to invade into their surrounding environment.[6,7] Extensive work has been carried out demonstrating that cell/nucleus shape and mechanics have been correlated with the invasiveness of the tumor cells using assays involving single tumor cells plated on a 2D substrate or single tumor cells embedded within a 3D extracellular matrix (ECM).[8–12] *In vivo*, solid tumors often are in the form of cell aggregates or compact cell mass containing many cell types including normal epithelial, endothelial, immune and tumor cells.[13] The architecture of breast tumor is thought to evolve with time, from an early stage DCIS with a simple core shell structure to a late stage tumor with a complex architecture involving vasculature. There is an extensive literature in understanding the roles of the tumor microenvironment in promoting tumor invasion.[6,14–18] However, less is known about how tissue architecture within the bulk of the tumor influences tumor invasion.[19,20]

Accordingly, we developed a co-culture tumor spheroid assay for studies of tumor architecture in tumor invasion that is compatible with optical microscopic imaging. We used a 1:1 ratio of metastatic and non-tumorigenic epithelial cells to create co-culture tumor spheroids and a type I collagen matrix as the surrounding 3D ECM. The invasion behavior of the tumor spheroids was followed by time lapse microscopy. Our work demonstrated that tumor architecture critically regulated tumor cell invasion.

2. Results and Discussion

2.1. *Formation of co-culture tumor spheroids*

Co-culture spheroids were formed within an array microwell platform previously developed in our labs.[21,22] 1:1 cell number ratio of metastatic MDA-MB-231 expressing EGFP (green) and non-tumorigenic MCF-10A cells expressing dTomato (red) fluorescent protein were placed in the microwells. Upon seeding, the cells of two types mixed uniformly as seen in Fig. 1(a). With time, cells of both types moved, proliferated, and re-organized within the microwell, and tumor spheroids started to form after overnight incubation. Interestingly, we found that the spatial organization of the two cell types within the co-culture spheroids evolved over time. On day 2, the non-tumorigenic cells (red MCF-10A cells) formed a cluster in the center with metastatic cancer cells (green MDA-MB-231 cells) on the periphery (top panels of Figs. 1 (b) and 1(c)). On day 4, this spatial organization was reversed with MCF-10A cells outside and MDA-MB-231 cells inside (low panels of Figs. 1(b) and 1(c)). The degree

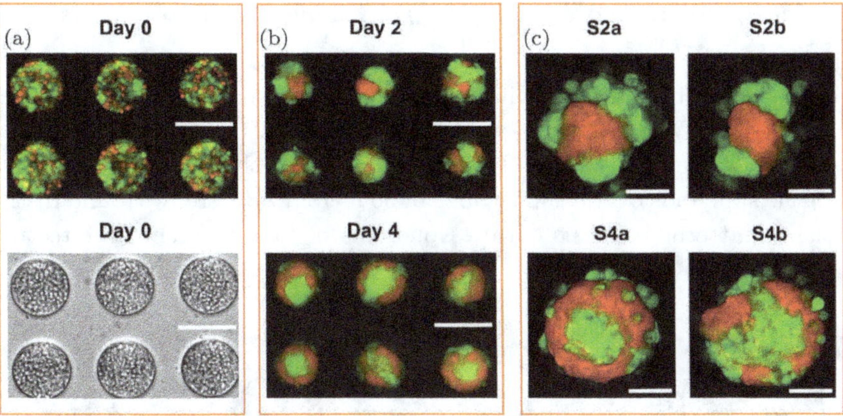

Fig. 1. Different architectures of co-culture spheroids formed within an array microwell platform. Co-culture tumor spheroids were formed by mixing two cell types in an array microwell. (a) Fluorescence (top panel) and bright field (bottom panel) micrographs of 1:1 ratio of two cell types in microwells at day 0 using an epi-fluorescence microscope. (b) Florescence image of co-culture spheroids at day 2 and day 4. Images were taken at the mid-Z-plane of the spheroids using an epi-fluorescence microscope. (c) Close-up confocal images of four types of spheroid architecture observed at day 2 and day 4. Images were reconstructed as projection of Z-planes of the entire spheroids. Spheroids on day 2 exhibited a morphology with metastatic MDA-MB-231 cells outside and nontumorigenic MCF-10A cells inside (S2a and S2b). Spheroids on day 4 presented a reversed morphology with MCF-10A cells surrounding MDA-MB-231 cells, with a few loose MDA-MB-231 cells attaching to the periphery (S4a and S4b). In S2a and 4a, one cell type were enclosed by the other cell type. In S2b and 4b, one cell type only partially surrounded by the other cell type. Green cells were malignant breast tumor cell line (MDA MB-231) expressing green fluorescent protein; red cells were a non-tumorigenic breast epithelial cell line (MCF-10A) expressing dTomato. Scale bar: 200 μm in (a) and (b) and 50 μm in (c).

of enclosure of one cell type over the other was not uniform from well to well, and varied as indicated in Fig. 1(c).

The phenomena of cell segregation has been studied extensively using differential cell–cell adhesion theory,[23,24] however, spheroid architecture inversion was discovered only recently and was poorly understood. We know that the malignant MDA-MB-231 cells are transformed cells, and have essentially no or very low expression of the cell–cell adhesion molecule E-cadherin. MCF-10A cells, on the other hand, have high E-cadherin expression.[21,24] During spheroid formation, cells of two different types can migrate and re-organize to keep the total free energy at the lowest level. Here, the free energy is computed using cell–cell adhesion. In addition, cells of both types proliferate and have different growth rates. In these conditions, MCF-10A cells have a higher proliferation rate than MDA-MB-231 cells. Our previous studies showed that the differential growth rate was an important factor for the spheroid architecture inversion. No architecture inversion was observed when cell growth was inhibited.[21] In this work, we use this inversion phenomena to generate spheroids of different architecture, further experiments will be needed to understand the underlying mechanism responsible for the inversion.

2.2. *The architecture of co-culture spheroid regulated tumor invasion*

To examine how the architecture of the co-culture spheroids affects tumor cell invasion, we embedded the spheroids within a 3D matrix at a concentration of 1.5 mg/mL type I collagen. We placed the spheroid embedded ECM into a microfluidic platform developed previously in the lab[25,26] and followed the spatial and temporal dynamics of the spheroids over a time course of 36 h (Fig. 2). Here, we found distinct tumor invasion patterns using co-culture spheroids of four different architectures.

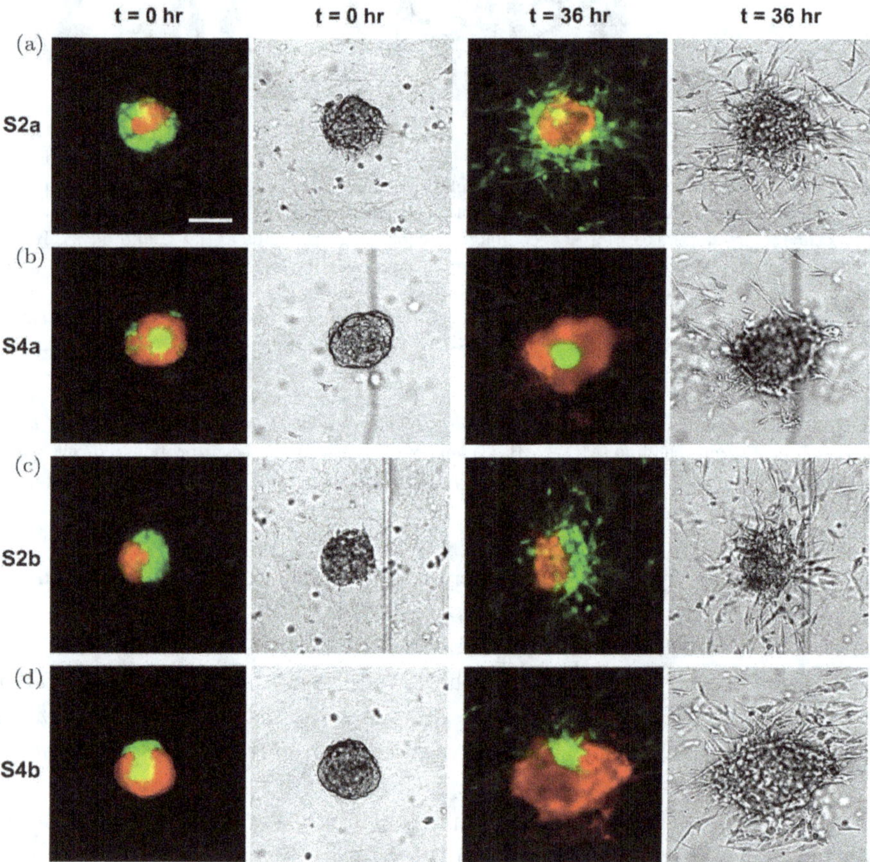

Fig. 2. Architecture of the co-culture spheroids regulated tumor invasion in 3D collagen matrices. Fluorescence (left) and brightfield (right) images of tumor spheroids at time points $t = 0$ and 36 h. (a) In the spheroids with MDA-MB-231 cells outside of a MCF-10A cluster (S2a), significant numbers of the MDA-MB-231 cells invaded out of the spheroid. (b) In the spheroids with MCF-10A cells outside of MDA-MB-231 cell cluster (S4a), only the peripheral MDA-MB-231 cells invaded out of the spheroid while the majority malignant MDA-MB-231 cells enclosed within the MCF-10A cells remained inside. (c) In the spheroids with two cell types side by side (S2b), significant numbers of MDA-MB-231 cells invaded out of the spheroid from one side of the spheroid. (d) In the spheroids with MCF-10A surrounding MDA-MB-231 cells with an opening (S4b), MDA-MB-231 cells invaded out through the opening. The collagen concentration is 1.5 mg/mL. Images were captured from the middle Z-plane of the spheroids. Scale bar: 100 μm. The scale bar is the same for (a)–(d).

In architecture S2a, where MDA-MB-231 cells surrounded the MCF-10A cells, a significant population of MDA-MB-231 cells was observed to detach from the spheroid (Fig. 2(a) and Movie S1). In contrast, in the spheroids with a reversed architecture (S4a), only a few MDA-MB-231 cells at the periphery were able to invade out, but majority of the MDA-MB-231 cells remained in the spheroid core enclosed by a MCF-10A shell (Fig. 2(b) and Movie S2). In architectures S2b and S4b, where part of the MDA-MB-231 cells was at the periphery and in direct contact with collagen matrix initially, majority of the tumor cells invaded out from the side of the spheroid away from the MCF-10A aggregate ((Figs. 2(c) and 2(d), Movies S3 and S4). In particular, in the spheroids with architecture S4b where MDA-MB-231 cells were in contact with collagen through a small opening of the MCF-10A shell, MDA-MB-231 cells streamed out through the opening in a directional way (Fig. 2(d)). Figure 2 clearly demonstrated that the initial cell–cell spatial arrangements within the co-culture spheroid significantly influenced the subsequent invasion.

2.3. *Malignant tumor cells detached from the spheroid more readily in day 2 spheroids than day 4 spheroids*

The first step of tumor cell invasion is for the malignant tumor cells to detach from the tumor spheroids. Using the time sequence images of tumor invasion, we quantified the number of malignant cells detached from the spheroid by using the fluorescence of the cells in the image. We assumed that the total fluorescence intensity of the tumor cells was proportional to the number of cells in this calculation. We note that cell proliferation was not observed during the 36 h of invasion. We found that MDA-MB-231 cells more readily detached from the spheroids harvested at day 2 than those from spheroids harvested at day 4 (Fig. 3) regardless of their initial architecture. At $t = 36$ h, our result demonstrated that spheroids S2a and S2b had the high percentages of detaching cells of $55.0 \pm 1.19\%$ and $50.5 \pm 0.88\%$. In contrast, the

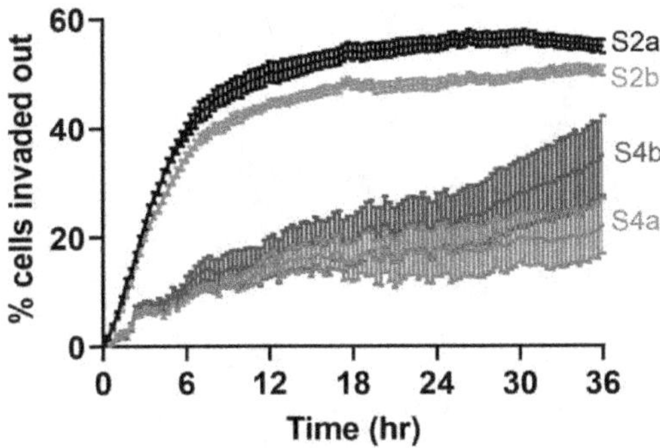

Fig. 3.   Tumor cell detachment from spheroids. Percentage of tumor cells (MDA-MB-231 cells) detached from the co-culture spheroids of four different architectures. Time lapse images were used to calculate the fluorescence of the MDA-MB-231 cells. Percentages of detaching cells for spheroids S2a, S2b, S4a, and S4b are $55.0 \pm 1.19\%$, $50.5 \pm 0.88\%$, $22.1 \pm 5.07\%$, and $34.9 \pm 7.34\%$.

percentages of detached cells from spheroids S4a and S4b were $22.1 \pm 5.07\%$ and $34.9 \pm 7.34\%$. This is consistent with the understanding that spheroids harvested at day 2 were not as compact as those from day 4. We know that cells secrete matrices and adhesion molecules during spheroid formation which may be the reason for spheroid compaction with time. It is interesting note that the percentage of cells invading out for S4b architecture will likely to increase and may reach the same value as S2a and S2b in longer experiments. However, for S4a architecture, the percentage of cells invading out may remain the same because the non-tumorigenic cells at the periphery secured the metastatic tumor cell cluster within the spheroid core and hence blocked their invasion.

2.4. *Invasion characteristics of malignant tumor cells from co-culture spheroids*

An important characteristic for tumor invasion is how far tumor cells migrate away from the primary tumor within a given time. To quantify the motility of tumor cells from co-culture spheroids of different architectures, we tracked individual MDA-MB-231 malignant breast tumor cells that invaded out from the spheroid. The trajectories of 60 cells for each of the four different spheroid architectures are shown in Fig. 4. Here, the starting point of the track was the time when the cell detached from tumor spheroid, and end point of the track was marked with a dot.

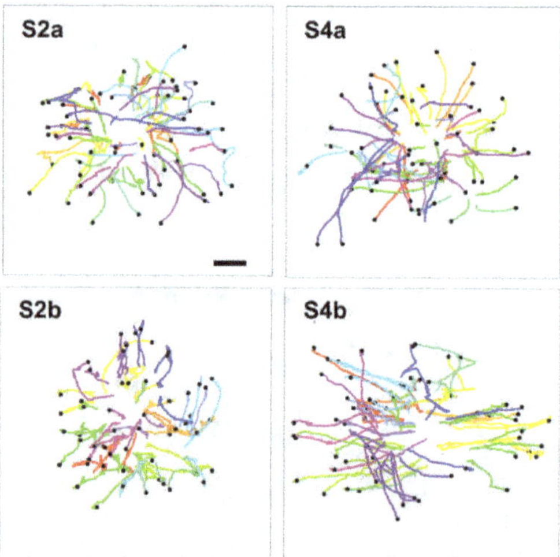

Fig. 4.   Trajectories of MDA-MB-231 cells invading out of the tumor spheroids. 60 tumor cell trajectories were presented for each spheroid architecture. Each colored line represents one cell trajectory. The start point of each trajectory was at the x and y coordinate for the cell position when it started to migrate away from the spheroid, and the end point is presented as a black dot. The time durations of the trajectories range from 14 to 36 h (average of 28 h) for S2a, 10 to 36 h (average of 29 h) for S4a, 12 to 36 h (average of 29 h) for S2b and 6 to 36 h (average of 23 h) for S4b. Scale bar: 100 μm. The scale bar is the same for the four trajectory plots.

Using the cell trajectories, we computed the cell migration speed, persistence, and mean squared displacements (MSDs) of the MDA-MB-231 cells under these four conditions (See Fig. 5(b)). We first compared the cell motilities between S2a (MCF-10A cell cluster surrounded by MDA-MB-231 cells) and S4a (the reversed architecture of S2a, with loosely associated MDA-MB-231 cells around the MCF-10A spheroid shell) architectures. Surprisingly, there was no significant difference between the cell migration speeds. The cell migration speed for S2a is $0.166 \pm 0.006\,\mu\mathrm{m/min}$, while the average cell migration speed for S4a is $0.164 \pm 0.007\,\mu\mathrm{m/min}$. However, the persistence of the tumor cells in S4a is greater than those in S2a, with an average persistence of 0.574 ± 0.022 for S4a in contrast to 0.491 ± 0.021 for S2a, or a 16.9% increase (see Fig. 5(b)). Accordingly, there is a slight increase at later time points in the MSDs measured for S4a spheroids than S2a spheroids (see Fig. 5(c)). We then compared the cell motilities between S4b (MDA-MB-231 cells with an exit route from full enclosure by MCF-10A cells) and S2b (MDA-MB-231 and MCF-10A cluster side by side) architectures. Interestingly, the S4b architecture significantly enhanced cell migration speed, persistence, and MSD. The average cell migration speed is $0.231 \pm 0.013\,\mu\mathrm{m/min}$ for S4b in contrast to $0.160 \pm 0.007\,\mu\mathrm{m/min}$ for S2b, which is a 44.2% difference. The average persistence is 0.638 ± 0.024 for S4b in contrast to 0.403 ± 0.021 for S2b, or a 58.3% difference. When we compared the speed, persistence and MSD of all four cases, it was clear that spheroids with architecture S4b had highest speed, persistence and MSD.

Taken together, we found that MDA-MB-231 cells from co-culture spheroids (see Fig. 5(a)) of S4b architecture to be the most invasive. This invasiveness was largely due to the architectural arrangement of the two cells types, where the malignant MDA-MB-231 cells were surrounded by the non-tumorigenic MCF-10A cells with an

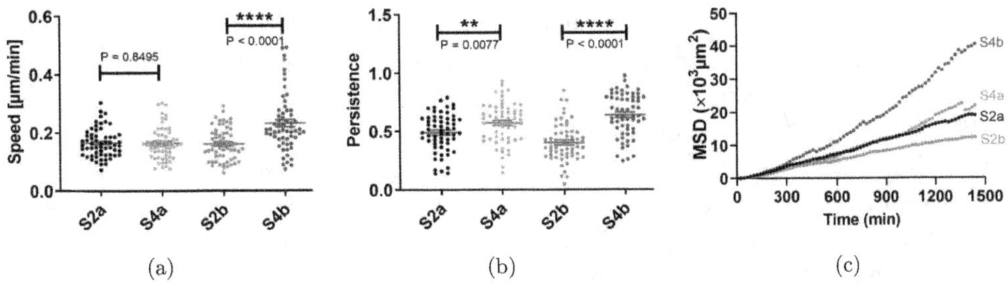

Fig. 5. Architecture of spheroids modulated speed, persistence, and MSDs of MDA-MB-231 malignant breast tumor cells. (a) Migration speed (in μm/min) of invading MDA-MB-231 tumor cells. S4b has a significant increase in cell migration speed in comparison to that of S2b. (b) Persistence of invading MDA-MB-231 tumor cells. Persistence of invading cells from day 4 spheroids is higher than those from day 2. One-way ANOVA test was used for both speed and persistence comparison for all four architecture with $p < 0.0001$. (c) MSDs of invading MDA-MB-231 tumor cells. Tumor cells from S4b have significantly greater MSD than those of the other architectures in the time duration of 24 h. To compute speed and persistence length in A and B, we used 60 cell tracks. To compute MSD, we used 30 cell tracks, and each of the track was longer than 1500 min.

opening at the periphery. This opening enabled the MDA-MB-231 cells to migrate much more directionally than other spheroid architectures. It is interesting to note that although malignant tumor cells in day 4 spheroids were not readily detached from the spheroids initially (in comparison to cells from spheroids of day 2), they were much more invasive once they invaded into the ECM.

3. Conclusion and Future Perspectives

Here, we showed that the architectural arrangement of the malignant tumor cells and the non-tumorigenic epithelial cells critically regulated tumor cell invasiveness. One important finding was that the non-tumorigenic epithelial shell can prevent the invasion of the malignant tumor cells when they were fully surrounded by the non-tumorigenic cells. Among the four tumor architectures we created, we found that S4b, the case where tumor cells were surrounded by the non-tumorigenic cells with a small opening to the ECM facilitated a persistent and fast movement of the malignant cells. In all four spheroid architectural cases, the invading cells were those that had direct contact with the ECM.

We propose a 3D co-culture tumor spheroid invasion model here for the studies of cell–cell interactions in tumor invasion. Our work highlighted the importance of tissue architecture in tumor invasion. The tumor spheroid assay presented here is a step forward in comparison to the single cell invasion assays.[27] It provided a straightforward platform to include cell–cell interactions within the tumor bulk in tumor invasion. Future direction will be to include immune cells, and other stroma cells within the tumor environment. This work can also be easily adapted to other cancer types.

Looking forward, an important question is how the tumor microenvironment regulates tumor architecture and subsequent invasion. Clinically, the transition from ductal carcinoma *in situ* (DCIS) to invasive ductal carcinoma (IDC) for breast cancer patients is still poorly understood (Fig. S1(A)).[4] We note the architectural similarities between DCIS to S4a spheroids (Fig. S1(B)), and IDC to S4b spheroids (Fig. S1 (C)). We believe that the presented tumor spheroid assay can potentially be used for modeling different stages of breast cancer, which will facilitate an understanding of the transition from DCIS to IDC. An immediate question will be the roles of biophysical and biochemical cues within the tumor environment in facilitating the transition from DCIS to IDC and subsequently tumor invasion.

4. Materials and Methods

Cell culture: Metastatic breast cancer cells (MDA-MB-231 cell line expressing EGFP) and non-tumorigenic mammary epithelial cells (MCF-10A cell line expressing dTomato variants) were kind gifts from Dr. Joseph Aslan at the Oregon Health & Science University. MDA-MB-231 cells were cultured every 3 to 4 days from passage 2 to 20, and used at 50–70% confluency.[28] The growth medium for MDA-MB-231 cells was composed of DMEM high glucose medium (Gibco, Life Technologies Corporation, Grand Island, NY), 10% fetal bovine serum (Atlanta biologicals, Lawenceville, GA), and 1% antibiotics (100 units/mL penicillin and 100 μg/mL streptomycin). MCF-10A cells were cultured every 3 to 4 days from passage 2 to 10, and used at 70–90% confluency. The growth medium for MCF-10A cells was composed of DMEM/F-12 medium (Gibco), 5% donor horse serum (Atlanta biologicals), 20 ng/mL human EGF (Gibco), 0.5 μg/mL hydrocortisone (Sigma-Aldrich, St. Louis, MO), 100 ng/mL Cholera Toxin (Sigma-Aldrich), 10 μg/mL insulin (Sigma-Aldrich), and 1% antibiotics (Gibco).

Co-culture spheroid formation: Tumor spheroids were formed using microfabricated microwell arrays previously developed in our labs.[21,22] Each spheroid was formed within a 200 μm diameter and 220 μm height non-adherent microwell treated with 1% pluronic F-127 solution (Sigma-Aldrich). The microwell array was patterned on a thin PDMS membrane (Fig. S2(A)) and then glued to the bottom surface of one of the 12-well plates. Each microwell array contained 1296 microwells (Fig. S2(A)), and within each microwell array, 2 million cells (1:1 ratio of EGFP MDA-MB-231: dTomato MCF-10A) suspended in 2.5 mL medium (1:1 ratio of DMEM and DMEM/F12 growth media) were seeded. Cells were first allowed to settle down into all microwells for 30 min in the incubator (Fig. 1(a)) before the device was placed on a rocker (Boekel Scientific, Rocker II Model 260350). Co-cultured spheroids were formed within the microwells after overnight and cultured for 2 or 4 days to obtain different architectures (Fig. S2(B)), with medium change every 2–3 days. The diameter of the spheroids was approximately 100 μm to fit the invasion device height constraint of 200 μm.

Experimental procedure: A microfluidic platform previously developed in our lab was used to study the co-culture spheroid invasion in 3D collagen matrices.[25,26] Briefly, sterilized PDMS devices were treated with oxygen plasma (Harrick Plasma Cleaner PDC-001, Harrick Plasma, Ithaca, NY) for 1 min on high power mode and then activated with 1% Poly(ethyleneimine) (Sigma-Aldrich) for 10 min followed by a 0.1% Glutaraldehyde (Electron Microscopy Sciences, Hatfield, PA) treatment for 30 min. The PDMS device was then sandwiched between a 3 inch by 1 inch glass slide and a plastic manifold. A 0.6% of agarose solution was used to fill the void space around the PDMS device to prevent medium from evaporating during spheroid invasion experiment.

On the day of experiment, the spheroids were collected from the microwells and filtered by a Falcon® Cell Strainer (Corning) with 70 μm pores to ensure the uniformity of the spheroid size for each experiment. To prepare spheroid embedded collagen matrices, 60 μL Type I collagen from stock concentration of 5.0 mg/mL (Corning, Discovery Labware Inc., Bedford, MA) was first titrated with 1.32 μL 1N NaOH and 20 μL 10X M199 (Sigma-Aldrich) to yield a final pH of approximately 7.4.[29] The collagen was then mixed with the collected co-culture spheroids to a final volume of 200 μL. The final average spheroid concentration was approximately one spheroid per one mm^2 from the top view of the device and the final collagen concentration was 1.5 mg/mL.

Spheroid-embedded collagen solution was introduced to all the channels in each device on an ice block. The microfluidic device was then placed in an incubator to allow collagen polymerization for 45 min at 37°C. To prevent spheroids from gravitationally settling down to the bottom of the device, the microfluidic chips were positioned up-side-down for the first 10 min and then flipped three times more at time points 5, 15, and 15 min in the incubator. Following polymerization, the channels and reservoirs were hydrated with 37°C 1:1 ratio medium, then were plugged with PDMS filled gel loading tips. The microfluidic device was then transferred to the microscope stage enclosed by an environmental control chamber (WeatherStation, PrecisionControl LLC), which was kept at 37°C, 5% CO_2 and about 70% humidity).

Imaging and data analysis: An inverted microscope (IX81, Olympus America, Center Valley, PA, USA) with a CCD camera (Orca-ER, Hamamatsu Photonics, Japan) was used for all the invasion experiments. In a typical experiment, the middle z-plane of the spheroids in the channels were captured using a 10X objective (Olympus, NA = 0.3) in bright field mode and in green fluorescence mode (EX: 460–500 nm, EM: 510–560 nm) for EGFP-MDA-MB-231 cells and red fluorescence (EX: 510–560 nm, EM: 572.5–647.5 nm) mode for dTomato-MCF-10A cells. A sequence of 109 images was captured every 20 min for a total of 36 h. A Zeiss LSM 710 confocal microscope was used to image the fluorescent labeled spheroids formed at day 2 and day 4. A z-stack of average 25 slices with each slice thickness of 6.45 μm was taken for each co-culture spheroid using a 10X objective.

To quantify tumor spheroid invasion, fluorescent images of the GFP mode were used to quantify tumor cell invasion by calculating the percentage of cells detached from the spheroid. Here, fluorescence intensity of the image was used to represent the spatial cell density. To determine the percentage of MDA-MB-231 cells detached from the spheroid, an outline of the spheroid was drawn around the tumor edge in bright field image at $t = 0$ using ImageJ. Then the outline was superimposed to the GFP fluorescent image of the same spheroid (Fig. S3). The total intensity within the outline was measured using ImageJ at all time points in 36 h range. The percentage of cells detached from the spheroid was calculated by the ratio of the total fluorescence outside the initial spheroid outline divided by the total fluorescence of the spheroid at $t = 0$.

To quantify tumor cell motility, time-lapse images of tumor spheroid invasion were processed in ImageJ and in house Matlab programs. MDA-MB-231 tumor cells were tracked after they invaded out of the spheroid. Note that the starting time for each cell differed because each cell invaded out from the spheroid at different time points. A total of 60 tumor cells from four spheroids were tracked for each spheroid architecture and all the tracked trajectories were used to compute the cell migration speed, persistence (the total cell displacement in a given time duration over the length of the cell trajectory of the same time period), and the MSDs.[22,28]

Statistical Analysis: All the data were plotted using Matlab or Prism GraphPad software. Student's *t*-test and one-way ANOVA test were performed for two-group and four-group comparisons, respectively, using Prism and mean $\pm$ SEM were presented in all numerical results as well as the average line and error bars in the plots.

Acknowledgments

This work was mainly supported by a grant from the National Cancer Institute [Grant No. R01CA221346]; partially supported by the Cornell Center on the Microenvironment & Metastasis [Award No. U54CA143876 from the National Cancer Institute]; the Cornell NanoScale Science and Technology, and the Cornell BRC imaging facility. We thank Professor Mingling Ma's lab for sharing their protocols for making spheroids and Young Joon Suh for helpful discussions.

References

1. J. E. Sero *et al.*, *Mol. Syst. Biol.* **11**(3), 790 (2015).
2. H. J. Burstein *et al.*, *N. Engl. J. Med.* **350**(14), 1430 (2004).
3. R. J. Lee *et al.*, *Int. J. Surg. Oncol.* **2012**, 123549 (2012).
4. S. J. Schnitt, *Breast. Cancer Res.* **11**(1), 101 (2009).
5. C. F. Cowell *et al.*, *Mol. Oncol.* **7**(5), 859 (2013).
6. K. Wolf *et al.*, *J. Cell Biol.* **201**(7), 1069 (2013).
7. K. Wolf *et al.*, *Semin. Cell Dev. Biol.* **20**(8), 931 (2009).
8. P. Friedl, K. Wolf and J. Lammerding, *Curr. Opin. Cell Biol.* **23**(1), 55 (2011).
9. A. C. Rowat *et al.*, *Bioessays* **30**(3), 226 (2008).
10. C. M. Denais *et al.*, *Science*, **352**(6283), 353 (2016).
11. K. Wolf *et al.*, *J. Cell Biol.* **160**(2), 267 (2003).
12. E. Sahai and C. J. Marshall, *Nat. Cell Biol.* **5**(8), 711 (2003).
13. A. F. Chambers, A. C. Groom and I. C. MacDonald, *Nat. Rev. Cancer* **2**(8), 563 (2002).
14. Y. L. Huang, J. E. Segall and M. Wu, *Lab. Chip.* **17**(19), 3221 (2017).
15. M. Wu and M. A. Swartz, *J. Biomech. Eng.* **136**(2), 7 (2014).
16. B. J. Kim and M. Wu, *Microfluidics for Mammalian Cell Chemotaxis Ann. Biomed. Eng.* **40**(6), 1316 (2012).
17. M. J. Paszek *et al.*, *Cancer Cell* **8**(3), 241 (2005).
18. E. T. Roussos, J. S. Condeelis and A. Patsialou, *Nat. Rev. Cancer* **11**(8), 573 (2011).
19. M. J. Bissell, A. Rizki and I. S. Mian, *Curr. Opin. Cell Biol.* **15**(6), 753 (2003).
20. M. C. Adriance *et al.*, *Breast Cancer Res.* **7**(5), 190 (2005).
21. W. Song *et al.*, *Soft Matter* **12**(26), 5739 (2016).
22. Y. J. Suh *et al.*, *Integr. Biol.* **11**(3), 109 (2019).
23. M. S. Steinberg, *Science* **137**(3532), 762 (1962).
24. S. Pawlizak *et al.*, *New Journal of Physics* **17**(8), 083049 (2015).
25. C.-K. Tung *et al.*, *Lab on a Chip* **13**(19), 3876 (2013).
26. Y. L. Huang *et al.*, *Integr. Biol. (Camb.)* **7**(11), 1402 (2015).
27. E. J. Brock *et al.*, *J. Mammary Gland Biol. Neoplasia* **24**(1), 1 (2019).
28. B. J. Kim *et al.*, *PLoS One* **8**(7), 69422 (2013).
29. V. L. Cross *et al.*, *Biomaterials* **31**(33), 8596 (2010).

Chapter 10

Investigations of Interactions Between Altretamine and Model Membranes: Spectroscopic and Calorimetric Analysis

D. Bilge*, N. Civelek and Z. Özçelik Çetinel

Department of Physics
Faculty of Science
Ege University, Bornova-Izmir 35100, Turkey
**duygu.bilge@ege.edu.tr*

Altretamine (ALT) is a Food and Drug Administration (FDA) approved antineo-plastic drug particularly used for ovarian cancer. This study examined, at a molecular level, the interactions of the drug with model membranes composed of phospholipids with different acyl chain lengths and head group charges at varied ALT concentrations based on temperature. For this purpose, spectroscopic studies of the liposomes in multilamellar vesicles form were conducted by Fourier transform-infrared spectroscopy (FTIR) and their calorimetric studies were carried out by differential scanning calorimetry (DSC) techniques. The results of the study showed that ALT clearly interacted with lipids and that these interactions were more significant in multilamellar vesicles made up of short chain phospholipids. Moreover, the results suggested that ALT settled into the tail group region, in particular the region that formed the hydrophobic part of lipids, and this effects the whole section of the membranes including glycerol backbones and head groups. This study is expected to contribute, on molecular level, to the studies on the knowledge of the mechanism in cancer, which is still very much a dangerous disease, and its related treatment.

Keywords: Altretamine; model membranes; spectroscopy; phase transition.

1. Introduction

The natural membrane is a complicated system made up of a wide range of different constituents including carbohydrates, lipids and proteins.[1] The complication and dynamism of biological membranes proves very difficult in studies regarding the interplay between anticancer drugs and membranes.[2] The complication of biological membranes has encouraged a breakthrough of various simpler model systems, the size, geometry and composition of which can be very precisely controlled.[3] The automatic mergence process of micelles, vesicles or the amphiphilic molecules in the membranes plays an important role in a wide area from biological systems to technical applications.[4] Liposomes are greatly available for different biophysical parameters to assess and the most widely employed models for studying anticancer drug–membrane interactions.[2] Phospholipids are the structural components in liposomes in addition to being major factors of mammalian cell membranes.[5] The most abundant phospholipids in eucaryotic organisms are phosphatidylcholines (PCs) or lecithins.[6,7] Thus, PCs are the most employed phospholipids in model membrane research.[7,8] Dipalmitoylphosphatidylcholines (DPPCs) and Dimyristoylphosphatidylcholine (DMPCs) are zwitterionic structures and Dipalmitoylphosphatidylglycerols (DPPGs) are negatively charged membrane lipids. Both lipids are used to reveal the effects of charge in drug–membrane interactions.[9–12] Tah *et al.* chose DPPG due to its biocompatible nature. In addition, because of its small size, it is suitable for drug transport.[13] For this reason, these three lipids are chosen to investigate acyl chain lengths and charge effects.

Cancer, which is the second major cause of mortality across the world, is a serious problem that affects the wellbeing of all human societies. It is caused by a series of successive mutations in genes that change cell functions.[14] Cancer cells exhibit a variety of chemical, structural, metabolic and biophysical characteristics that are different from normal cells. The structural, biophysical and functional differences between cancer and normal cell membranes can induce or prevent drug activity and/ or toxicity. Furthermore, drugs tend to have different influences on membranes involving modifications in lipid conformation, surface charge, lipid domains and packing, membrane fluidity and curvature and finally cell function. Accordingly, comprehension of the interactions of chemotherapeutic compounds with biological membranes is vital as it is closely related to the therapeutic process and drugs toxicity.[2]

Hexamethylmelamine (HMM),[15] most commonly known as ALT, is an alkylating antineoplastic drug utilized for the therapy of progressive ovarian cancer.[16] According to the Biopharmaceutical Classification System (BCS), ALT is a class II drug with poor aqueous solubility and high permeability, which implies that its absorption is restricted by the former. It is commercially available in Hexalen® [50 mg] capsules, which are recommended to be ingested four times a day. It is an Food and Drug Administration (FDA)-approved drug, taken either alone or in combination with other anticancer medication.[15,17] The therapeutic activity of ALT is inhibited by its poor solubility, bioavailability and dose-induced toxicity such as gastrointestinal process, hematology and neurotoxicity.[18]

This study chose DPPC and DMPC lipids that had the same head groups but different chain lengths and DPPC and DPPG lipids with different head groups but the same chain lengths. No studies that spectroscopically and calorimetrically examined, on a molecular level, the interactions between ALT and different model membranes were found in the literature. The zwitterionic DPPC and DMPC model membranes and the anionic DPPG model membranes in multilamellar vesicle (MLV) forms were studied via two noninvasive Fourier transform-infrared spectroscopy (FTIR) and differential scanning calorimetry (DSC) techniques. This study is the first of its kind as it displayed the relative impacts of ALT on membrane order and fluidity, nature of hydrogen bonding and phase-transition behavior in distinct

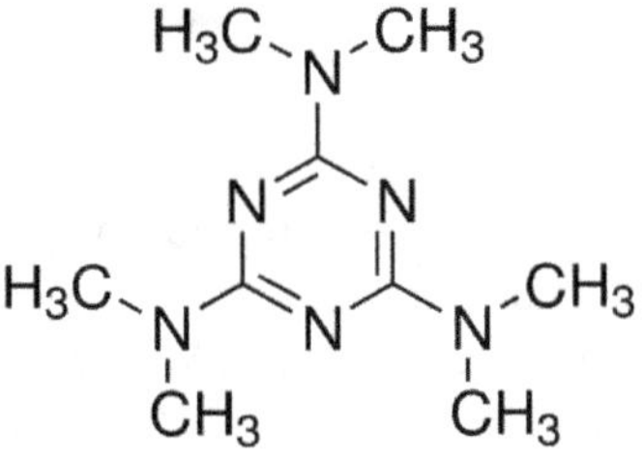

$$H_3C-N-CH_3$$

Fig. 1.   Chemical structure of ALT.

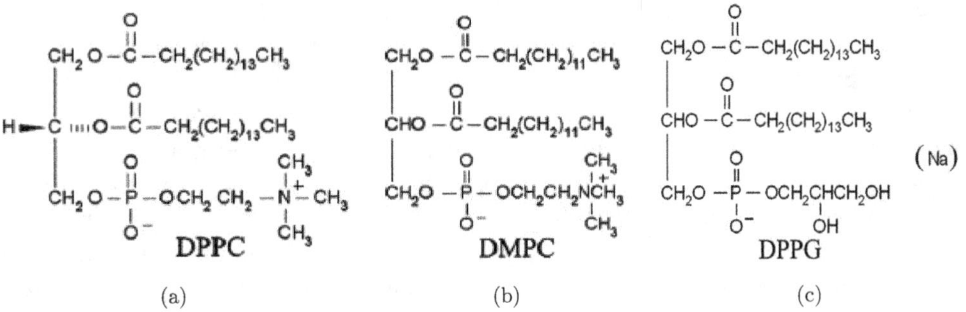

DPPC

$CH_2O-\overset{O}{\overset{\|}{C}}-CH_2(CH_2)_{13}CH_3$

$H-\overset{|}{C}-O-\overset{O}{\overset{\|}{C}}-CH_2(CH_2)_{13}CH_3$

$CH_2O-\overset{O}{\overset{\|}{P}}-OCH_2CH_2-\overset{CH_3}{\overset{|}{N^+}}-CH_3$
$\overset{|}{O^-}\qquad\overset{|}{CH_3}$

(a)

DMPC

$CH_2O-\overset{O}{\overset{\|}{C}}-CH_2(CH_2)_{11}CH_3$

$CHO-\overset{O}{\overset{\|}{C}}-CH_2(CH_2)_{11}CH_3$

$CH_2O-\overset{O}{\overset{\|}{P}}-OCH_2CH_2N\overset{CH_3}{\overset{|+}{}}CH_3$
$\overset{|}{O^-}\qquad\overset{|}{CH_3}$

(b)

DPPG

$CH_2O-\overset{O}{\overset{\|}{C}}-CH_2(CH_2)_{13}CH_3$

$CHO-\overset{O}{\overset{\|}{C}}-CH_2(CH_2)_{13}CH_3$

$CH_2O-\overset{O}{\overset{\|}{P}}-OCH_2CHCH_2OH$
$\overset{|}{O^-}\qquad\overset{|}{OH}$

(Na)

(c)

Fig. 2.   Chemical structures of (a) DPPC, (b) DMPC and (c) DPPG.

phospholipids in which different acyl chain lengths and charge groups existed including variable ALT concentrations. Figures 1 and 2 exhibit the chemical structures of the ALT and all of the lipids used.

2. Material and Methods

ALT, DPPC, DMPC and DPPG were purchased from Sigma (St. Louis, MO, USA) and used without further purification.

The pure phospholipid MLVs were prepared according to the method used by Severcan *et al.* for infrared measurements.[19] In order to prepare all the three lipid MLVs, 5 mg phospholipid was dissolved in chloroform in an Eppendorf tube. A dried lipid film was obtained after the samples were exposed to 3 h vacuum drying using a Christ LT-105 spin vac system. The addition of 25 μl of 10 mM phosphate buffer in 7.4 pH caused the lipid films to be hydrated. The samples were vortexed for 20 min at a temperature 20°C higher than the main phase transition temperature (T_m) to obtain homogeneous vesicles. In order to prepare the ALT-containing liposomes, firstly the required amount of ALT was placed into tubes at 3, 6, 9, 30, 35 and 40 mol% concentrations from the stock solution, in which ALT was dissolved in chloroform. To eliminate the excessive chloroform, the solutions in the tubes were exposed to a stream of nitrogen and 5 mg lipid was added. This procedure was followed to prepare all three liposomes. For the FTIR studies, sample suspensions of 20 μl were put between CaF_2 windows, the cell thicknesses of which were 12 mm. Infrared spectra were acquired by a PerkinElmer Frontier FTIR spectrometer. The interferograms were averaged by being scanned for 50 times at a resolution of $2\,cm^{-1}$. A Specac digital temperature controller unit was used to regulate temperature. The samples were incubated for 5 min at each temperature prior to the spectrum being scanned. The samples were scanned between 10°C and 40°C for DMPC liposomes and between 20°C and 65°C for DPPC and DPPG liposomes. The power spectrum of air was recorded as a background and automatically subtracted from the spectra of the samples by means of the PerkinElmer Spectrum v10.3.7 software previously utilized to analyze data. Experiments were repeated twice and the spectra of the samples were collected at the above temperatures. Buffer spectra were also collected at these temperatures and digitally subtracted from the sample spectra under the same circumstances. Subtraction was carried out until the bulk water signal at $2300\,cm^{-1}$ was flattened. The normalized average spectra were compared to demonstrate the changes. The band positions were measured according to the center gravity and the bandwidth measurements at 0.80 of the maximum band height. Moreover, qualitatively comparative outcomes were obtained using the bandwidth at half maximum of the absorption bands. The comprehensive analyses were made by means of the subtracted native spectra. The lipid mixture for the DSC measurements were prepared by following the same procedure. However, this time thin films were obtained by hydrating 2 mg of phospholipid with $50\,\mu l$ phosphate buffer. The samples were scanned over a temperature range of 10–40°C for DMPC and 25–55°C for DPPC and DPPG. A TA Q 100 DSC (TA Instruments Inc., New Castle, Delaware, USA) instrument was used with a heating rate of 1°C/min.

3. Results

3.1. *Calorimetric results*

This study calorimetrically examined the interactions of ALT with three different lipids, namely DPPC, DMPC and DPPG at low concentrations of 3, 6 and 9 mol% and at high concentrations of 30, 35 and 40 mol% using the DSC technique. Figures 3(a)–3(c) shows the DSC curves of the liposomes in the absence of ALT and at different concentrations of DPPC, DMPC and DPPG. Accordingly, Figs. 3(a)–3(c) exhibits pretransition (T_p) and main phase transition (T_m) for pure DPPC at 35.91°C and 41.90°C, for pure DMPC at 12.73°C and 23.33°C and for pure DPPG at 34.03°C and 41.13°C, respectively.

Tables 1(a)–1(c) presents main phase transition temperatures, width of the transition at half peak height $(\Delta T_{1/2})$ and enthalpy values for main phase transition (ΔH) of DPPC, DMPC and DPPG MLVs including ALT.

It was observed that the addition of ALT in concentrations of 3, 6, 9, 30, 35 and 40 mol% to pure DPPC liposomes caused the pretransition temperatures to shift to lower values and though the peak intensities decreased, pretransition did not

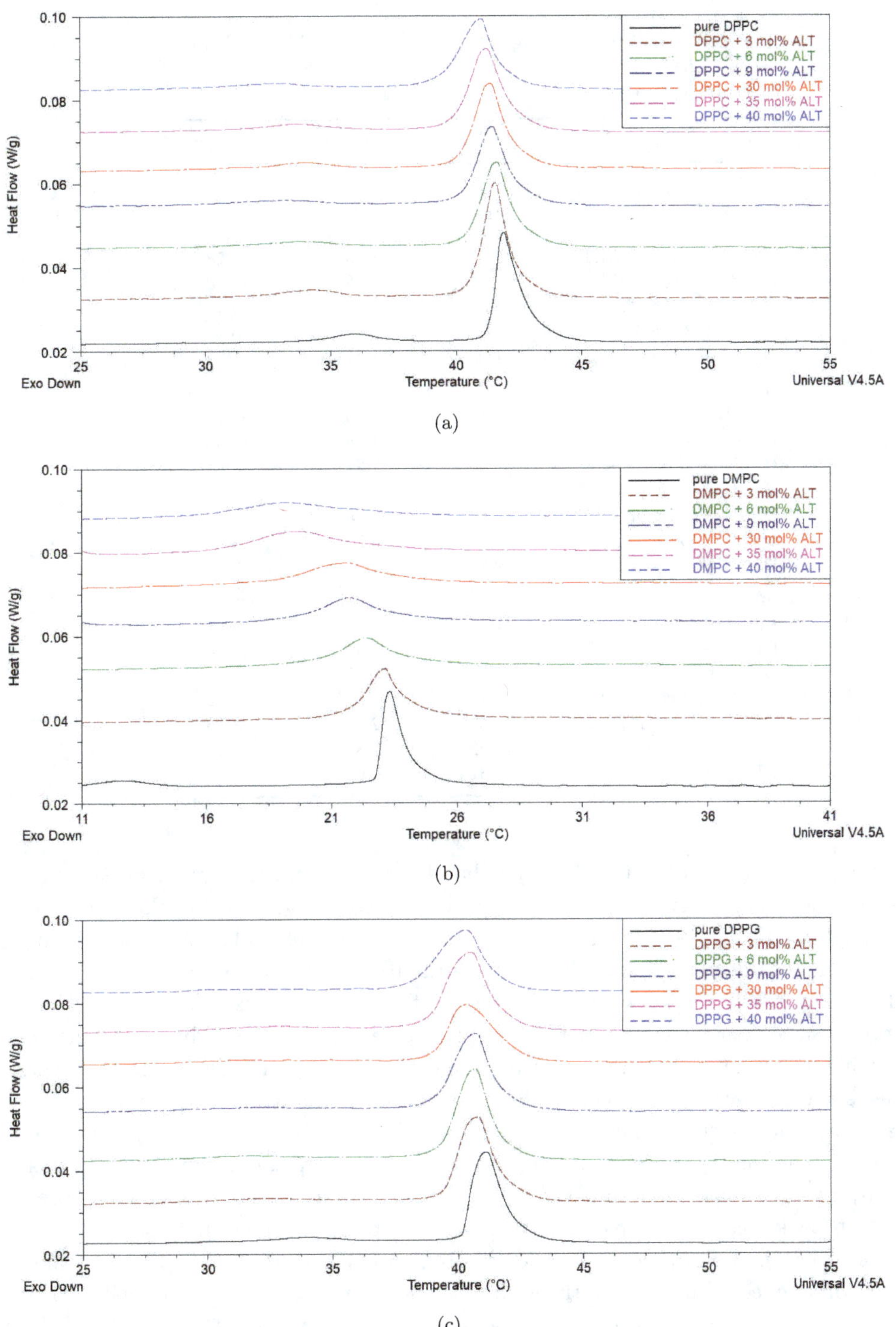

Fig. 3.   DSC curves for (a) DPPC, (b) DMPC and (c) DPPG at varying concentrations of ALT.

Table 1. Main phase transition temperature (T_m), width of the transition of at half peak height $(\Delta T_{1/2})$ and main phase enthalpy values (ΔH) for (a) DPPC (b) DMPC and (c) DPPG with increasing concentrations ALT.

	T_m (°C)	$\Delta T_{(1/2)}$ (°C)	ΔH (J/g)
(a)			
Pure DPPC	41.90	0.99	43.64
DPPC + 3 mol% ALT	41.58	1.02	43.48
DPPC + 6 mol% ALT	41.64	1.25	41.78
DPPC + 9 mol% ALT	41.47	1.28	40.48
DPPC + 30 mol% ALT	41.38	1.31	39.70
DPPC + 35 mol% ALT	41.23	1.43	38.40
DPPC + 40 mol% ALT	41.03	1.47	37.79
(b)			
Pure DMPC	23.33	0.89	34.13
DMPC + 3 mol% ALT	23.14	1.44	33.33
DMPC + 6 mol% ALT	22.40	2.27	31.31
DMPC + 9 mol% ALT	21.72	2.53	28.13
DMPC + 30 mol% ALT	21.61	3.35	30.63
DMPC + 35 mol% ALT	19.60	3.69	28.27
DMPC + 40 mol% ALT	19.22	5.01	26.62
(c)			
Pure DPPG	41.13	1.32	46.51
DPPG + 3 mol% ALT	40.78	1.54	46.14
DPPG + 6 mol% ALT	40.70	1.47	50.97
DPPG + 9 mol% ALT	40.73	1.62	46.52
DPPG + 30 mol% ALT	40.33	2.07	38.16
DPPG + 35 mol% ALT	40.52	1.82	43.13
DPPG + 40 mol% ALT	40.31	2.04	39.58

disappear even at 40 mol%. On the other hand, it was found that main phase temperatures moved to lower values of ~ 1°C, peaks slightly broadened thus the widths of the transitions at half peak height values increased slightly, while main phase transition enthalpies decreased slightly. The examination of the calorimetric results of DMPC bilayers with and without ALT at different concentrations showed that pretransition temperatures completely disappeared even with the addition of 3 mol% ALT while the main phase transition temperatures significantly decreased to lower values such as ~ 4°C and the peaks broadened to finally lose their intensity remarkably. Moreover, the main transition enthalpies of the liposomes considerably reduced. It was consequently seen that when the MLV group composed of DPPG was studied, the pretransition temperature shifted to lower values whereas the pretransition peak was not completely abolished even at higher concentrations. It was observed that the main phase transition shifted to lower values of ~ 1°C, the peaks slightly broadened, thus the width of the transition at half peak height values increased a little and the main phase enthalpies decreased insignificantly.

When the interactions of all three lipids with ALT were studied, the calorimetrically obtained results showed that with the addition of ALT, chosen as the anticancer agent to model membranes pretransition, did not disappear in the MLVs made up of DPPC and DPPG, while the one in those made up of DMPC was completely abolished. In addition, ALT slightly shifted the main phase transition of DPPC and DPPG, whereas it was more significantly moved to lower temperatures and the DSC peaks considerably broadened for DMPC.

3.2. *Spectroscopic results*

The Fourier transform-infrared (FTIR) technique was used for the spectroscopy of the study in order to examine the effects of the lipid acyl chain lengths and the net charge of their head groups on the drug–membrane interactions. As mentioned in the literature, the net charge of the head group of DPPC and DMPC remain in the zwitterionic lipid groups, while the net charge of the head group of DPPG remain in the anionic lipid groups. For this reason, ALT-DPPG and ALT-DPPC MLV systems were used to investigate the charge effect of the polar head group of lipids on ALT–membrane interactions and ALT-DPPC and ALT-DMPC MLV systems were studied to examine the effects of different acyl chain lengths using the FTIR technique based on concentration and temperature. Lipid systems including pure lipids and low and high ALT concentrations were used in the experiments. The calorimetric results of the study showed that the DSC peaks exhibited significant changes after $30 \, \text{mol}\%$ concentrations in DMPC MLVs in particular. To study how the drug–lipid interactions varied based on charge and acyl chain lengths, $30 \, \text{mol}\%$ was chosen as the high drug concentration and $3 \, \text{mol}\%$ was chosen as the low drug concentration. CH_2 antisymmetric and symmetric stretching modes were observed in the values of ~ 2920 and $2850 \, \text{cm}^{-1}$ wavenumbers, whereas C=O stretching bands of the ester groups were seen between ~ 1721–$1739 \, \text{cm}^{-1}$ and PO_2^- antisymmetric stretching bands between ~ 1200–$1260 \, \text{cm}^{-1}$ wavenumber regions. The CH_2 antisymmetric stretching band wavenumber was analyzed to obtain the states of order and disorder of the system while the values of the CH_2 antisymmetric stretching band width were examined to acquire its dynamics (fluidity). The analyses wavenumber of the C=O and PO_2^- antisymmetric double bond stretching bands provide information on whether or not there is a hydrogen bond around such functional groups. Figures 4(a)–4(c) shows the changes in the wavenumber values of the DPPC, DMPC and DPPG MLVs depending on temperature with and without ALT in gel and liquid crystalline phases.

It is clear from Fig. 4(b) that the phase transition for pure DMPC was completed at $\sim 2°C$ while the transitions for the high and low ALT concentrations were concluded at $\sim 5°C$. The higher the ALT concentrations, the lower the phase transition temperatures and the broader the phase transition gap. The wavenumber values increased at gel phase in accordance with higher concentration while at liquid crystalline phase no significant changes were observed at the low concentration and only

     D. Bilge, N. Civelek & Z. Özçelik Çetinel

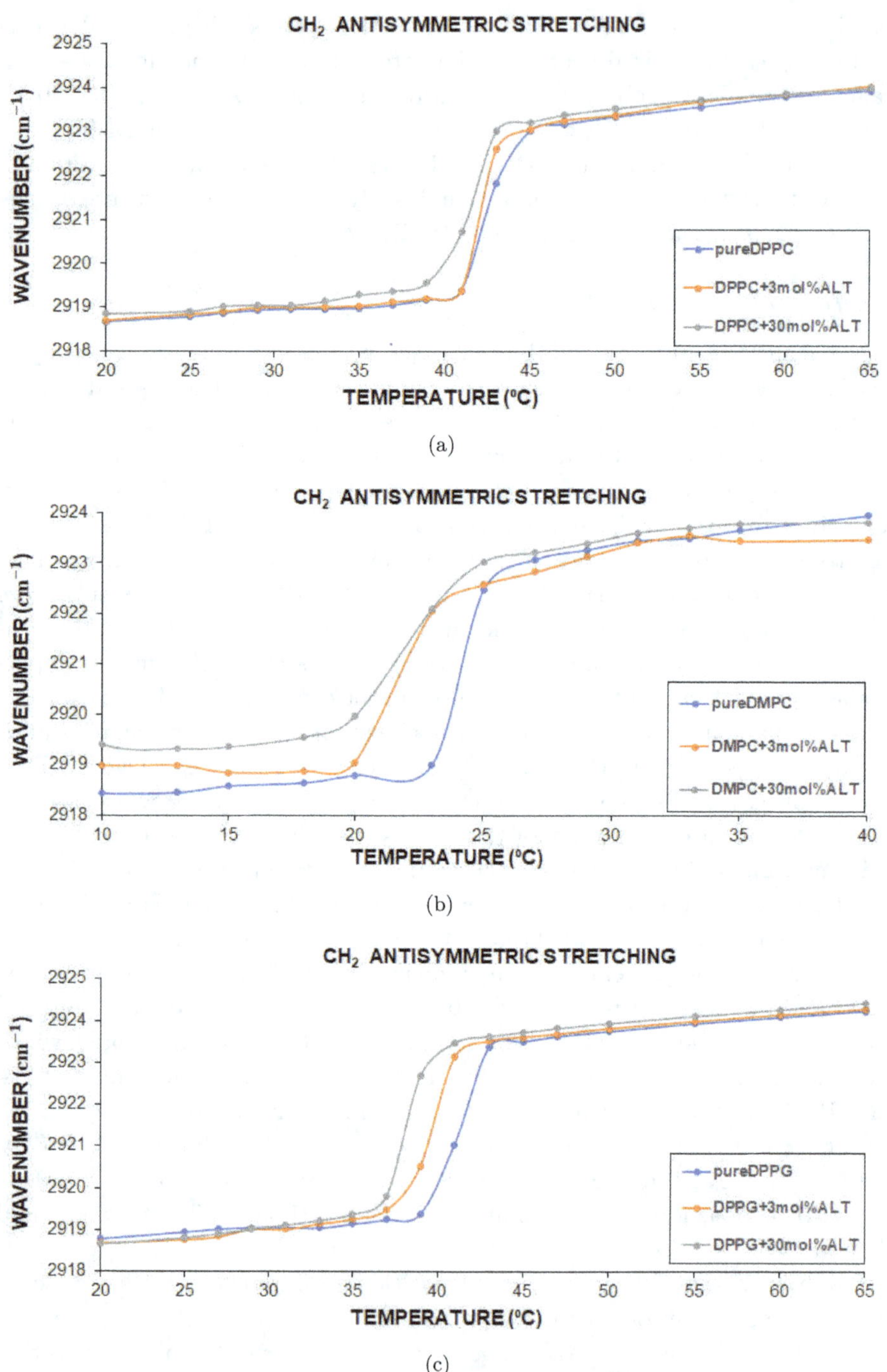

Fig. 4. The wavenumbers variations of the CH$_2$ antisymmetric stretching mode based on the temperature (a) DPPC, (b) DMPC and (c) DPPG liposomes in the absence and presence of 3 and 30 mol% concentration of ALT.

a slight increase was found to occur at the high concentration, which indicates that increase in ALT concentration at gel phase significantly disturbed the order of lipid bilayers and almost no effects occurred in the membrane order with 3 mol% drug concentration whereas a slight disturbance was seen with 30 mol% at the liquid crystalline phase. Figures 4(a) and 4(c) show that low drug concentration for DPPC/ DPPG-ALT systems did not have any considerable effects on the bilayer order at both phases while slightly disturbing effects emerged on the DPPC MLVs in the gel phase at the high concentration, although no significant effects were seen on DPPG MLVs. On the other hand, a slightly disturbing effect was determined on the membrane order of both lipids at the high concentration at the liquid crystalline phase. Figures 5(a)–5(c) exhibit the values of CH_2 antisymmetric stretching band width change based on temperature at low and high concentrations of ALT for the DPPC, DMPC and DPPG liposomes.

As can be seen from Figs. 5(a)–5(c), the band width values increased in accordance with an increase in temperature at both phases for the DPPC, DMPC and DPPG MLVs at both concentrations. In other words, higher ALT concentration increased the fluidity of lipid bilayers. Interaction between the glycerol backbone of the DPPC, DMPC and DPPG liposomes and ALT could be found by analyzing the C=O stretching band. Figures 6(a)–6(c) present the changes of the wavenumber values of the C=O stretching band based on concentration at both the gel and liquid crystalline phases.

An increase was seen in the wavenumber values at both phases and at both concentrations in the DPPC and DMPC lipids, implying that ALT accounted for the dehydration around the functional group in the glycerol backbone of the lipid membranes. An adverse effect took place on the DPPG-ALT system at gel phase, where the wavenumber decreased at the low concentration and increased at the high concentration. Accordingly, hydrogen bonding occurred around the functional group at the low concentration, while dehydration occurred at the high concentration. However, no significant effects appeared at low concentrations while dehydration ensued at the high concentration at the liquid crystalline phase. The PO_2^- antisymmetric double bond stretching band was analyzed to obtain knowledge on the head groups of the phospholipids, in which change in wavenumber implied that hydrogen bonding took place between the hydrogen atoms of the water molecules and phosphate group. Figures 7(a)–7(c) show the temperature-dependent changes of the wavenumber values of the PO_2^- antisymmetric double bond stretching band for the DPPC, DMPC and DPPG liposomes at ALT concentrations of 3 and 30 mol%. It is clear to see that dehydration occurred based on the increase in the wavenumber at the low concentration at the gel phase in the DPPC-ALT system. No significant effect were found to occur at the high concentration.

However, the liquid crystalline phase exhibited an adverse effect at both concentrations as dehydration occurred at the low concentration and hydrogen bonding resulted at the high concentration. In the DMPC liposomes at the gel phase, a significant effect was not observed with 3 mol% ALT, however the wavenumber

 D. Bilge, N. Civelek & Z. Özçelik Çetinel

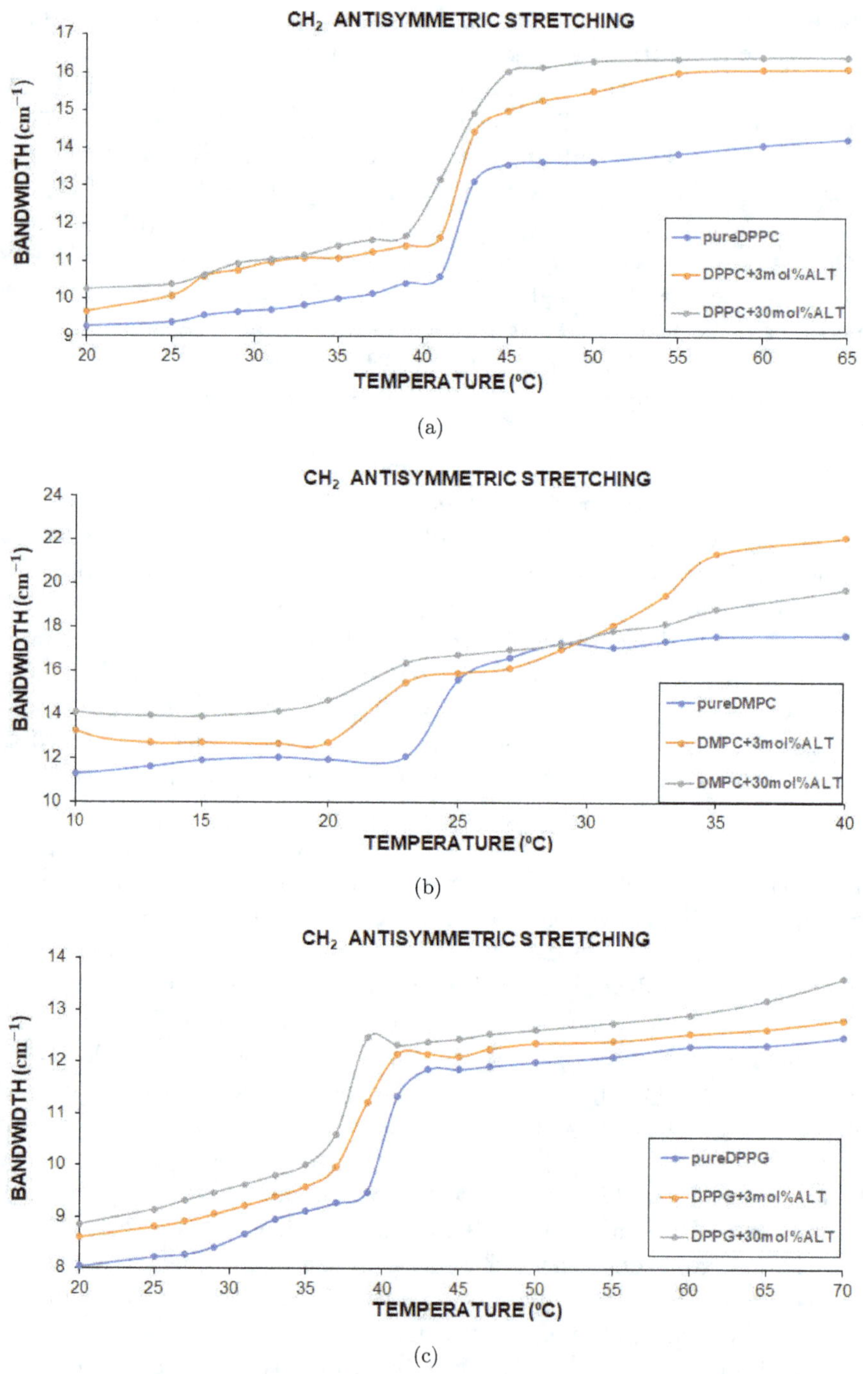

Fig. 5.   The bandwidths variations of the CH_2 antisymmetric stretching mode based on the temperature (a) DPPC, (b) DMPC and (c) DPPG liposomes in the absence and presence of 3 and 30 mol% concentration of ALT.

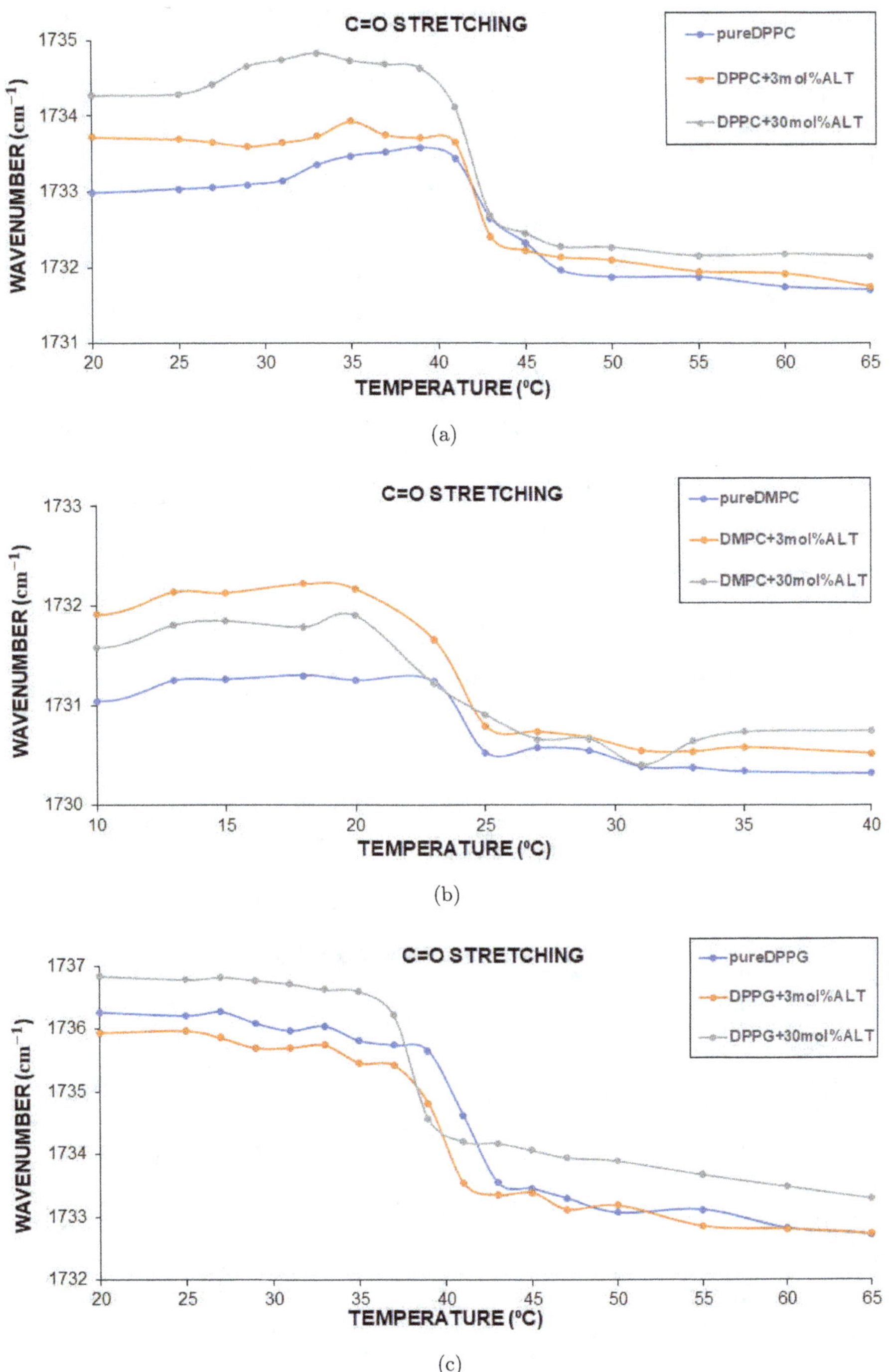

Fig. 6. The wavenumbers variations of the C=O antisymmetric stretching mode based on the temperature (a) DPPC, (b) DMPC and (c) DPPG liposomes in the absence and presence of 3 and 30 mol% concentration of ALT.

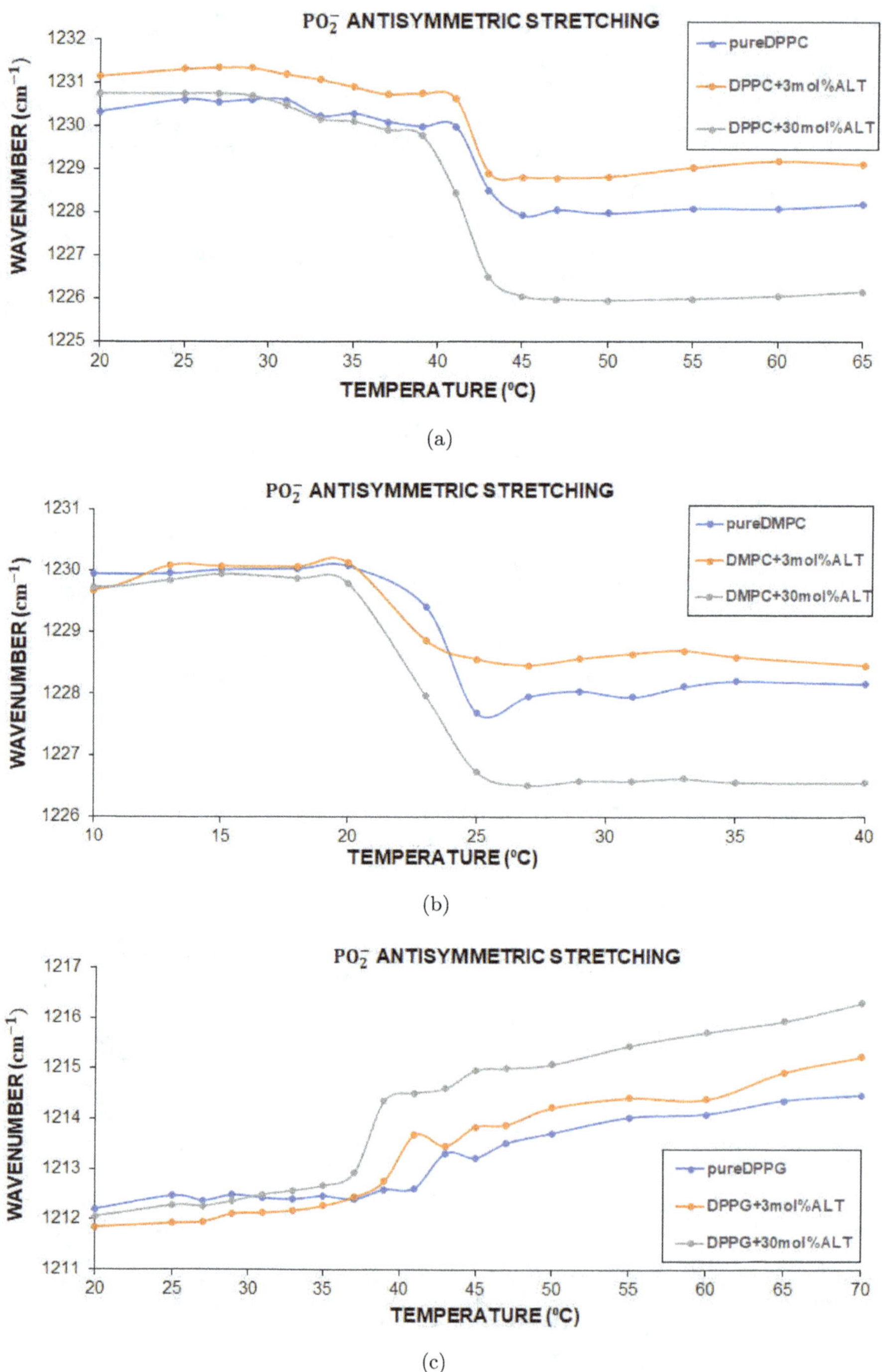

Fig. 7. The wavenumbers variations of the PO_2^- antisymmetric double bond stretching mode based on the temperature (a) DPPC, (b) DMPC and (c) DPPG liposomes in the absence and presence of 3 and 30 mol% concentration of ALT.

decreased a little with 30 mol%. A similar effect was found with the DPPC liposomes at the liquid crystalline phase. Hydration occurred in the DPPG MLVs at the low concentration, whereas no significant effect was found with the high concentration at the gel phase. Dehydration occurred in the liquid crystalline phases at both low and high concentrations.

4. Discussion

The reason why membranes are extremely complicated structures is because they contain lipid domains, lipid asymmetry and the phases and diversity in lipid composition. Their complexity is further increased by the fact that they are associated with proteins and carbohydrates, which makes the biophysical interactions with bioactive molecules very difficult to investigate under actual circumstances. Therefore, simplified artificial membrane systems that resemble the natural bilayer lipid membrane have been developed.[20] Liposomes are extremely crucial in drug transportation,[11] therefore the interaction between a drug molecule and lipid bilayers is vitally related to its pharmaceutical process.[21] In terms of cellular processes and diseases, lipid order and fluidity are extremely important parameters for the existing functions of natural membranes.[11] For instance, the membrane fluidity of neoplasms is different from that of normal cells. Although no common pattern has been found, a series of studies have proven a significant association between plasma membrane fluidity and malignance due to cancer.[22] Molecular research involving the dynamics and fluidity of membranes in drug–membrane interactions are therefore very important. Phospholipid polar head groups and acyl chains are known to have a considerable impact on lipid bilayers. This study calorimetrically and spectroscopically examined the interactions between ALT and model membranes that have the same chain length but different head groups, namely DPPC and DPPG. In addition, the interactions between ALT and model membranes that have the same head groups but different chain lengths, namely DPPC and DMPC were also investigated calorimetrically and spectroscopically. In light of the DSC analyses, Fig. 3 shows that pretransition was abolished for DMPC with the addition of 3 mol% ALT, while it appeared at even high concentrations of ALT for DPPC and DPPG. As pretransition is quite sensitive to changes in the packing of lipid acyl chains, its disturbance and/or disappearance suggests that lipid bilayer structures can be modified by the drug at gel phase.[23] Table 1 demonstrates the significant decrease in the values of the main phase transition temperature for DMPC compared to DPPC and DPPG. It is known that the physical consistence of liposomes is a function of the acyl chain lengths of phospholipids.[11,24] Phospholipids with long chain lengths and high- transition temperatures are more stable than those with short chain lengths and low-transition temperatures.[11,25,26] Accordingly, the fact that the intermolecular interactions of DPPC and DPPG, the hydrophobic groups of which are more packed, are stronger suggests that ALT could be diffused into both lipids less than the DMPC MLVs. Similarly, the study by Ali *et al.* indicated that as a result of the addition of Taxane prodrug to lipid bilayers with varied chain lengths, differences between the behaviors of the liposomes of DPPC and DMPC and those of DSPC were observed. They explained these differences by the fact that DSPC has stronger intermolecular hydrophobic chain interactions compared to DMPC and DPPC.[27] Figures 3(a)–3(c) shows how peaks broadened while the transition from gel to liquid crystalline lamellar phase shifted to low temperatures at increased ALT concentrations. The broadening of the phase transition curves showed strong van der Waals interactions between the hydrophobic chains to be decayed and that ALT could further penetrate into the hydrophobic section of the bilayers.[10,28,29] Such broadening in the phase transition curves also indicated a decrease in cooperation in the phase transition profile. Moreover, the decrease in the phase transition temperature and the broadening of peaks suggested that the drug interacted with the C2–C8 methylene region of the hydrocarbon chains in particular and that it could be located in the hydrocarbon core of the bilayer.[10,30]

From Fig. 3, it is clear that the most significant peak broadening emerged in the DMPC MLVs. The broadening of the peaks and the decrease in their density is indicative of a breakdown in the interactions between the phospholipid acyl chains.[31] On the other hand, the influence on the DMPC liposomes could be caused by the fact that the van der Waals interactions between the drug–lipid are stronger than those in the DPPC-ALT and DPPG-ALT systems. Ali *et al.* observed that the addition of the prodrug taxane to DMPC liposomes caused the transition from gel to liquid crystalline lamellar phase to more significantly shift to low temperatures, hence, causing the peaks to further broaden compared to those in the DPPC or DSPC. They also put forward that the addition of taxane acyl chains into phospholipid bilayers maximized the van der Waals interactions between drug–lipid.[27] The variation of lipid bilayer enthalpies could be associated with the changes in the van der Waals interactions due to the interplay of the drug with the hydrophobic part of the liposomes.[10,27,32–36] Enthalpy change is related to the thermal energy input to the lipid system and the phase transitions of the phospholipids are the result of the decreases in the van der Waals interactions between the lipid acyl chains.[37] When the enthalpy values of main phase transition with increasing ALT concentrations are considered, it can be said that there was a separate decrease in each lipid, as can be seen from Tables 1(a)–1(c). The decrease in the ΔH values suggests that the interplays increased between the lipids and ALT and the van der Waals interactions decreased among the lipids, which in turn could increase the disorder in the membrane.[27,38] The reason for the decrease in the main phase transition enthalpy could be that some liposomes do not undergo phase transition due to drug–lipid interactions.[39–42] The study by Ergun *et al.* reported that the phase transition of various DPPG liposomes was similarly inhibited and thus phase transition enthalpy decreased because of the interaction between DPPG and Agomelatine.[10]

The states of order and disorder in the lipid bilayers could be subtracted from the change in the frequency values of the CH_2 antisymmetric stretching band obtained from the FTIR spectrum. The decrease in the frequency shows an increase in the order of lipids, while the increase in the frequency indicates a decrease in the order of lipids. Variations in the band width of the CH_2 antisymmetric stretching band reflected the changes in the system's dynamics, i.e., the changes in the fluidity of the system. An increase in the band width indicated that the fluidity of the system also increased. On the other hand, changes in the CH_2 antisymmetric stretching band width reflect those in the dynamics of the system.[10,43] From Figs. 4(a)–4(c), it is clear that the increase in the frequency values of the DMPC lipids are more significant than those in the DPPC and DPPG lipids at both concentrations. In other words, the drug used decreased the order of the DMPC lipids more than it did in the DPPC and DPPG lipids. Moreover, ALT increased the fluidity in the gel and liquid crystalline phases for all three lipids (Figs. 5(a)–5(c)). The increase in the wavenumber was consistent with the increase in the Gauche conformers of the hydrocarbon chains, which showed that the order of the system decreased.[44] According to the obtained results, there was a higher increase in the trans/gauche isomerizations in the acyl chains of the DMPC lipids than those in the DPPC and DPPG and therefore a further increase in the trans/gauche rotameric energy of the acyl chains in the DMPC lipid could be cited. This result was very consistent with the results from the DSC curves in Figs. 3(a)–3(c), where the broadening in the phase transition gap based on analyses of CH_2 antisymmetric stretching band was higher in the DMPC lipids and the related phase transition temperature shifted to lower values than those in the DPPC and DPPG MLVs. Based on such calorimetric and spectroscopic data, the fact that the most significant drug–lipid interaction was in the DMPC instead of in the DPPC and DPPG MLVs suggests that the drug could interact with the lipids of the shorter acyl chain or generally with the hydrophobic chain groups of the lipids. From monitoring the frequency of the $C{=}O$ stretching and the PO_2^- symmetric stretching bands, it is clear that the hydration states of the glycerol molecules and head groups of phospholipids existed.

Under the empirical principles, the decreased frequency values exhibited higher strength in the existing hydrogen bonding or in the formation of new hydrogen bonding between the components. The addition of ALT to the DPPC and DMPC bilayers caused dehydration to occur at the two phases and at both concentrations, which indicated that the new hydrogen bonds did not emerge in the glycerol backbone of phospholipids. A contrasting effect was observed at the gel and liquid crystalline phases in the DPPG MLVs (Figs. 6(a)–6(c)). From Figs. 7(a)–7(c), which shows the state of the hydration around phosphate head groups of the bilayers, it can be seen that the DPPC and DMPC lipids with the same head groups, at the liquid crystalline phase in particular exhibited a similar effect, whereas the DPPG lipids with different head groups had a different effect. ALT has 0H donor and 6H acceptors, which could cause changes to appear in the capacity of the hydrogen bonding of all lipids used in this study. These hydrogen bonds could be composed of the head group of the phospholipids and/or other ALT molecules. The difference in the hydrogen bonding of the DPPG head group in the gel and liquid crystalline phases could be accounted for by a variation in the permeability of the membranes between both phases. The bilayers are tightly packed and have a low permeability in the gel phase while those in the liquid crystalline phase are loosely packed and show a high permeability. The highest permeability occurred around T_m at which both the gel and liquid crystalline phases remained altogether. Therefore, a mismatch that produces defects through which molecules can permeate is advisable in the lipid packing between the two phases.[10] As can be seen from the analyses of the C=O stretching and PO_2^- antisymmetric double stretching bands, ALT exhibited similar behavior in the DMPC and DPPC lipids with the same head groups whereas it showed different behavior in the DPPG lipid with different head groups.

Briefly, in light of all of the calorimetric and spectroscopic experiments conducted within the scope of this study, ALT was found to interact with the tail groups of the phospholipids and that such interaction could influence all the membrane structures including the glycerol backbone and the phosphate head groups in general and as a whole by destabilizing the membranes. This means that ALT could change the structure and the physicochemical properties of liposomes. Cancer is a serious disease that can lead to death. Recently, investigations into the reasons for and the effects of cancer have been highly prevalent. Drugs can lead to variations in lipid conformation, lipid packing, membrane fluidity, etc. Therefore, the analysis of interactions between biological membranes and agents is extremely important.[2] ALT is a drug that is preferred in the treatment of ovarian cancer.[17] The comprehensive results of this study could be guiding in terms of a potential molecular basis for clinical practices and other similar studies conducted in the future as there is no available literature on ALT–model membrane interactions.

Acknowledgment

This work was supported by the Ege University Research Fund 2016/FEN/032.

References

1. J. G. Shapter, I. Köper, D. K. Suhendro, J. L. Zieleniecki and J. Knobloch, Membrane–drug interactions studied using model membrane systems, *Saudi J. Biol. Sci.* **22**, 714–718 (2015), https://doi:10.1016/j.sjbs.2015.03.007.
2. A. C. Alves, D. Ribeiro, C. Nunes and S. Reis, Biophysics in cancer: The relevance of drug-membrane interaction studies, *Biochim. Biophys. Acta - Biomembr.* **1858**, 2231–2244 (2016), https://doi:10.1016/j.bbamem.2016.06.025.
3. Y. H. M. Chan and S. G. Boxer, Model membrane systems and their applications, *Curr. Opin. Chem. Biol.* **11**, 581–587 (2007), https://doi.org/10.1016/j.cbpa.2007.09.020.
4. O. Masalci, N. Kazanci, I. Orujalipoor and S. Ide, Nano-structural analysis of a lyotropic liquid crystal (TTAB + water + decanol ternary) system by salt (NH_4Br) addition, *Mol. Cryst. Liq. Cryst.* **609**, 70–79 (2015), https://doi:10.1080/15421406.2014.957476.
5. B. Roy, P. Guha, R. Bhattarai, P. Nahak, G. Karmakar, P. Chettri and A. K. Panda, Influence of lipid composition, pH, and temperature on physicochemical properties of liposomes with curcumin as model drug, *J. Oleo Sci.* **65**, 399–411 (2016), https://doi.org/10.5650/jos.ess15229.

6. M. Caffrey and J. Hogan, LIPIDAT: A database of lipid phase transition temperatures and enthalpy changes. DMPC data subset analysis, *Chem. Phys. Lipids.* **61**, 1–109 (1992), https://doi:10.1016/0009-3084(92)90002-7.

7. D. Bilge, I. Sahin, N. Kazanci and F. Severcan, Molecular and interactions of tamoxifen with distearoyl phosphatidylcholine multilamellar vesicles: FTIR and DSC studies, *Spectrochim. Acta A. Mol. Biomol. Spectrosc.* **130**, 250–256 (2014), https://doi:10.1016/j.saa.2014.04.027.

8. P. Neven and H. Vernaeve, Guidelines for monitoring patients taking tamoxifen treatment, *Drug Saf.* **22**, 1–11 (2000), https://doi:10.2165/00002018-200022010-00001.

9. G. Cakmak Arslan and F. Severcan, The effects of radioprotectant and potential antioxidant agent amifostine on the structure and dynamics of DPPC and DPPG liposomes, *Biochim. Biophys. Acta - Biomembr.* **1861**, 1240–1251 (2019), https://doi:10.1016/j.bbamem.2019.04.009.

10. S. Ergun, P. Demir, T. Uzbay and F. Severcan, Agomelatine strongly interacts with zwitterionic DPPC and charged DPPG membranes, *Biochim. Biophys. Acta - Biomembr.* **1838**, 2798–2806 (2014), https://doi:10.1016/j.bbamem.2014.07.025.

11. D. Bilge, N. Kazanci and F. Severcan, Journal of molecular structure acyl chain length and charge effect on tamoxifen – lipid model membrane interactions, *J. Mol. Struct.* **1040**, 75–82 (2013), https://doi:10.1016/j.molstruc.2013.02.031.

12. C. B. Fox and J. M. Harris, Confocal Raman microscopy for simultaneous monitoring of partitioning and disordering of tricyclic antidepressants in phospholipid vesicle membranes, *J. Raman Spectrosc.* **41**, 498–507 (2010), https://doi:10.1002/jrs.2483.

13. B. Tah, P. Pal, S. Mishra and G. B. Talapatra, Interaction of insulin with anionic phospholipid (DPPG) vesicles, *Phys. Chem. Chem. Phys.* **16**, 21657–21663 (2014), https://doi:10.1039/c4cp03028a.

14. S. H. Hassanpour and M. Dehghani, Review of cancer from perspective of molecular, *J. Cancer Res. Pract.* **4**, 127–129 (2017), https://doi:10.1016/j.jcrpr.2017.07.001.

15. B. Gidwani and A. Vyas, The potentials of nanotechnology-based drug delivery system for treatment of ovarian cancer, *Artif. Cells. Nanomed. Biotechnol.* **43**, 291–297 (2015), https://doi:10.3109/21691401.2013.853179.

16. L. A. Hansen and T. E. Hughes, Altretamine, *DICP* **25**, 146–152 (1991), https://doi.org/10.1177/106002809102500209.

17. B. Gidwani and A. Vyas, Designing and evaluation of extended release matrix tablet containing altretamine-HP-β-CD inclusion complex, *J. Incl. Phenom. Macrocycl. Chem.* **83**, 401–409 (2015), https://doi:10.1007/s10847-015-0569-9.

18. B. Gidwani and A. Vyas, Pharmacokinetic study of solid-lipid-nanoparticles of altretamine complexed epichlorohydrin-β-cyclodextrin for enhanced solubility and oral bioavailability, *Int. J. Biol. Macromol.* **101**, 24–31 (2017), https://doi:10.1016/j.ijbiomac.2017.03.047.

19. F. Severcan, I. Sahin and N. Kazanci, Melatonin strongly interacts with zwitterionic model membranes-evidence from Fourier transform infrared spectroscopy and differential scanning calorimetry, *Biochim. Biophys. Acta - Biomembr.* **1668**, 215–222 (2005), https://doi.org/10.1016/j.bbamem.2004.12.009.

20. M. Deleu, J. M. Crowet, M. N. Nasir and L. Lins, Complementary biophysical tools to investigate lipid specificity in the interaction between bioactive molecules and the plasma membrane: A review, *Biochim. Biophys. Acta - Biomembr.* **1838**, 3171–3190 (2014), https://doi:10.1016/j.bbamem.2014.08.023.

21. T. T. T. Do, U. P. N. Dao, H. T. Bui and T. T. Nguyen, Effect of electrostatic interaction between fluoxetine and lipid membranes on the partitioning of fluoxetine investigated

using second derivative spectrophotometry and FTIR, *Chem. Phys. Lipids.* **207**, 10–23 (2017), https://doi:10.1016/j.chemphyslip.2017.07.001.

22. A. C. Alves, D. Ribeiro, M. Horta, J. L. F. C. Lima, C. Nunes and S. Reis, The daunorubicin interplay with mimetic model membranes of cancer cells: A biophysical interpretation, *Biochim. Biophys. Acta - Biomembr.* **1859**, 941–948 (2017), https://doi:10.1016/j.bbamem.2017.01.034.

23. A. B. Hendrich, O. Wesolowska and K. Michalak, Trifluoperazine induces domain formation in zwitterionic phosphatidylcholine but not in charged phosphatidylglycerol bilayers, *Biochim. Biophys. Acta - Biomembr.* **1510**, 414–425 (2001), https://doi:10.1016/S0005-2736(00)00373-4.

24. M. Anderson and A. Omri, The effect of different lipid components on the in vitro stability and release kinetics of liposome formulations, *Drug Deliv.* **11**, 33–39 (2004), https://doi:10.1080/10717540490265243.

25. G. Gregoriadis, Overview of liposomes, *J. Antimicrob. Chemother.* **28**, 39–48 (2012), https://doi:10.1093/jac/28.suppl_b.39.

26. D. J. A. Crommelin and H. Schreier, Liposomes, in *Colloidal Drug Delivery Systems*, ed. J. Kreuter, Vol. **66** (Marcel Dekker, New York, 1994).

27. S. Ali, S. Minchey, A. Janoff and E. Mayhew, A differential scanning calorimetry study of phosphocholines mixed with paclitaxel and its bromoacylated taxanes, *Biophys. J.* **78**, 246–256 (2000), https://doi:10.1016/S0006-3495(00)76588-X.

28. N. Kazanci, N. Toyran, P. I. Haris and F. Severcan, Vitamin D2 at high and low concentrations exert opposing effects on molecular order and dynamics of dipalmitoyl phosphatidylcholine membranes, *Spectroscopy*, **15**, 47–55 (2001), https://doi:10.1155/2001/890975.

29. F. Korkmaz and F. Severcan, Effect of progesterone on DPPC membrane: Evidence for lateral phase separation and inverse action in lipid dynamics, *Arch. Biochem. Biophys.* **440**, 141–147 (2005), https://doi:10.1016/j.abb.2005.06.013.

30. M. Biol, L. Bilayer and H. Sciences, Effect of small molecules on the dipalmitoyl lecithin liposomal bilayer, *Transition*, **201**, 157–201 (1977), https://link.springer.com/content/pdf/10.1007%2FBF01870299.pdf.

31. O. Wesolowska, Presence of anionic phospholipids rules the membrane localization of phenothiazine type multidrug resistance modulator, *Biophys. Chem.* **109**, 399–412 (2004), https://doi.org/10.1016/j.bpc.2003.11.004.

32. C. Potamitis, P. Chatzigeorgiou, E. Siapi, K. Viras, T. Mavromoustakos, A. Hodzic, G. Pabst, F. Cacho-Nerin, P. Laggner and M. Rappolt, Interactions of the AT1 antagonist valsartan with dipalmitoyl-phosphatidylcholine bilayers, *Biochim. Biophys. Acta - Biomembr.* **1808**, 1753–1763 (2011), https://doi:10.1016/j.bbamem.2011.02.002.

33. C. Fotakis, D. Christodouleas, P. Zoumpoulakis, E. Kritsi, N. P. Benetis, T. Mavromoustakos, H. Reis, A. Gili, M. G. Papadopoulos and M. Zervou, Comparative biophysical studies of sartan class drug molecules losartan and candesartan (CV-11974) with membrane bilayers, *J. Phys. Chem. B.* **115**, 6180–6192 (2011), https://doi:10.1021/jp110371k.

34. A. Hasanovic, C. Hollick, K. Fischinger and C. Valenta, Improvement in physicochemical parameters of DPPC liposomes and increase in skin permeation of aciclovir and minoxidil by the addition of cationic polymers, *Eur. J. Pharm. Biopharm.* **75**, 148–153 (2010), https://doi:10.1016/j.ejpb.2010.03.014.

35. F. Martinez and A. Gómez, Thermodynamics of partitioning of some sulfonamides in 1-octanol-buffer and liposome systems, *J. Phys. Org. Chem.* **15**, 874–880 (2002), https://doi:10.1002/poc.564.

36. J. J. Chicano, A. Ortiz, J. A. Teruel and F. J. Aranda, Organotin compounds alter the physical organization of phosphatidylcholine membranes, *Biochim. Biophys. Acta - Biomembr.* **1510**, 330–341 (2001), https://doi:10.1016/S0005-2736(00)00365-5.

37. C. Demetzos, Differential Scanning Calorimetry (DSC): A tool to study the thermal behavior of lipid bilayers and liposomal stability, *J. Liposome Res.* **18**, 159–173 (2008), https://doi:10.1080/08982100802310261.

38. H. Maswadeh, C. Demetzos, I. Daliani, I. Kyrikou, T. Mavromoustakos, A. Tsortos and G. Nounesis, A molecular basis explanation of the dynamic and thermal effects of vinblastine sulfate upon dipalmitoylphosphatidylcholine bilayer membranes, *Biochim. Biophys. Acta - Biomembr.* **1567**, 49–55 (2002), https://doi:10.1016/S0005-2736(02)00564-3.

39. A. M. S. Cardoso, S. Trabulo, A. L. Cardoso, A. Lorents, C. M. Morais, P. Gomes, C. Nunes, M. Lúcio, S. Reis, K. Padari, M. Pooga, M. C. Pedroso De Lima and A. S. Jurado, S4(13)-PV cell-penetrating peptide induces physical and morphological changes in membrane-mimetic lipid systems and cell membranes: Implications for cell internalization, *Biochim. Biophys. Acta - Biomembr.* **1818**, 877–888 (2012), https://doi:10.1016/j.bbamem.2011.12.022.

40. A. W. Shaw, M. A. McLean and S. G. Sligar, Phospholipid phase transitions in homogeneous nanometer scale bilayer discs, *FEBS Lett.* **556**, 260–264 (2004), https://doi.org/10.1016/S0014-5793(03)01400-5.

41. A. B. Hendrich, R. Malon, A. Pola, Y. Shirataki, N. Motohashi and K. Michalak, Differential interaction of Sophora isoflavonoids with lipid bilayers, *Eur. J. Pharm. Sci.* **16**, 201–208 (2002), https://doi:10.1016/S0928-0987(02)00106-9.

42. T. Tsunoda, T. Imura, M. Kadota, T. Yamazaki, H. Yamauchi, K. O. Kwon, S. Yokoyama, H. Sakai and M. Abe, Effects of lysozyme and bovine serum albumin on membrane characteristics of dipalmitoylphosphatidylglycerol liposomes, *Colloids Surf. B. Biointerfaces*, **20**, 155–163 (2001), https://doi.org/10.1016/S0927-7765(00)00188-0.

43. I. Sahin, D. Bilge, N. Kazanci and F. Severcan, Concentration-dependent effect of melatonin on DSPC membrane, *J. Mol. Struct.* **1052**, 183–188 (2013), https://doi.org/10.1016/j.molstruc.2013.08.060.

44. N. Ezer, I. Sahin and N. Kazanci, Alliin interacts with DMPC model membranes to modify the membrane dynamics: FTIR and DSC studies, *Vib. Spectrosc.* **89**, 1–8 (2017), https://doi:10.1016/j.vibspec.2016.12.006.

Chapter 11

Kinetic Aspects of the Interplay of Cancer and the Immune System

Vladimir P. Zhdanov

Section of Biological Physics, Department of Physics
Chalmers University of Technology, Göteborg, Sweden

Boreskov Institute of Catalysis, Russian Academy of Sciences
Novosibirsk, Russia
zhdanov@catalysis.ru

The understanding of the interplay between cancer and the immune system is still limited. Herein, I focus on two aspects of this interplay. First, I propose a kinetic model describing the likely role of the immune system in the lifetime risk of cancer at the level of the whole human population. For each tissue, the risk is predicted to be influenced by the heterogeneity of the population and to depend exponentially on time. The expression for the risk does not, however, depend explicitly on the total number of divisions of the corresponding stem cells. For this reason, the correlation with the latter number can only be indirect. Second, using another kinetic framework, I describe how the growth of a few tumors can depend on their interaction via the immune system. The analysis shows that depending on specific details, the tumors of different sizes tend either to reach the same size or remain to be of different sizes.

Keywords: Cancer; immune system; initiation; lifetime risk; tumor growth.

1. Introduction

Cancer is a complex disease occurring on different lengths and time scales (reviewed in Refs. 1–6; for general introduction, see Ref. 7). Phenomenologically, it can be divided into three interconnected stages including initiation, tumor growth and propagation of metastases. During these stages, cancer development depends or may depend on the response of the immune system.[4,8,9] Eventually, activation of oncogenic pathways in tumor cells is known to reduce induction of a local antitumor immune response, and the tumor growth becomes faster.[4] Recent progress in the understanding of the effect of the immune system on the late stage of cancer has resulted in the development of antibody-based immunotherapies that modulate immune responses against tumors.[10]

Due to its importance and complexity, cancer has long attracted the attention of mathematicians and physicists working in biophysics and statistical physics. The kinetic models of various stages of cancer can be traced across decades and are now numerous (reviewed in Refs. 11–15; for a few recent related papers published in

                              V. P. Zhdanov

Biophysical Reviews and Letters, see Refs. 16–18). The goals and standards of various models are very diverse, and, in general, there is a tendency to construct and use the so-called multiscale models (see e.g., Refs. 19–23). Despite the apparent abundance of the kinetic models, the understanding of the corresponding mechanistic details is still incomplete in general and far from complete if one is interested in the interplay between cancer and the immune system. In this work, I focus on two aspects of this interplay.

First, I discuss the lifetime risk of cancer (Sec. 2). How to interpret the corresponding experimental data is still open for debate. My analysis is focused on the likely role of the immune system from this perspective. The very assumption that the lifetime risk can be primarily determined by the immune system is natural because this system is known to play the key role in the suppression of tumors.[4,8,9] How this effect can be mathematically scrutinized and explained is, however, not clear now.

Secondly, I describe how the growth of a few tumors can depend on their interaction via the immune system (Sec. 3). In general, one can expect that the tumors of different sizes can tend either to reach the same size or to remain to be of different sizes. Herein, I clarify the conditions of realization of these kinetic regimes.

The two corresponding models, introduced in Secs. 2 and 3, are independent of each other, each has its own parameters and designations, and accordingly Secs. 2 and 3 can be read separately (in particular, each section has its own brief summary in the end).

2. Lifetime Risk of Cancer

2.1. *General remarks*

The experimental studies of the very first phase of cancer initiation are challenging because the process begins long before the appearance of clinically detectable morphological lesions in the tissue.[1] Some related general features of the illness can, however, be clarified by comparing the large-scale statistics of distribution of cancer among different tissues during the human life span, T, with the tissue properties. This was done by Tomasetti and Vogelstein by employing the statistics for USA[24] and 69 countries, representing more than one-half of the world's population.[25] Referring to the well-known facts that (i) some tissue types give rise to cancers much more often than others, and (ii) the endogenous mutation rate of conventional cell types appears to be nearly identical, they argued that the DNA mutations resulting in cancer initiation should take place primarily during cell division, and at the level of the whole population, there should be a correlation between the lifetime number of divisions, $D(T)$, among a particular class of cells or, more specifically, stem cells within each organ and the lifetime risk, $R(T)$, of cancer initiation in that organ. The data they presented confirm this conjecture. Looking at the array of the data points presented for various tissues (e.g., Fig. 1 in Ref. 24), one can conclude that the correlation is roughly of the power-law type,

$$R(T) \propto D^{\alpha}(T), \tag{1}$$

where $\alpha \simeq 0.68$.

The results reported by Tomasetti and Vogelstein[24,25] were widely debated at the levels of concepts and conventional models of cancer initiation (see e.g., Refs. 26 and 27 and references therein). In such models (e.g., Refs. 27–33; reviewed in Ref. 12), the initiation of cancer is commonly viewed as the probabilistic growth of a cell population and mathematically described as a branching process. In addition, the cellular events (replication, mutation of DNA and death) are assumed to not influence each other and the corresponding rate constants are considered to be literally constant, i.e., to be independent of time. The lifetime risk of cancer is usually expressed as[26]

$$R(T) = \int_0^T \int_0^{T-t} g(\tau)d\tau f(t)dt, \tag{2}$$

where $f(t)$ and $g(\tau)$ are the probability density functions that the first cancer stem cell arises at age t, and that cancer progresses to full disease after some further time τ. Taking into account that the maximum time scale of the cancer progression, $\tau_{\max}$, is much shorter that T, expression (7) can be factorized[26] as

$$R(T) = \int_0^{\tau_{\max}} g(\tau)d\tau \int_0^T f(t)dt \equiv QP(T), \tag{3}$$

where $P(T)$ is the lifetime probability of cancer initiation, and Q is the probability of cancer progression ($\tau_{\max}$ is the corresponding time span). To get a weak dependence of R on T, one should admit that the number of mutations resulting in cancer initiation in a cell should be small. For one and two mutations, one can elementarily get[26]

$$R(T) \propto P(T) \propto NrT \text{ or } R(T) \propto D(T), \tag{4}$$

and

$$R(T) \propto P(T) \propto Nr^2T^2 \text{ or } R(T) \propto D^2(T)/N, \tag{5}$$

where N is the number of stem cells in the tissue of interest, r is the rate constant of division of these cells, and $D(T) = NrT$ is the total number of their divisions. Such expressions based on the conventional models predict that even for minimal numbers of mutations, 1 or 2 (usually, this number is considered to be larger[28]), the dependence of the lifetime risk of cancer initiation on the lifetime number of stem-cell divisions is stronger compared to that identified in Refs. 24 and 25. Taking this divergence into account, it is instructive to look once again through the key assumptions employed in the conventional models, to modify them including new biologically reasonable ingredients, and to scrutinize the corresponding kinetic scenarios. As already noted in Sec. 1, one of the ways to extend the conceptual basis here is to include the immune system into the interplay. Below, I follow this line.

2.2. *Kinetic model*

In the context under consideration, one can notice that there are indications that during the early stages of tumor development, cytotoxic immune cells (e.g., natural killer (NK) and CD8$^+$ T cells) are able to recognize and eliminate some of the subpopulations of cancer cells.[9,34] The detailed data on the effect of the immune system on the very first events of cancer initiation are, however, still lacking.

The available kinetic models describing the interplay of the immune system and cancer are aimed at late stages of the disease with emphasis on the immunotherapies (see, e.g., Ref. 35–37 and references therein). In such models, the equation for the population of the tumor cells is usually represented as (see, e.g., Ref. 35)

$$dn_t/dt = rn_t(1 - n_t/n_t^*) - qn_e n_t, \tag{6}$$

where r and n_t^* are the rate constants of division and the maximum population of these cells, respectively, n_e is the population of the effector cells, and q is the rate constant of elimination of the former cells by the latter cells. In turn, the equation for the population of the effector cells includes the positive and negative feedbacks between the two populations. All the rate constants in such equations are considered to be independent of time.

Focusing on cancer initiation, I consider that on the life-span scale, a few mutations resulting in conversion of a stem cell to a cancer stem cell occur rapidly. The immune system is assumed to not act on the preconverted cells, and accordingly, this conversion is considered to be described by using the conventional probabilistic approach and not treated here explicitly. The elimination of the newly formed cancer stem cells by the immune system is admitted, and the evolution of these cells after formation of a single cell is described by the simplest equation,

$$dn_s/dt = rn_s - qn_s, \tag{7}$$

with the corresponding initial condition

$$n_s(0) = 1, \tag{8}$$

where r and q are the rate constant of division and elimination. This equation can be viewed to be a reduced form of Eq. (6). In particular, Eq. (7) does not take into account that the growth is restricted [via the existence of the maximum population as in the first term on the right-hand side of Eq. (6)] because n_s is small. For the same reason, the effect of cancer stem cell on the immune system is negligible and accordingly n_e [in Eq. (6)] can be considered to be constant and included into q. With these simplification, Eq. (6) is indeed reduced to (7).

According to Eq. (7), the growth of the population of cancer stem cells is exponential if $r > q$, while at $r < q$ this population goes extinct. In other words, this means that the immune system cannot eliminate cancer cells in the former case and can do this in the latter case.

On the life-span scale, the time scales, $1/r$ and $1/q$, characterizing the cell growth and elimination are short, and Eq. (7) employed with the fixed values of r and q predicts the all-or-none transition and accordingly is insufficient in order to interpret the global statistics. To use Eq. (7) on the life-span scale for the whole human population, I take into account that the efficiency of the immune system of different men or in different organ of a man is different and deteriorates with age. Phenomenologically, this can be described by introducing the time-dependent q distribution, $f(q, t)$, normalized as

$$\int_0^\infty f(q, t)dq = 1. \tag{9}$$

This distribution is considered to correspond to a given organ of the whole population. With increasing time, the rate constant of division of cancer stem cells can in principle also vary, but such variations are expected to be less important compared to those in q in relation with deterioration of the immune system and accordingly are neglected. Then, taking into account that Eq. (7) employed with the fixed values of r and q predicts the all-or-none transition, the risk of cancer can be represented as

$$R(t) = \int_0^r f(q, t)dq. \tag{10}$$

To be specific, I represent the distribution of q in the Gaussian form,

$$f(q, t) = (\alpha(t)/\pi)^{1/2}\exp[-\alpha(t)(q - \bar{q}(t))^2], \tag{11}$$

where $\bar{q}(t)$ and $\alpha(t)$ are the corresponding parameters defining the mean and determining the variance, respectively. The use of this expression implies that the distribution is narrow ($\alpha^{1/2}(t) \gg \bar{q}(t)$) and accordingly the normalization defined by (9) is fulfilled with good accuracy. The deterioration of the immune system means that the whole distribution of q shifts downwards with increasing time. At the following simplest level adopted, this can be described by considering α to be constant and assuming the mean to decrease as

$$\bar{q}(t) = \bar{q}_\circ - \beta t, \tag{12}$$

where β $(\beta < \bar{q}_\circ/T)$ is the decrease rate.

Substituting (11) into (10) and employing (12) yield

$$R(t) = (\alpha/\pi)^{1/2} \int_0^r \exp[-\alpha(\bar{q}_\circ - q - \beta t)^2]dq. \tag{13}$$

Taking into account that $\alpha^{1/2}(t) \gg \bar{q}(t)$, the integral in this expression can easily be calculated analytically as

$$R(t) \simeq \frac{\exp[-\alpha(\bar{q}_\circ - r - \beta t)^2]}{2(\alpha\pi)^{1/2}(\bar{q}_\circ - r - \beta t)}. \tag{14}$$

In the latter expression, the dependence of the exponential term in the numerator on time is more important than that of the denominator. Ignoring the latter dependence or, more specifically, replacing $\bar{q}_{\circ} - r - \beta t$ there by $\bar{q}_{\circ} - r$ results in

$$R(t) \simeq \frac{\exp[-\alpha(\bar{q}_{\circ} - r - \beta t)^2]}{2(\alpha\pi)^{1/2}(\bar{q}_{\circ} - r)}. \tag{15}$$

The lifetime risk of cancer is accordingly given by

$$R(T) \simeq \frac{\exp[-\alpha(\bar{q}_{\circ} - r - \beta T)^2]}{2(\alpha\pi)^{1/2}(\bar{q}_{\circ} - r)}. \tag{16}$$

Taken together, Eqs. (7)–(16) specify mathematically the model under consideration.

2.3. *Summary*

The outcome of the analysis presented is as follows:

(i) To describe the effect of the immune response on the lifetime risk of cancer, one should take the heterogeneity of the population into account [Eqs. (9) and (10)]. Another important aspect is that the function of the immune system deteriorates with age. The extent of this deterioration depends on various factors, and accordingly the heterogeneity of the immune response increases with age and is expected to be appreciable at the second part of the life span.

(ii) The risk of cancer is predicted to depend exponentially on time [Eq. (16)]. Mathematically, this dependence is different compared to the power-law dependences [e.g., (4) and (5)] predicted by the conventional model. Qualitatively, the results are, however, similar because the increase of the risk with increasing time is anyway fast.

(iii) The lifetime risk of cancer is predicted to depend exponentially on T as well. The corresponding expression [Eq. (16)] does not, however, explicitly contain $D(T)$, and accordingly cannot be directly used to interpret the results reported by Tomasetti and Vogelstein.[24,25] To move in this direction, one needs to clarify the extent of the difference in the effect of the immune system on cancer stem cells in different organs, or, more specifically, how different are α, β and $\bar{q}_{\circ}$ for different organs. Such data are now lacking.

Finally, I note that focusing on the scenario of cancer initiation with emphasis on the role of the immune system in this process does not exclude other scenarios.

3. Interaction of Tumours Via the Immune System

3.1. *General remarks*

The inflammatory response of the immune system to cancer development has long been recognized to influence the growth of tumours.[4,8,9] The corresponding kinetic models are typically focused on the interplay between a single tumor and the immune system (for recent and earlier studies, see e.g., Refs. 35–37 and 38–41, respectively; for an individual-based approach, see Ref. 42). The growth of cancer cells in a tumor is usually described by the first-order equation with the "logistic" correction restricting the population [Eq. (6)]. In this section, I propose two models focused on the interaction of a few tumors via the immune system. The difference between the models is in the expressions used to describe the birth and death of cancer cells. The goal of this analysis is to show to what extent this difference can modify the type of the growth kinetics.

The interaction between the immune system and tumors is complex and involves various cells including NK and effector CD8$^+$ T cells.[4,8,9] In the available models (e.g., Refs. 35–41), the focus is usually on one or two types of cells of the immune system. The models I use are of the same category and mimic the function of NK cells. These cells are known to mediate the tumor killing mainly by releasing cytotoxic perforin and granzyme, eliminating tumor cells and also triggering apoptotic pathways in tumor cells through the production of TNFα or via direct cell–cell contact through activation of the TRAIL and FASL pathways (reviewed in Ref. 9). Mechanistically, the corresponding minimal model scheme includes proliferation of tumour cells,

$$T \to 2T, \tag{17}$$

birth and conventional death of NK cells,

$$K_1 \to 2K_1, \quad K_1 \to \varnothing, \tag{18}$$

conversion of NK cells into the active form, K_2, via interaction with tumor cells, related death of NK cells, and conventional death of active cells,

$$K_1 + T \to K_2 + T, \quad K_1 + T \to T, \quad K_2 \to \varnothing, \tag{19}$$

formation of perforin or granzyme, M, by active cells and its conventional degradation,

$$K_2 \to K_2 + M, \quad M \to \varnothing, \tag{20}$$

and elimination of tumor cells and related degradation of perforin or granzyme,

$$T + M \to M, \quad T + M \to T, \quad \text{or} \quad T + M \to \varnothing. \tag{21}$$

In the models below, I consider that there are m tumors and operate with the corresponding populations of tumour cells, n_i ($1 \leq i \leq m$), and with K_1, K_2 and M concentrations, C_1, C_2, and c. One of the tumors, e.g., with the population n_1, can be considered as the initial one, while other $m - 1$ tumors can be identified with metastases. As already noticed, the focus is on the immune-system-mediated interplay between tumors. The formation of new tumors is not included.

3.2. *Model 1*

The first model is based on the assumption that the steps including cancer cells occur near or are related primarily to the interface between a tumor and healthy tissue. The rates of these steps are phenomenologically considered to be proportional to the interface area. A tumor is viewed to be spherically shaped, its volume is proportional to n_i, and accordingly its radius and interface area are considered to be proportional to $n_i^{1/3}$ and $n_i^{2/3}$, respectively. In this case, the kinetic equations corresponding to steps (17)–(21) are represented as

$$dn_i/dt = kn_i^{2/3} - rn_i^{2/3}c, \tag{22}$$

$$dC_1/dt = w - \kappa_1 C_1 - r_1 C_1 \sum_{i=1}^{m} n_i^{2/3}, \tag{23}$$

$$dC_2/dt = qC_1 \sum_{i=1}^{m} n_i^{2/3} - \kappa_2 C_2, \tag{24}$$

$$dc/dt = uC_2 - \kappa_* c - r_* c \sum_{i=1}^{m} n_i^{2/3}. \tag{25}$$

The attribution of the rate constants used in these equations to the mechanistic steps under consideration is as follows: k is for (17); r is for $T + M \to M$ and $T + M \to \varnothing$; w and κ_1 are for (18); $r_1 \equiv q + r_\star$, where q is for $K_1 + T \to K_2 + T$, and $r_\star$ is for $K_1 + T \to T$; κ_2 is for $K_2 \to \varnothing$; u and κ_* are for (20); and r_* is for $T + M \to T$ and $T + M \to \varnothing$.

To analyze the model, one can take into account that the evolution of tumor cells is usually slower compared to other step, and accordingly, Eqs. (23)–(25) can be solved in the steady-state approximation, i.e.,

$$C_1 = \frac{w}{\kappa_1 + r_1 \sum_i n_i^{2/3}}, \tag{26}$$

$$C_2 = \frac{qw \sum_i n_i^{2/3}}{\kappa_2 \left(\kappa_1 + r_1 \sum_i N_i^{2/3}\right)}, \tag{27}$$

$$c = \frac{quw \sum_i n_i^{2/3}}{\kappa_2 \left(\kappa_1 + r_1 \sum_i n_i^{2/3}\right)\left(\kappa_* + r_* \sum_i n_i^{2/3}\right)}. \tag{28}$$

Substituting then (28) into (22) yields

$$\frac{dn_i}{dt} = kn_i^{2/3} - \frac{rquwn_i^{2/3} \sum_j n_j^{2/3}}{\kappa_2 \left(\kappa_1 + r_1 \sum_j n_j^{2/3}\right)\left(\kappa_* + r_* \sum_j n_j^{2/3}\right)}. \tag{29}$$

This equation can be rewritten in a more compact form as

$$\frac{dn_i}{dt} = kn_i^{2/3} - \frac{\mu n_i^{2/3} \sum_j n_j^{2/3}}{\left[1 + \sum_j (n_j/K)^{2/3}\right]\left[1 + \sum_j (n_j/K_*)^{2/3}\right]}, \tag{30}$$

where

$$\mu \equiv rquw/(\kappa_1\kappa_*\kappa_2), \quad K \equiv (\kappa_1/r_1)^{3/2}, \quad \text{and} \quad K_* \equiv (\kappa_*/r_*)^{3/2}. \tag{31}$$

An important feature of Eqs. (22), (29), and (30) is that both terms on their right-hand parts are proportional to $n_i^{2/3}$. For this reason, these equations can be factorized. In particular, Eq. (30) can be represented as

$$\frac{dn_i}{dt} = n_i^{2/3}\left(k - \frac{\mu\sum_j n_j^{2/3}}{\left[1 + \sum_j (n_j/K)^{2/3}\right]\left[1 + \sum_j (n_j/K_*)^{2/3}\right]}\right). \tag{32}$$

This means that for arbitrary i, the steady-state solution of this equation can be reached simultaneously provided

$$k - \frac{\mu\sum_j n_j^{2/3}}{\left[1 + \sum_j (n_j/K)^{2/3}\right]\left[1 + \sum_j (n_j/K_*)^{2/3}\right]} = 0. \tag{33}$$

This condition can be fulfilled for different combinations of n_j. Practically, this means that despite the interaction via the immune system, the tumors of different sizes can coexist under steady-state conditions.

To illustrate explicitly the kinetics predicted by the model, I set $K_* \equiv K$, use the normalized rate constants, $\tilde{k} \equiv k/K^{1/3}$ and $\tilde{\mu} \equiv \mu K^{1/3}$, and populations, $\tilde{n}_i = n_i/K$, and rewrite Eq. (33) as

$$\frac{d\tilde{n}_i}{dt} = \tilde{k}\tilde{n}_i^{2/3} - \frac{\tilde{\mu}\tilde{n}_i^{2/3}\sum_j \tilde{n}_j^{2/3}}{\left[1 + \sum_j \tilde{n}_j^{2/3}\right]^2}. \tag{34}$$

For a single tumor, the subscripts and summation sign can in Eqs. (34) be removed, and they are reduced to

$$\frac{d\tilde{n}}{dt} = \tilde{k}\tilde{n}^{2/3} - \frac{\tilde{\mu}\tilde{n}^{4/3}}{[1 + \tilde{n}^{2/3}]^2}. \tag{35}$$

The right-hand part of this equation contains two terms representing the birth and death rates of tumor cells. The comparison of these rates (Fig. 1) indicates that the type of the solution predicted by the equation depends on the ratio of the corresponding rate constants, $\tilde{\mu}/\tilde{k}$. If this ratio is sufficiently large, there are stable and unstable steady-state solutions with relatively low and high population of tumor cells, respectively, and with increasing time depending on the initial population the tumor either reaches a stable steady state or its growth is not limited. If this ratio is

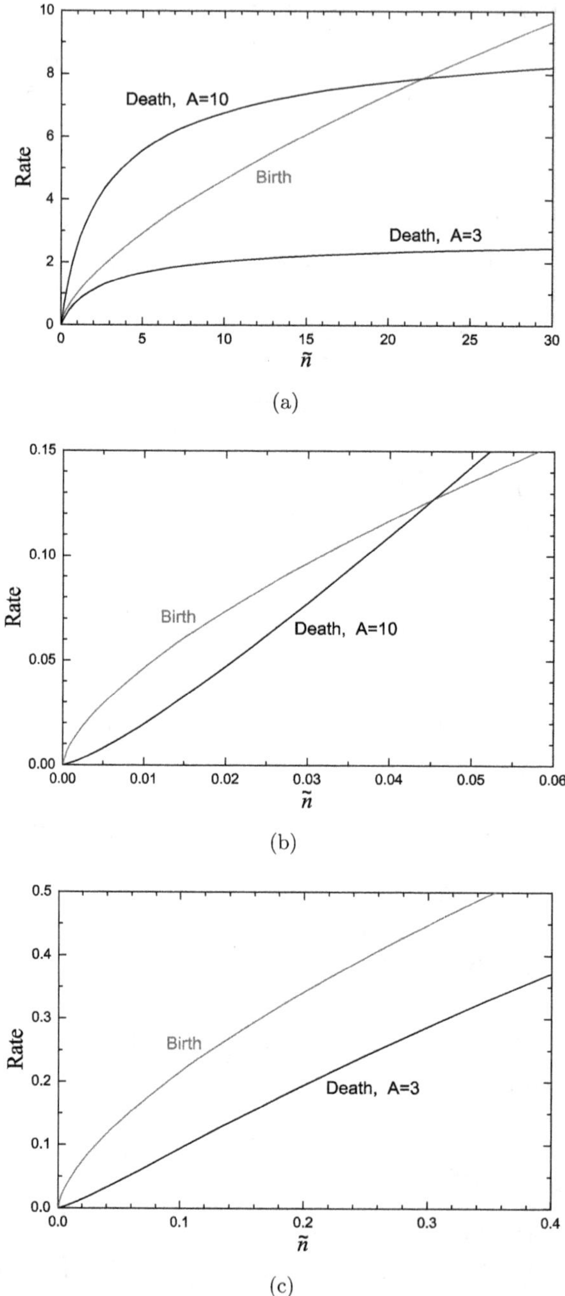

Fig. 1. (a) Birth and death rates of cells in a single tumor as a function of the normalized population of these cells according to Eq. (35). Both rates are normalized to $\tilde{k}$. The death rate is shown for $A \equiv \tilde{\mu}/\tilde{k} = 10$ and 3. In the former case (with $A = 10$), there are stable and unstable steady-state solutions of Eq. (35). The stable solution is nearly not visible, because the corresponding $\tilde{n}$ is small. The unstable solution with $\tilde{n} \simeq 22$ is well visible. In the latter case (with $A = 3$), there are no steady-state solutions. [(b) and (c)] The same rates at small values of $\tilde{n}$.

sufficiently small, there are no steady-state solutions, and the tumor growth is not limited.

Typical transition kinetics, calculated by using Eqs. (34) for two tumors in the practically interesting case when initially one tumor is large and close to that predicted under steady-state conditions in the absence of the other tumor while the other tumor is relatively small, are shown in Fig. 2. With increasing time, the tumors are seen to remain large and small, respectively.

3.3. *Model 2*

The second model of the phenomenon under consideration is similar to the first one. The only difference is that the rate of birth of cells in a tumor, $kn_i^{2/3}$ in Eq. (22), is now approximated as $k_\circ + kn_i$, where $k_\circ$ and k are constants. In both these phenomenological expressions, the birth rate per cell slightly decreases with increasing n_i. Concerning the latter expression, one can notice that the rate remains finite at $n_i \to 0$, and accordingly can hardly be accepted in this limit. In the following analysis, this is not a problem because the cell populations in cells are considered to be not too small.

With the modification proposed, Eq. (22) is replaced by

$$dn_i/dt = k_\circ + kn_i - rn_i^{2/3}c, \tag{36}$$

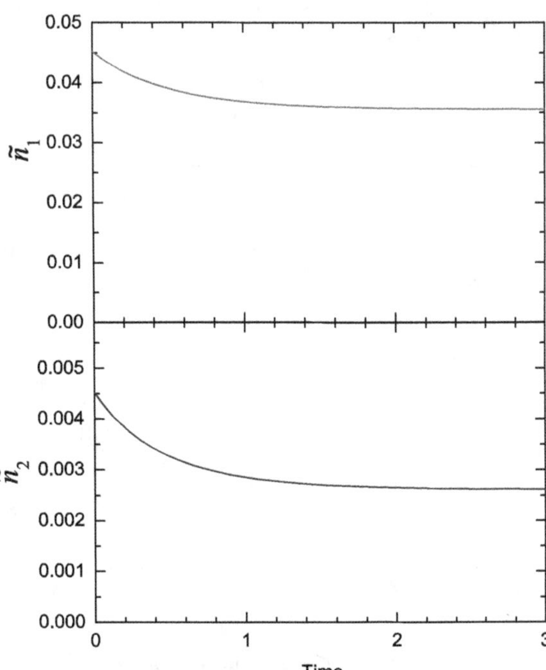

Fig. 2. Normalized populations of cells in two tumors as a function of the dimensionless time, $\tilde{k}t$, according to Eqs. (34) with $\tilde{\mu}/\tilde{k} = 10$.

240   *V. P. Zhdanov*

while the other equations [(23)–(25)] defining the model remain the same. The latter means that expressions (26)–(28) remain to be applicable, and Eq. (22) can be rewritten as

$$\frac{dn_i}{dt} = k_o + k n_i - \frac{\mu n_i^{2/3} \sum_j n_j^{2/3}}{\left[1 + \sum_j (n_j/K)^{2/3}\right]\left[1 + \sum_j (n_j/K_*)^{2/3}\right]}. \tag{37}$$

In contrast with Eqs. (30), the right-hand part of the latter equations cannot be factorized. For this reason, the asymptotic behaviors of Eqs. (30) and (37) are different. In particular, Eqs. (37) predict that eventually the tumors reach the same size.

To illustrate the kinetics predicted by this model, I again set $K_* \equiv K$, use the normalized rate constant, $\tilde{\mu} \equiv \mu K^{1/3}$, and populations, $\tilde{n}_i = n_i/K$, and then rewrite Eqs. (37) as

$$\frac{d\tilde{n}_i}{dt} = k_o + k\tilde{n}_i - \frac{\tilde{\mu} \tilde{n}_i^{2/3} \sum_j \tilde{n}_j^{2/3}}{\left[1 + \sum_j \tilde{n}_j^{2/3}\right]^2}. \tag{38}$$

For a single tumor, these equations are reduced to

$$\frac{d\tilde{n}}{dt} = k_o + k\tilde{n} - \frac{\tilde{\mu} \tilde{n}^{4/3}}{[1 + \tilde{n}^{2/3}]^2}. \tag{39}$$

The kinetics predicted by the latter equation are qualitatively similar to those predicted by Eq. (35) (cf. Figs. 1 and 3). For two tumors, Eqs. (38) predict that with increasing time, the sizes of tumors become equal (Fig. 4).

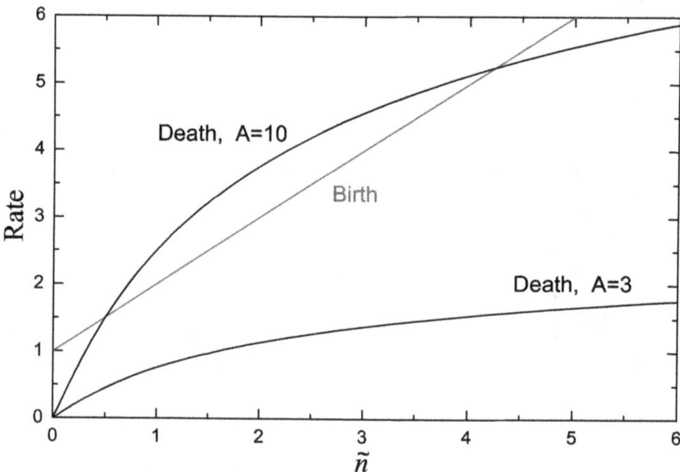

Fig. 3.   Birth and death rates of cells in a single tumor as a function of the normalized population of these cells according to Eqs. (39). Both rates are normalized to k. The birth and death rates are shown for $k_o/k = 1$ and $A \equiv \tilde{\mu}/k = 10$ and 3, respectively. In the former case (with $A = 10$), there are stable and unstable steady-state solutions of Eq. (39). In the latter case (with $A = 3$), there are no steady-state solutions.

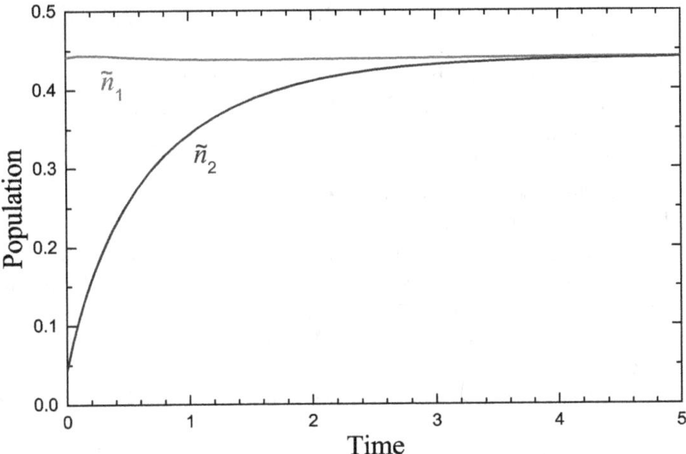

Fig. 4.　Normalized populations of cells in two tumors as a function of the dimensionless time, kt, according to Eq. (39) with $k_o/k = 1$ and $\tilde{\mu}/k = 10$.

3.4. *Summary*

The models proposed in this section show that the kinetics of the interplay of tumors and the immune system may be different depending on the specifics of the birth and death processes. In general, there is either unlimited growth or transition to a stable steady state. In the latter case, tumors of different sizes tend either to reach the same size or to remain to be of different sizes.

Finally, I notice that stable steady states are predicted by various models of tumor growth. From the perspectives of reality, the stability of such states can, of course, be maintained only for a while because eventually in the absence of suitable treatment, the growth of cancer cells is unlimited anyway owing to the processes (e.g., formation of numerous metastases and new tumors or global deterioration of the immune system (cf. Sec. 2)) which where not taken into account.

P.S. Among new related studies, I may mention review 43, experiments 44, and theoretical analysis 45.

References

1.　K. Curtius, N. A. Wright and T. A. Graham, *Nat. Rev. Cancer* **18**, 19 (2017).
2.　S. Maman and I. P. Witz, *Nat. Rev. Cancer* **18**, 359 (2018).
3.　T. Shibue and R. A. Weinberg, *Nat. Rev. Clin. Oncol.* **14**, 611 (2017).
4.　S. Spranger and T. F. Gajewski, *Nat. Rev. Cancer* **18**, 139 (2018).
5.　H. Liu, A. A. Kiseleva and E. A. Golemis, *Nat. Rev. Cancer* **18**, 511 (2018).
6.　B. T. Finicle, V. Jayashankar and A. L. Edinger, *Nat. Rev. Cancer* **18**, 619 (2018).
7.　R. A. Weinberg, *The Biology of Cancer* (Garland Science, New York, 2014).
8.　S. I. Grivennikov, F. R. Greten and M. Karin, *Cell* **140**, 883 (2010).
9.　H. Gonzalez, C. Hagerling and Z. Werb, *Gen. Dev.* **32**, 1267 (2018).
10.　M. Binnewies *et al.*, *Nat. Med.* **24**, 541 (2018).

11. D. Wodarz and N. L. Komarova, *Dynamics of Cancer: Mathematical Foundations of Oncology* (World Scientific, Singapore, 2014).

12. P. M. Altrock, L. L. Liu and F. Michor, *Nat. Rev. Cancer* **15**, 730 (2015).

13. A. R. A. Anderson and P. K. Maini, *Bull. Math. Biol.* **80**, 945 (2018).

14. P. Dogra *et al.*, *Biomed. Microdev.* **21**, 40 (2019).

15. R. C. Rockne *et al.*, *Phys. Biol.* **16**, 041005 (2019).

16. S. Singh, P. Sharma and P. Singh, *Biophys. Rev. Lett.* **12**, 69 (2017).

17. S. Khajanchi, *Biophys. Rev. Lett.* **12**, 187 (2017).

18. D. H. Margarit, L. Romanelli and A. J. Fendrik, *Biophys. Rev. Lett.* **13**, 19 (2018).

19. D. Burini, E. De Angelis and M. Lachowicz, *Commun. Appl. Ind. Math.* **9**, 106 (2018).

20. W. O. Oduola and X. Li, *Cancer Inf.* **17**, 1 (2018).

21. Z. Szymanska, M. Cytowski, E. Mitchell, C. K. Macnamara and M. A. J. Chaplain, *Bull. Math. Biol.* **80**, 1366 (2018).

22. I. Granata, E. Troiano, M. Sangiovanni and M. R. Guarracino, *BMC Bioinform.* **20**, 162 (2019).

23. W. Liang, Y. Zheng, J. Zhang and X. Sun, *BMC Bioinform.* **20**, 203 (2019).

24. C. Tomasetti and B. Vogelstein, *Science* **347**, 78 (2015).

25. C. Tomasetti and B. Vogelstein, *Science* **355**, 1330 (2017).

26. M. A. Nowak and B. Waclaw, *Nature* **355**, 1266 (2017).

27. X. Zhang, H. Fröhlich, D. Grigoriev, S. Vakulenko, J. Zimmermann and A. G. Weber, *Sci. Rep.* **8**, 3388 (2018).

28. I. Bozic *et al.*, *Proc. Natl. Acad. Sci. USA* **107**, 18545 (2010).

29. T. Antal and P. L. Krapivsky, *J. Stat. Mech.-Theory Exp.* P08018 (2011).

30. J. Foo, K. Leder and F. Michor, *Phys. Biol.* **8**, 015002 (2011).

31. B. Werner, D. Dingli and A. Traulsen, *J. R. Soc. Interf.* **10**, 0349 (2013).

32. B. Bauer, R. Siebert and A. Traulsen, *J. Theor. Biol.* **358**, 52 (2014).

33. C. D. McFarland, L. A. Mirny and K. S. Korolev, *Proc. Natl. Acad. Sci. USA* **111**, 15138 (2014).

34. M. W. L. Teng, J. Galon, W.-H. Fridman and M. J. Smyth, *Clin. Invest.* **125**, 3338 (2015).

35. A. Talkington, C. Dantoin and R. Durrett, *Bull. Math. Biol.* **80**, 1059 (2018).

36. A. Besse, G. D. Clapp, S. Bernard, F. E. Nicolini, D. Levy and T. Lepoutre, *Bull. Math. Biol.* **80**, 1084 (2018).

37. M. Qomlaqi, F. Bahrami, M. Ajami and J. Hajati, *Math. Biosci.* **292**, 1 (2017).

38. V. A. Kuznetsov, I. A. Makalkin, M. A. Taylor and A. S. Perleson, *Bull. Math. Biol.* **56**, 295 (1994).

39. D. Kirschner and J. C. Panetta, *J. Math. Biol.* **37**, 235 (1998).

40. H. Moorea and N. K. Li, *J. Theor. Biol.* **227**, 513 (2004).

41. Y. Dong, R. Miyazaki and Y. Takeuchi, *Discrete Continuous Dyn. Syst. B* **19**, 55 (2014).

42. F. R. Macfarlane, T. Lorenzi and M. A. J. Chaplain, *Bull. Math. Biol.* **80**, 1539 (2018).

43. G. E. Mahlbacher, K. C. Reihmer and H. B. Frieboes, *J. Theor. Biol.* **469**, 47 (2019).

44. C. E. DeSantis *et al.*, *CA Cancer J. Clin.*, in press, doi.org/10.3322/caac.21577.

45. R. F. Alvarez, J. M. Barbuto and R. Venegeroles, *J. Theor. Biol.* **471**, 42 (2019).

Chapter 12

Comparison of the Atavistic Model of Cancer to Somatic Mutation Theory: Phylostratigraphic Analyses Support the Atavistic Model

Charles H. Lineweaver

Research School of Astronomy and Astrophysics
Research School of Earth Sciences
Australian National University
Canberra, ACT, Australia
charley.lineweaver@anu.edu.au

Paul C.W. Davies

Beyond Center, Arizona State University
Tempe, Arizona 85281, USA
Paul.Davies@asu.edu

We review the atavistic model of cancer[1] and compare it with the leading model: Somatic Mutation Theory (SMT). We identify their differences and describe specific predictions of the atavistic model that make it more easily tested than SMT. The increasingly dense phylogenetic tree of all life on Earth has permitted a range of phylostratigraphic analyses which support the atavistic model.

Keywords: Cancer; hallmarks; somatic mutation theory; atavistic model.

1. Hallmarks of Cancer

1.1. *Descriptions of crime do little to prevent crime*

Police and prosecutors spend much time asking: Who did it? Who was the victim? Where and when did it happen? What kind of gun was used? Was the shooter right- or left-handed? How did they get into the house? Did they steal any money? Were there accomplices?

Such a detailed description of a crime doesn't contribute much to crime prevention. Whatever the crime, the perpetrator started out life as an innocent baby. How did an innocent baby become a murderer? Relatively little time is spent addressing this more basic issue. Strong correlations between crime and poverty, abusive upbringing, addiction and racism, are somehow considered too theoretical, too broad and too difficult to address.[2,3] After finding the defendant guilty, the crime is usually ascribed to the "evilness" of the defendant. The real causes of crime are ignored. And yet we like to think we are doing our best to prevent crime.

The conventional approach to preventing cancer resembles the approach to preventing crime. Like crime, cancer ruins many lives.[4] Like the legal system, oncologists usually focus on the excruciatingly complex molecular details of cancer[5,6]: How did cancer get in? How did mammary CAFs recruit endothelial precursor cells? When did CDK inhibitors regulate cyclin-CDK complexes? When and where did TGFβ control EMT? How was EMT induced by the NF-κB signalling pathway? How did STAT3/5 and E2A TF act as signal intermediaries for interleukin-5 receptors? And so on. These technical descriptions are incredibly detailed, but like the legal system they sidestep the ultimate causes of cancer.

By contrast, our atavistic model[1] addresses the foundational questions: What is cancer? Why does it exist? What is its place in the great sweep of evolutionary history? We do this by tracing the deep evolutionary roots of cancer back billions of years. Doing so not only provides a new conceptual framework for understanding cancer as a biological phenomenon, but has very specific implications for therapy.[7] Above all, the theory is readily testable in a quantitative way in aspects where the dominant Somatic Mutation Theory (SMT) is silent. Before describing the essentials of the atavism theory, we first review the hallmarks of cancer and the basics of the prevailing SMT.

1.2. *Hallmarks: Symptoms or causes?*

Although the vast majority of cancer research focuses on technical specifics, a helpful over-arching framework has been provided in terms of cancer 'hallmarks'.[8–11] These include:

(1) Insensitivity to immune destruction.
(2) Angiogenesis; cancer cells acquire a sustained ability to create blood vessels, allowing the tumor to grow beyond the limitations of passive nutrient diffusion.
(3) Unregulated cell proliferation; cancer cells acquire a self-sufficiency in growth signals, an insensitivity to anti-growth signals, an ability to evade apoptosis and to evade signals to senesce. This leads to unregulated proliferation surpassing Hayflick's limit.
(4) Invasion and metastasis; cancer cells acquire the ability to grow independently of tissue anchoring, to invade neighbouring tissues, and to metastasize to distant sites.
(5) Reprogramming metabolism; cancer cells shift their metabolism from oxidative phosphorylation to glycolysis even when oxygen is available: the Warburg Effect.

However, Hanahan and Weinberg[9] acknowledge the limited nature of this list:

> "... *it is now clear that the biology of tumors can no longer be understood simply by enumerating the traits of the cancer cells, but instead must encompass the contributions of the tumor microenvironment to tumorigenesis.*"

In spite of this big picture characterization, almost all cancer research continues to focus on molecular minutiae – much of it based on massive sequencing efforts of uncertain significance. This huge effort to define cancer in ever-more elaborate molecular terms has not led to substantial improvements in outcomes for most cancer patients.[12,13] Similarly, detailed descriptions of crimes, crime scenes, murder weapons, and accomplices have not led to substantial improvements in crime prevention.

In addition, caution is necessary when it comes to the causal interpretation of the enumerated traits. The hallmarks might be caused by cancer, or simply facilitate its progress, or be incidental common features or just *be* what cancer is. Likewise, long, complex lists of molecular correlations cannot distinguish symptoms from causes. Consider, for example, cancer aneuploidy. Salmina *et al.*[14] write:

> *"This is followed by autokaryogamy and a homologous pairing preceding a bi-looped endoprophase. The associated RAD51 and DMC1/γ-H2AX double-strand break repair foci are tandemly situated on the AURKB/ REC8/kinetochore doublets along replicated chromosome loops, indicative of recombination events. MOS-associated REC8-positive peri-nucleolar centromere cluster organises a monopolar spindle."*

Are these complicated molecular features causes or symptoms of cancer? Is aneuploidy in general a cause or a symptom of cancer? Nobody knows. The biggest obstacle to preventing or treating cancer is not a lack of detailed information about the molecular mechanisms, it is our poor theoretical understanding of what cancer is as a biological phenomenon.[15,16] In our view, fine-grained mechanistic accounts of cancer are not very useful without a valid interpretational framework. Given that cancer is widespread in multicellular life, that framework will necessarily be based on evolutionary biology. As Dobzhansky famously remarked,[17] *". . . nothing makes sense in biology except in the light of evolution. . ."* Evolutionary biologists recognize the need to distinguish proximate from ultimate causes:[18]

> *"As pointed out by ethologist Nikolaas Tinbergen and evolutionary biologist Ernst Mayr, there exist two distinct mechanisms of life: proximate and ultimate (Tinbergen 1963, Mayr 1961). Proximate mechanisms refer to what organisms are, studied by molecular, systems and behavioral biology; and ultimate mechanisms refer to why organisms have come to be as they are, the subject in evolutionary biology. Fields on the two sides have been developing in parallel with little crosstalk."*

Advances in oncology are severely hampered by a dearth of crosstalk between those who study the proximate molecular mechanisms and the more theoretically-minded with interests in the ultimate cause of cancer. Comparing the atavistic model with SMT can help promote such crosstalk.

2. The Dominant Theory: Somatic Mutation Theory (SMT)

The Somatic Mutation Theory (SMT), developed over the past five decades, postulates that cancer is the outcome of an evolutionary process in the host organism.

According to Nowell[19]: "*tumors... develop over time through a series of stepwise somatic mutations followed by repeated selection...*" Cairns[20] argued that '*cancer can be viewed as the operation of Darwinian selection among competing populations of dividing cells.*' More recent papers develop SMT further.[21–28] A concise account of SMT is provided by Stratton *et al.*[25]:

> "*All cancers are thought to share a common pathogenesis. Each is the outcome of a process of Darwinian evolution occurring among cell populations within the microenvironments provided by the tissues of a multicellular organism. Analogous to Darwinian evolution occurring in the origins of species, cancer development is based on two constituent processes, the continuous acquisition of heritable genetic variation in individual cells by more-or-less random mutation and natural selection acting on the resultant phenotypic diversity. ... Occasionally, however, a single cell acquires a set of sufficiently advantageous mutations that allows it to proliferate autonomously, invade tissues and metastasize... The merit of an increased somatic mutation rate with respect to the development of cancer is that it increases the DNA sequence diversity on which selection can act...*"

About a decade ago, SMT emerged as the consensus model of cancer, as summarized in Pepper *et al.*[29]: "*somatic (within-body) cellular selection and evolution is the fundamental process by which neoplasms arise, acquire malignancy, and evade therapeutic interventions*". Today, SMT remains the dominant model.[30,31] In a "Consensus Statement" Maley *et al.*[32] describe the SMT point of view: "*the ecology of the microenvironment of a neoplastic cell determines which changes provide adaptive benefits.*"

Godfrey-Smith[33] refers to the type of somatic cell evolution invoked by SMT as "within-organism evolution" because it occurs inside the body. Such "internal Darwinism" involves competition between somatic cells, not between individual multicellular organisms. Thus, 'evolutionary processes' in SMT contrast sharply with efforts to trace the evolution of cancer mechanisms over billions of years via conventional Darwinism – an approach that is the foundation of the atavism theory. These two uses of the term 'evolution' are often conflated and need to be distinguished.

3. The Atavistic Model of Cancer

3.1. *Brief summary of the atavistic model*

The notion that cancer is a sort of throwback, or reversion to an ancestral phenotype, is neither new nor particularly contentious. Early precursors to what we term the atavistic model of cancer can be found as far back as the nineteenth and early twentieth centuries,[34–37] and have been discussed in more modern versions by Soto &

Sonnenschein[38] and Vincent.[39] In Refs. 1 & 7, we proposed our own variant of the atavistic model of cancer. The main ingredients are:

(1) Pre-existence: the hallmark capabilities of cancer are postulated to be pre-existing in that they are based on latent functions already in the genomes of normal human cells. The capabilities acquired by cancer therefore cannot be the result of *newly* evolved traits as postulated by the internal Darwinism of SMT. The latent functions appropriated by cancer must be of some use to modern organisms, for example, during embryogenesis, development, wound healing, tissue protection and maintenance, or in the immune system, otherwise they would not already be in our genomes.[40] This point is summarized in Table 1, where we divide cancer hallmarks into loss-of-function, gain-of-function and latent functions. Functions lost are obviously pre-existing. We contend that the same is true of functions gained. In our theory 'gain' does not imply novelty; rather, it results from a reversion to a deeply embedded, pre-existing function.

(2) Reversion: The emergence of multicellularity (between approximately 1.5 to 0.5 billion years ago) involved the evolution and regulation of cell differentiation and cooperation. During this transition, previously free-living, independently-reproducing cells differentiated into germ cells and into the many types of somatic cells in multicellular bodies.[41] During this evolution of multicellularity many pre-multicellular capabilities (that had evolved between ~ 4 and 1.5 billion years ago) became regulated and suppressed, resulting in controlled proliferation and cellular cooperation. The relatively new genes responsible for this regulation include what are now known as tumor suppressors. Damage to these genes results in loss-of-function. The atavistic theory readily explains the correlation between loss-of-function and gain-of-function in cancer (Table 1, columns 2 and 3). The gains-of-function resemble atavisms in the sense that they are reversions to latent functions. These functions were, however, not latent ~ 4 to 1.5 billion years ago.

(3) The age differences ~ 4 to 1.5 billion years for functions gained and ~ 1.5 to 0.5 billion years ago for functions lost, is responsible for a differing degree of vulnerability to carcinogens. The more recently-evolved regulators are lost, allowing the older, less-constrained proliferation to reappear. The unicellular functions are more integrated and robust and should be considered a major strength of cancer. Thus, antimitotic therapies are a 'target-the-strength' therapy: life has had 4 billion years to evolve defenses against attacks on its proliferative prowess (which is, after all, the most basic feature of life). Not surprisingly therefore, target-the-strength therapies produce profound side-effects in normal cells in which the usually-latent capabilities are needed. Interpreted within the atavistic model, loss-of-function can provide the basis for a 'target-the-weakness' therapeutic strategy.[7]

 C. H. Lineweaver & P. C. W. Davies

Table 1. Deconstruction of the Hallmarks of Cancer into Functions Lost, Gained and Latent[a]

1	2	3	4
Hallmarks of Cancer[8-11]	**Lost**	**Gained**	**Latent**
	regulatory functions of normal cells lost in cancer; these functions regulate cell differentiation and enforce cooperative behaviour	functions gained ('de-repressed') in cancerous cells	non-cancer origin of functions "gained" in cancerous cells
Insensitivity to immune destruction	adaptive immunity, complete maturation of hematopoietic cells, including the full suite of cells active in the adaptive immune system	freedom from the threat of destruction by adaptive immunity, reversion to innate immunity	innate-only immunity, fetal immunity before adaptive immunity becomes functional
Angiogenesis; cancer cells acquire a sustained ability to create blood vessels, allowing the tumor to grow beyond the limitations of passive nutrient diffusion.	normal well-regulated angiogenesis	unregulated, rapid and aggressive angiogenesis	placental implantation, embryogenesis, organ formation, wound healing
Unregulated cell proliferation; cancer cells acquire a self-sufficiency in growth signals, an insensitivity to anti-growth signals, an ability to evade apoptosis and evade signals to senesce; this leads to unlimited replication surpassing Hayflick's limit.[42]	well-regulated cell proliferation, Hayflick limit, p53, cell cycle checkpoints, signalling to control proliferation and bring an end to wound healing	unregulated cell proliferation, relief from curfew and checkpoints, no Hayflick limit, wound healing that doesn't stop stem cell like behaviour	tissues where rapid cell proliferation is needed,[43] embryogenesis,[40] stem cells, wound healing, tissue maintenance
Invasion and metastasis; cancer cells acquire the ability to grow independently of tissue anchoring, to invade neighbouring tissues, and to metastasize to distant sites	regulated release and adhesion to neighboring cells, E-cadherin, signalling to maintain adhesion	unregulated EMT migration, aggressive invasion and metastasis	EMT needed for tissue maintenance, wound healing, normal cell migration during embryogenesis,[5] e.g. migration of neural crest cells, placentation, tissue invasion and displacement to distant sites are normal properties of leukocytes
Reprogramming metabolism; cancer cells shift their metabolism from oxidative phosphorylation to glycolysis even when oxygen is available: the Warburg Effect.[44,45]	facultative switching between oxidative phosphorylation and anaerobic glycolysis	aerobic glycolysis (loss of switch means gain of aerobic glycolysis)	glycolysis needed in hypoxic environments during embryogenesis and during extreme physical exertion
Stemness; Stem cell-like behaviour[9,46-49]	normal well-regulated maturation and cell differentiation, loss of terminal differentiation in maturational cascades	unregulated and truncated cell-differentiation cascades, maturation blocks, stem-cell like behaviour	maintenance of stem-like cell behaviour is needed to replace damaged cells in wound healing

[a]The capabilities listed here, and many of the correspondences between them, are supported by a vast amount of literature.[5,33,40,43,50-57]

3.2. *Pre-existence of hallmark capabilities*

In the atavistic model[1,7] there are no newly evolved hallmark capabilities,[11] only reversion to pre-existing hallmark capabilities. Care is therefore needed when describing cancer as an 'evolutionary process'. SMT uses the language of genomic instability, random mutations and *"Darwinian selection among competing populations of dividing cells."*[20] Random mutations provide the raw material which selection sculpts into something that gives cancer cells an advantage over normal cells. By analogy with Darwinian organismal evolution, one might expect the SMT to predict that populations of cancer cells diverge both genotypically and phenotypically as cancer progresses. Such is not observed to be the case. Although, genotypic divergence (in mostly passenger mutations) is seen,[58] at the same time we see phenotypic convergence as cancers acquire the same hallmark capabilities. To explain this oddity in the context of SMT, appeal is made to the concept of convergent evolution.[59–63] This explanation requires the mutations in different patients to significantly overlap, and selection pressures to be strong enough and similar enough for the cancer in the different patients to display the common hallmark capabilities.[8–10] However, Martincorena *et al.*[64] find no evidence for the negative selection expected in SMT.[19]

There cannot be convergent evolution unless there is first divergence from a common ancestor. The language of SMT – genomic instability, random mutations and Darwinian evolution - implies that the genetic alterations and mutations provide the necessary divergence to allow selection to produce convergent outcomes. However, there is no evidence to support this supposition. Indeed, if the SMT explanation were correct, we might expect to see *more* phenotypic divergence between different kinds of cancer as cancer progresses. Instead we see the opposite: a pattern of cancer cells becoming more similar and stem-cell-like.

In the atavistic model, the similarity of the hallmarks in different cancer patients is due to the similarity of the genomes of all humans. Cancer is understood as the result of deep homology (shared features among all humans) rather than convergent evolution. The distinctions between the two theoretical models is summarized in Fig. 1.

Several other features of cancer seem paradoxical in the framework of SMT but expected in the atavistic model. For example, quite generally, mutations "are not constructive but destructive in their action".[65] Mutations are a wrecking ball. Many SMT-inspired sequencing projects are predicated on the idea that there are meaningful patterns hidden in the wreckage; such efforts are in our view futile. Another feature hard to reconcile with the SMT is the "progression puzzle": the extremely low probability of simultaneously evolving the combination of the several abilities required for metastasis – and yet that is what cancers often do.[66,67] In the atavistic framework there is no mystery; for example, a cell's ability to move in and out of the vascular system and around the body is needed for wound healing, and evolved over a long period hundreds of millions of years ago.

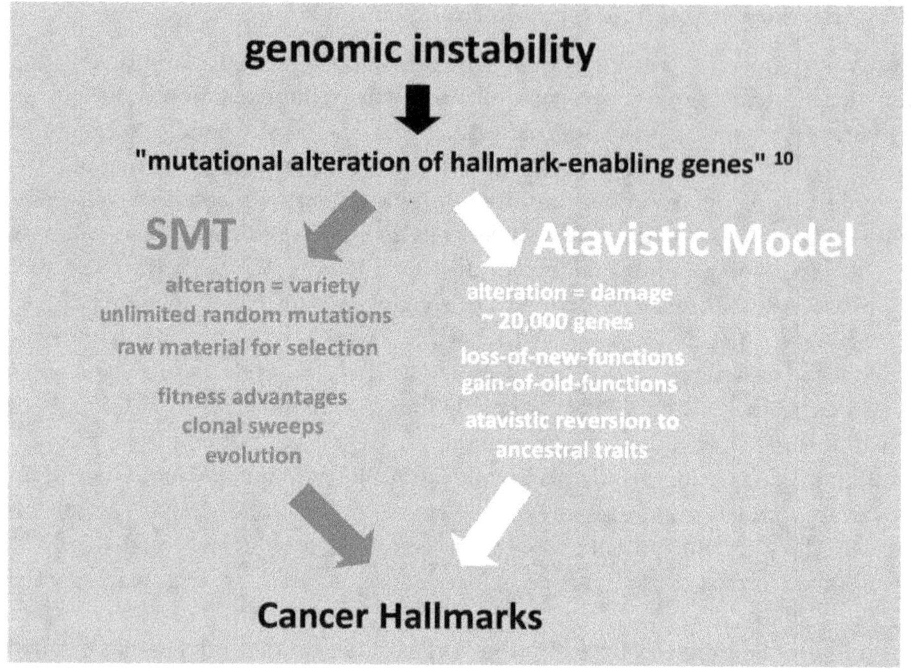

Fig. 1. The SMT (grey) and atavistic model (white) have different explanations for how genomic instability (top) leads to the acquisition of the hallmarks of cancer (bottom). SMT emphasizes an evolutionary process based on selection of cells with random, potentially unlimited genomic alterations. The atavistic model emphasizes the limitation of evolution in cancer to losses-of-function and gains-of-function. The gains are inevitably reversions to pre-existing capabilities. Thus, the atavistic model has to explain the hallmarks with less than $\sim 20,000$ genes, while SMT has an unlimited supply of randomly mutated genes from which to select and produce the hallmarks.

The "aneuploid paradox"[68,14] is also puzzling. Aneuploidy in normal cells decreases proliferation, but is a hallmark of hyper-proliferative cancer cells. How can such deviation from normal diploidy contribute to making cancer cells more fit than normal cells? An explanation offered by SMT proponents is that selection for proliferation might be accelerated by aneuploidy-induced genomic instability.[68] The atavistic explanation is that there must be good adaptive evolutionary reasons for the phenomena of chromothripsis and aneuploidy (seen for example in unicellular eukaryotes) which have been conserved until today, possibly to help manage mitotic failures or shuffle genes.[69,70]

3.3. *Correspondences between lost, gained and hallmark functions in Table 1*

In this section we provide a detailed discussion of Table 1. The hallmarks of cancer[8–10] (column 1) are often recognized as combinations of two processes: loss-of-function (column 2) and gain-of-function (column 3). The functions lost in cancer are the functions of normal cells necessary for the regulation of differentiation and cooperation of cells in a multicellular organism. As cancer progresses, the genetic wiring of these normal regulatory functions is first partially corrupted, then severely damaged (by the many types of carcinogen: UV radiation, smoke, viruses, etc) and then eventually lost. This is often called the loss-of-function of tumor suppressors. In normal cells, column 3 functions are usually suppressed and regulated by column 2 functions. As the regulators in column 2 disappear, column 3 functions appear. The functions gained (column 3) are the functions that the tumor suppressors in column 2 can no longer effectively suppress.

Column 4 lists the special times, places or processes in normal cells during development in which these usually latent molecular mechanisms are useful. This is often during embryogenesis and wound healing when rapid cell proliferation is required. These same latent molecular mechanisms (when unregulated in cancer cells by loss of column 2 functions) are responsible for cancer's gain-of-functions. In other words, the genetic networks underlying column 3 functions are *already* part of the human genome.

3.4. *Atavistic interpretation of the correspondences in Table 1*

The basic hypothesis of the atavistic theory is that what is reverted to during cancer must be already present in normal cells. Most biologists agree that *"evolution is mostly the reutilization of essentially constituted genomes"*[71] since it is obviously easier to co-opt or revert to existing pathways than to evolve new ones. The lost functions (column 2) are obviously based on normal cell functions. It is not as obvious that the gained functions – the functions driven by oncogenes (column 3) – are also based on latent pre-existing functions (column 4). So, evidence of this offers support for the atavism theory. Such evidence is in fact available. It comes from studies demonstrating that the functions gained (column 3) are indeed based on pre-existing ancient gene regulatory networks (GRN),[5,40,56] and suppressed in normal differentiated cells. In order for such ancient capabilities to be conserved until today, they must currently be serving some fundamental role, such as in early embryogenesis, wound healing and tissue maintenance.[40]

The atavistic model is based on (and therefore consistent with) the one-to-one correspondences listed in Table 1. These include correspondences between loss and gain (columns 2 and 3), between gain and pre-existing latent functions (columns 3 and 4) and between gain, latent and hallmarks (3, 4 and 1). These correspondences form an integral part of the atavistic model. Therefore, as we find out more about the molecular mechanisms of cancer, embryogenesis, regeneration and wound healing, the atavistic model predicts the uncovering of even closer, more comprehensive correspondences linking the gain-of-function hallmarks of cancer (column 3) to latent capabilities of normal cells (column 4). By contrast, the SMT makes no such predictions.

To further test the atavistic interpretation, more needs to be done to check how close and comprehensive the correspondences in Table 1 are. For example, if apparent differences are found between columns 3 and 4, we need to answer the question: Are those differences incidental or are they significant? Significant differences between functions gained in cancer (column 3) and the latent, normally-suppressed pre-existing functions of normal cells (column 4), would disfavour the atavistic model and lend support to SMT.

3.5. *Anabolic evolution and catabolic evolution*

It is 'a truth universally acknowledged'[72] that it takes much longer to construct a building than to knock it down. Cars, computers, machines of any kind, genetic regulatory networks, signalling pathways; all are easier to break than to make. Construction is iterative, cumulative and time-consuming. This is a recognition of the universal applicability of the second law of thermodynamics. Through predator/prey competition over many generations and millions of years, complex adaptations like sharper teeth, thicker skin, stealth, speed and tangled banks 'have been and are being evolved'.[73] In contrast, destruction can be rapid and simple. This, asymmetry between breaking and making is genetically identifiable and can be used to distinguish two different kinds of evolutionary processes (1) slow cumulative selection for an advantage by constructing something vs (2) quick selection for breaking, disabling or removing something. Biochemists make a similar distinction between the catabolism (breakdown) and anabolism (build up) of molecules. We borrow their terminology and call (1) 'anabolic evolution' (evolution that makes things) and (2) 'catabolic evolution' (evolution that breaks and removes things).

According to the atavistic model, the kind of evolution that occurs during tumorigenesis and cancer progression is catabolic. In cancer, the signalling pathways that control genetic regulatory networks necessary for the cooperation of cells in a multicellular organism, as well as the late stages of cell differentiation and maturation cascades, become damaged or broken. Whatever advantages result from this damage are those conferred by latent genetic networks, modules and signalling pathways that had been regulated by undamaged column 2 functions in Table 1. Germain[11] provides the most thorough analysis of the anabolic-catabolic asymmetry in the context of cancer:

> "... *most adaptations (and arguably the most important adaptations) in cancer cells are not complex adaptations, in other words they are not the result of cumulative evolution, but rather like one-step changes. For this reason, natural selection is not the essential component in their explanations. Instead, it is the pre-existing wiring of the cell which best accounts for these features. This is not to deny that the changes are selected for, but simply to say that the most enlightening explanatory material is already inside the cell, akin to architectural constraints. The healthy cells—their structure, possible states, pathways, and weak spots—already contain the resources to be drawn upon and developed by cancer cells. It dictates their evolution to a large extent... explanations based on the original molecular*

> *architecture of the cell do much more explanatory work than invoking natural selection. . ."*

During the catabolic evolution of tumorigenesis, cancer cells that have reverted to the newly unencumbered functions (column 3) can out-proliferate normal cells. The distinction between anabolic and catabolic evolution provides a new tool that can be used by modern analyses of cancer genomes[25,74–76] to test the atavistic model.

4. Phylostratigraphic Tests of the Atavism Theory

4.1. *Cancer's connection to the origin of multicellularity*

Despite the fact that cancer is the most studied phenomenon in biology, and that evolution is the central organizing principle around which the subject can be understood, remarkably few oncologists or cancer researchers pay much attention to evolution. Mostly this is simple oversight, but in some cases it reflects intentional neglect. For example, although Germain[11] acknowledges cancer's dependence on pre-existing wiring, he is not concerned about the *evolution* of this wiring[11]:

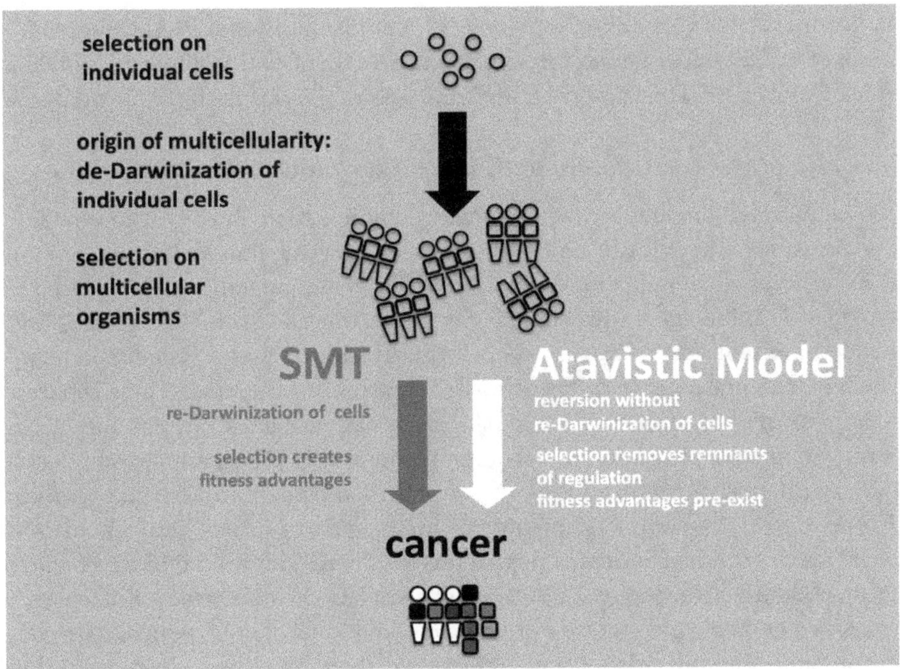

Fig. 2. Multicellular organisms evolved by the de-Darwinization[33] of single cells as they became somatic cells. SMT emphasizes that cancer is a reversal of this process; cancer is the re-Darwinization of somatic cells. The atavistic model resists the classification of cancer cells as fully re-Darwinized. The extent to which they are partially re-Darwinized can be debated. Germain[11] and Godfrey-Smith[33] argue that cells in a tumour might be too integrated to fully compete with one another, and therefore do not qualify as 'paradigmatic Darwinian populations'.

> *"By speaking of cancer development, I explicitly wish to avoid discussions of cancer in light of the evolution of multicellular life... What I am interested in is the evolution of cancer cells within a given tumour."*

In the atavistic model, the problem with this avoidance is that *"the evolution of cancer cells within a given tumour"* cannot be explained without invoking the evolution of multicellularity. The corruption and failure of both cellular differentiation and cooperation is a feature of cancer, and is readily interpreted as the atavistic reversion to less differentiated and less cooperative cellular behaviour. The functions that are lost are the regulatory capabilities that emerged with the evolution of cooperation necessary for multicellularity.[33,51–56] The genes responsible for cellular cooperation in multicellular organisms are precisely the genes that are corrupted in cancer and lead to loss of regulatory function. As Rokhsar[77] expresses it:

> *"If you are a cell in a multicellular organism, you have to cooperate with other cells in your body, making sure that you divide when you are supposed to as part of the team. The genes that regulate this cooperation are also the ones whose disruption can cause cells to behave selfishly and grow in uncontrolled ways to the detriment of the organism."*

Additional evidence for this link is provided by Aktipis and Nesse,[78] Aktipis *et al.*[79] and Moczek *et al.*[80] Support for cancer's link to the origin of multicellularity, which is not predicted by the SMT, is now substantial, based on phylostratigraphic analyses.

4.2. *Genomic phylostratigraphy with molecular clock-based gene ages*

Now that the human genome has been sequenced along with those of thousands of other organisms, we have a fairly accurate picture of how the human lineage and our genes fit into the phylogenetic tree of all life. In Fig. 3, extant humans are placed in the upper right. Our lineage is the thick diagonal line: our ancestors lived on this line. Groups of other extant species are listed horizontally along the top. The further they are to the left, the more genetically distant they are from humans. Their separate evolutionary lineages are represented by the thin diagonal lines. Such a line intersecting our own ancestral line defines a common ancestor.[81] For example, our closest cousins shown in this plot are reptiles ('node' 16). Dates can be assigned to these nodes. For example, the common ancestor of humans and reptiles lived about 0.31 billion years ago. Our lineage became metazoans ('animals') about a billion years ago (starting at node 31 with Sponges). A few thousand known species of amoeba are represented by one line which meets our lineage about 1.48 billion years ago at node 37. Our common ancestor with Eubacteria lived about 4 billion years ago. It is labelled "46" and is the deepest, most ancient node at the bottom of the plot. The nodes and estimated dates are taken from a compilation of molecular clock dating analyses.[82]

A caveat: this is not a comprehensive tree of life. We have deliberately drawn the tree to be human-centric inasmuch as our lineage is the only one displayed showing

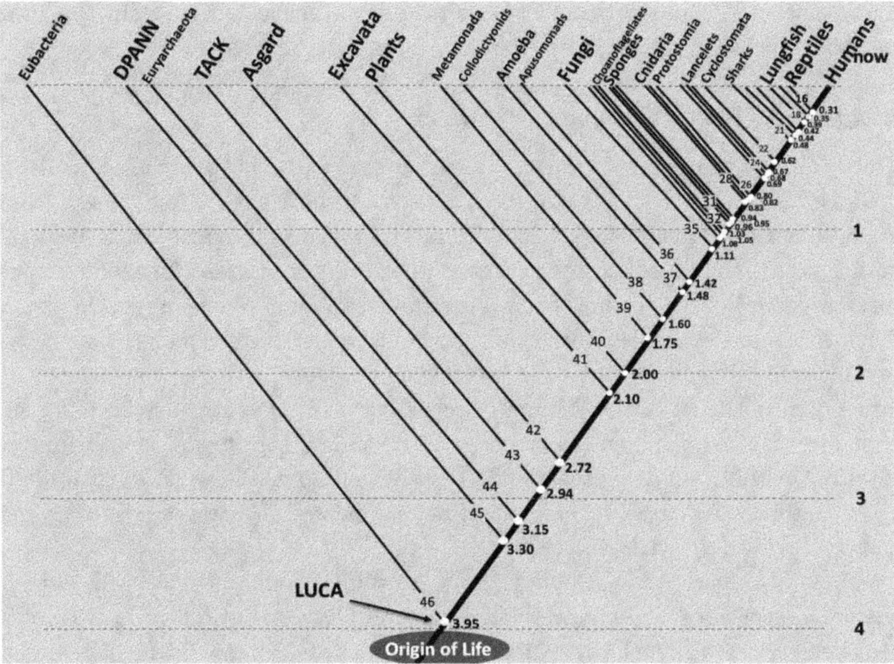

Fig. 3.   Phylogenetic tree of life on Earth showing 4 billion years of evolution. We humans are at the top right. The thick diagonal line traces the evolution of our lineage. Names of the groups of our living cousins are listed across the top at "now". Their lineages are represented by the thin diagonal lines. Their lineages diverged from our lineage at nodes labelled 16 – 46 and marked with a white dot. Age estimates of the time of divergence are given in billions of years next to the white dot. For example, humans and reptiles have a common ancestor that lived 0.31 billion years ago. The label "16" next to the node indicates this is the 16th node in the compilation of Lineweaver and Chopra.[82] Since the evolution of our multicellularity was largely complete by ∼ 300 million years ago, we have omitted the first 15 nodes and only show 30 of the nodes spanning the age range 0.31 to 4.0 billion years ago.

all the divergences from it. Once a species has diverged from our lineage, Fig. 3 ignores all subsequent divergences. For example, approximately 20,000 species of reptiles and birds are represented simply by *one* line.[81] Without this simplification, the diagram would be unintelligibly complicated.

The dates attached to the nodes of the tree allow us to estimate an age for each human gene. We omit the technical details here, but the general idea is as follows. Choose a human gene. Do a sequence similarity search for homologous genes in the tree (aka BLAST[83]) of fully sequenced genomes, starting with species that are most closely related to humans (other apes, primates, mammals). Continue the search among the genomes of the next most distant relatives (reptiles, amphibians, lungfish, coelacanths...). Keep searching for homologs of the gene among the full genomes of extant species that have common ancestors with us increasingly deep into the past. When you can no longer find a homolog (above a certain defined similarity threshold) that node becomes the age estimate of that human gene. A more detailed account is provided by Domazet-Lošo & Tautz,[84] Capra *et al.*,[85] Domazet-Lošo *et al.*[86] and

Moyers and Zhang.[87] The thirty nodes shown here were selected to be in the time range relevant to the origin of multicellularity about a billion years ago.

4.3. *Phylostratigraphic analyses of cancer*

Phylostratigraphic analyses relevant to cancer began with the pioneering work of Domazet-Lošo & Tautz.[88] They assigned ages to 'cancer-associated genes' from several cancer gene compilations by searching for sequence similarity in a 19-node stratigraphy (their Fig. 1). They found two peaks of cancer-associated-gene ages over-represented compared to the age distribution of all human gene ages. One peak was in their first pre-eukaryote unicellular node. The other peak correlated with the origin of multicellularity and metazoa in their 5th node. Despite these peaks, the most significant signal in the data was the under-representation of cancer-associated-gene ages more recent than the metazoan peak. Particularly significant is the under-representation in the most recent nodes (13-19) which span the period 400 million years ago until today. Apparently, genes that have appeared since our ancestors came on land, have little to do with cancer.

To test the atavistic model Chen *et al.*[89] examined human-breast-cell-derived xenograft tumors in mice and characterized the complete evolutionary history of a tumor. In their stratigraphy, they used the full genomes of 37 species, which yielded a dozen nodes for their age resolution. The expression profiles were found to evolve towards that of embryonic stem cells – the cell type resembling unicellular life. The most highly mutated and consistently downregulated genes in the metastatic samples were enriched in functions related to multicellularity – consistent with the previous results of Domazet- Lošo & Tautz[88] and the atavistic model. The Wu et al.[90] results from multiple myeloma cancer cells in a microfabricated ecology also *"support the model that cancer represents a revision back to ancient forms of life"*.

Cisneros *et al.*[91] used phylostratigraphy to estimate the ages of all human genes. Several different age bins were used. Based on gene ages, it was found that genes causally implicated in cancer are under-represented among genes younger than 0.5 billion years, again confirming earlier results of Domazet-Lošo & Tautz.[88] Another significant result is that some recessive cancer genes older than 900 million years are homologs of the ancient genes in bacteria that turn up mutation rates when the cells are stressed–suggesting that the well-known genomic instability of cancer can be interpreted as a reversion to an ancient prokaryotic stress response.

An Australian group led by David Goode has published three papers[92–94] based on phylostratigraphic analyses of RNA transcript sequencing data from The Cancer Genome Atlas for seven solid cancers. Their stratigraphy differentiated between "unicellular and multicellular genes" using 16 nodes. Only 6 of their nodes are earlier than 300 million years ago. Consistent with Domazet-Lošo & Tautz,[88] they found significant patterns in the relationship between gene age and expression levels in cancer. They found that genes of unicellular evolutionary origin are over-expressed in human cancers, whereas genes appearing at multicellular stages are down-regulated.

The over-expression of unicellular genes was associated with major dysregulation of the control structures imposed on unicellular processes during the evolution of multicellularity. Significantly, Trigos *et al.*[92–94] found that the atavistic signature was not a simple re-primitivisation to unicellularity. Rather it is a rewiring of the coupling between the gene networks that control unicellular processes from those that control multicellular processes. Their results thus provide evidence that cancer is a nuanced reorganization of the relationship between the unicellular and multicellular domains. Determining whether this reorganization is a reversion that can be seen during early embryogenesis or is a more recent adaptation will help decide which model, SMT or atavistic, yields a better understanding of cancer.

Phylostratigraphic support for the atavism theory also comes from the work of Chen *et al.*,[95] who looked at the phylogenetic tree of Tyrosine Kinases (TKs) which compose a major portion of oncogenes (abnormal activation of a TK gene often contributes to uncontrolled proliferation). Their results likewise *"demonstrated a general trend of atavism in tumorigenesis"*.

Zhou *et al.*[96] examined transcriptomes to compare differential gene expression between normal and cancer cells, and between embryonic and mature eptithelial cells. Starting with 11 phylostrata, they made 5 age bins: LUCA, Eukaryota, Metazoa, Vertebrata, Primata, and then reduced these to two bins: pre-metazoan and post-metazoan. They found that compared to normal cells, cancer cells were enhanced in pre-metazoan and depleted in post-metazoan gene expression. Compared to embryonic cells, differentiated epithelial cells were enriched in post-metazoan expression. Their summary: *"These findings support the atavism theory that cancer cells manifest the reactivation of an ancient ancestral state featuring unicellular modalities."*

Thus, several independent phylostratigrahic analyses of cancer support the atavistic model in various ways.

5. Summary and Discussion

We describe and compare the atavistic model and the prevailing SMT. The systematic emergence of cancer's hallmark capabilities and the correspondences presented in Table 1, have very different explanations in these two models. In contrast to SMT, the atavistic model makes many predictions about gene ages and cancer progression based on our best understanding of the origin and evolution of multicellularity within our lineage. In the atavistic model all of the many capabilities of cancer pre-exist in normal human cells, while in SMT they are interpreted as novel adaptations. We introduce the distinction between anabolic evolution (making things) and catabolic evolution (breaking things) to help distinguish the two models. In the atavistic model, the internal Darwinism of SMT, is not rejected outright, but given a secondary role. It could clear away the remnants of lost regulatory functions and deleterious mutations produced by genetic instability. We describe the technique of phylostratigraphy and review the results of recent independent genome and transcriptome phylostratigraphic analyses. They all support the atavistic model.

To paraphrase Carlo Maley (quoted in Barras[97]), the atavistic model is a source of interesting ideas, but even if it is on the right track and can predict features of cancer that SMT cannot, the human genome (upon which the predictions are based) is so complicated that there is much work to be done. We haven't found the complexity limit of cancer even if that complexity is limited to pre-existing features.

Acknowledgments

Research reported in this publication was supported by the National Cancer Institute of the National Institutes of Health under Award Number U54CA217376. The content is solely the responsibility of the authors and does not necessarily represent the official views of the National Institutes of Health.

References

1. P. C. W. Davies and C. H. Lineweaver, *Phys. Biol.* **8**, 1–7 (2007).
2. R. Poulton, T. E. Moffitt and P. A. Silva, *Soc. Psychiatry Epidemiol.* **50**(5), 679–693 (2015).
3. B. Western, *Homeward: Life in the Year After Prison*, (Russell Sage Foundation, 2018).
4. A. Jemal, F. Bray, M. M. Center, J. Ferlay, E. Ward and D. Forman, *Canadian Cancer Journal* **61**, 2, (2011).
5. R. A. Weinberg, *The Biology of Cancer*, (New York: Garland Science, 2007).
6. E. P. Gelman, C. L. Sawyers and F. J. Rauscher IIIEdts. *Molecular Oncology: Causes of Cancer and Targets for Treatment*, (Cambridge Univ. Press, Cambridge, UK, 2014).
7. C. H. Lineweaver, P. C. W. Davies and M. Vincent, *Bioessays* **36**, 827–835 (2014).
8. D. Hanahan and R. A. Weinberg, *Cell* **100**, 57–70 (2000).
9. D. Hanahan and R. A. Weinberg, *Cell* **144**, 646–674 (2011).
10. D. Hanahan and R. A. Weinberg, Chapter 2: Hallmarks of Cancer: an organizing principle of cancer medicine, in *Cancer: Principles & Practice of Oncology.* 10th Edn, Chapter 2, (Lippincott Williams & Wilkins Health Library, Wolters Kluwer, 2015).
11. P.-L. Germain, *Biol. Philos.* **27**, 785–810 (2012).
12. H. H. Q. Heng, J. B. Stevens, S. W. Bremer, K. J. Ye, G. Liu and C. J. Ye, *Cellular Biochemistry* **109**, 1072–1084 (2010).
13. R. A. Weinberg, *Cell* **157**(1) 267–271 (2014).
14. K. Salmina, A. Huna, M. Kalejs, D. Pjanova, H. Scherthan, M. S. Cragg and J. Erenpreisa, *Genes* **10**, 83 (2019), doi: 10.3390/genes10020083.
15. R. Gatenby, *Nature* **491**, 7425 S55 (2012).
16. Y. Cao, *Cell & Bioscience* **7**, 61 (2017), doi: 10.1186/s13578-017-0188-9.
17. T. Dobzhansky, *Am. Zoo.* **4**, 443–52 (1964).
18. X. Yi, *Synth. Syst. Biotechnol.* Dec: **2**, 4253–258 (2017).
19. P. C. Nowell, *Science* **194** 23–28 (1976).
20. J. Cairns, *Cancer Science and Society* (Freeman and Co, San Francisco, 1978).
21. F. Michor, Y. Iwasa and M. A. Nowak, *Nat. Rev. Cancer* **3**, 197–205 (2004).
22. L. M. F. Merlo, J. W. Pepper, B. J. Reid and C. C. Maley, *Nat. Rev. Cancer* **6**, 924–35 (2006).

23. A. R. Anderson, A. M. Weaver, P. T. Cummings and V. Quaranta, *V. Cell* **127**, 905–915 (2006).
24. J. W. Pepper, K. Sprouffske and C. C. Maley, *PLoS Comput. Biol.* **2**, 12 (2007).
25. M. R. Stratton, P. J. Campbell and P. A. Futreal, *Nature* **458**, 719–24 (2009).
26. M. Greaves and C. C. Maley, *Nature* **481** 306–313 (2012).
27. M. Gerlinger, N. McGranahan, S. M. Dewhurst, R. A. Burrell, I. Tomlinson and C. Swanton, *Annu. Rev. Genet.* **48**, 215–36 (2014).
28. M. Greaves, *Cancer Discov.* Aug; **5**(8), 806–20 (2015), doi: 10.1158/2159-8290.
29. J. W. Pepper, C. S. Findlay, R. Kassen, S. L. Spencer and C. C. Maley, *Evolutionary Applications* ISSN 1752-4571, 62–70 (2009).
30. C. C. Maley and M. Greaves edt, *Frontiers in Cancer Research: Evolutionary Foundations, Revolutionary Directions*, (Springer, 2016)
31. K. Curtius, N. A. Wright and T. A. Graham, *Nature Reviews Cancer*, **18**, 19–32 (2018).
32. C. C. Maley, A. Aktipis, T. A. Graham, A. Sottoriva, A. M. Boddy, M. Janiszewska, A. S. Silva, M. Gerlinger, Y. Yuan, K. J. Pienta, K. S. Anderson, R. Gatenby, C. Swanton, D. Posada, C.-I. Wu, J. D. Schiffman, E. S. Hwang, K. Polyak, A. R. A. Anderson, J. S. Brown, M. Greaves and D. Shibata, *Nature Reviews Cancer* **17**, 605–619 October (2017).
33. P. Godfrey-Smith, *Darwinian populations and natural selection* (Oxford University Press, Oxford UK, 2009).
34. H. Snow, *Cancers and the Cancer Process* (London: J & A Churchill Publishers, 1893).
35. M. Roberts, *Malignancy and Evolution* (Grayson and Grayson Publishers, London, 1926).
36. T. Boveri, *The Origins of Malignant Tumors* (Williams & Wilkins, Baltimore, 1929).
37. L. Israel, *J. Theor. Biol.* **178**, 375–80 (1996).
38. A. M. Soto and C. Sonnenschein, *BioEssays* **26**, 10, 1097–107, doi: 10.1002/bies.20087 (October 2004).
39. M. D. Vincent, *BioEssays* **34**, 72–82 (2012).
40. K. Naxerova, C. J. Bult, A. Peaston, K. Fancher, B. B. Knowles, S. Kasif and I. S. Kohane, *Genome Biol.* **9**, R108 (2008).
41. A. Weismann, *Das Keimplasma: eine Theorie der Vererbung [The Germ Plasm: A theory of inheritance]* (Jena: Fischer, 1892).
42. L. Hayflick, *Exp. Cell Res.* **37**(3), 614–636 (1965).
43. C. Tomasetti and B. Vogenstein, *Science*, **347**, 78-2 Jan (2015).
44. O. Warburg, *Science*, **123**, 3191 (1956).
45. M. G. Vander Heiden, L.C. Cantley and C. B. Thompson, *Science* **324**, 1029–1033 (2009).
46. T. Reya, S. J. Morrison, M. F. Clarke and I. L. Weissman, *Nature* **414**, 105–111 (2001).
47. D. L. Vaux, *Bioessays* **33**, 341–343, (2011).
48. D. Friedmann-Morvinski and I. M. Verma, *EMBO Rep.* **15**(3), 244–53 (2014).
49. K. E. Liu, *Biological Theory* **13**, 228–242 (2018).
50. Z. Zhang, A. Lei, L. Xu, L. Chen, Y. Chen, X. Zhang, Y. Gao, X. Yang, M. Zhang and Y. Cao, *J. Biol. Chem.* 2017; **292**(31), 12842–59 (2017).
51. D. M. Berman, S. S. Karhadkar, A. R. Hallahan, J. I. Pritchard, C. G. Eberhart, D. N. Watkins, J. K. Chen, M. K. Cooper, J. Taipale, J. M. Olson and P. A. Beachy, *Science* **297**, 1559–1561 (2002).
52. K. Xie and J. L. Abbruzzese, *Cancer Cell* **4**(4), 245–247 (2003).
53. F. Ragtke and H. Clevers, *Science* **307**(5717), 1904–1909 (2005).
54. L. L. Rubin and F. J. de Sauvage, *Nature Reviews Drug Discovery* **5**, 1026–1033 (2006).
55. M. Srivastava, O. Simankov, J. Chapman, B. Fahey, M. E. A. Gauthier, T. Mitros, G. S. Richards, C. Conaco, M. Dacre, U. Hellsten, C. Larroux, N. H. Putnam, M. Stanke, M. Adamska, A. Darling, S. M. Degnan, T. H. Oakley, D. C. Plachetzki, Y. Zhai, M. Adamski, A. Calcino, S. F. Cummins, D. M. Goodstein, C. Harris, D. J. Jackson, S. P.

Leys, S. Shu, B. J. Woodcroft, M. Bervoort, K. S. Kosik, G. Manning, B. M. Degnan and D. S. Rokhsar, *Nature* **466**, 5 Aug, (2010).

56. Z. Song, W. Yue, B. Wei, N. Wang, T. Li, L. Guan, S. Shi, Q. Zeng, X. Pei and L. Chen, *PLOS/ONE* doi.org/10.1371/journal.pone.0017687 March 4, (2011).

57. D. Tautz and T. Domazet-Lošo, *Nat. Rev. Genet.* **12**, 692–702 (2011).

58. C. D. McFarland, K. S. Korolev, G. V. Kryukov, S. R. Sunyaev and L. A. Mirny, *PNAS* Feb 19, **110**(8), 2910–2915 (2013).

59. S. Conway Morris, *Life's Solution: Inevitable Humans in a Lonely Universe* (CUP, UK, 2004).

60. S. Conway Morris, *The Runes of Evolution: How the Universe became Self-Aware* (Templeton Press, 2015).

61. G. R. McGhee Jr, *Convergent Evolution: Limited Forms Most Beautiful* (MIT Press, Cambridge, MA, 2011).

62. G. R. McGhee Jr, *Convergent Evolution on Earth: Lessons for the Search for Extraterrestrial Life* (MIT Press, Cambridge, MA, 2019).

63. F. Thomas, B. Ujvari, F. Renaud and M. Vincent, *BioEssays* **39**, 1700039 (2017).

64. Martincorena *et al.*, Universal Patterns of Selection in Cancer and Somatic Tissues, *Cell* **171**, 1029–1041 (2017).

65. S. Huang, *Seminars in Cancer Biology* **21**, 183–199 (2011).

66. R. Bernards and R. A. Weinberg, *Nature* **418**, 823 (2002).

67. J. Massagué and A. C. Obenauf. *Nature* **529**, 298–306 (2016), https://doi.org/10.1038/nature17038.

68. J. M. Sheltzer and A. Amon, *Trends in Genetics* **27**(11), 446–453 (2011).

69. E. C. Swart, J. R. Bracht, V. Magrini, P. Minx, X. Chen, Y. Zhou, J. S. Khurana, A. D. Goldman, M. Nowacki, K. Schotanus, S. Jung, R. S. Fulton, A. Ly, S. McGrath, K. Haub, J. L. Wiggins, D. Storton, J. C. Matese, L. Parsons, W.-J. Chang, M. S. Bowen, N. A. Stover, T. A. Jones, S. R. Eddy, G. A. Herrick, T. G. Doak, R. K. Wilson, E. R. Mardis and L. F. Landweber, *PLoS Biol.* **11**(1), e1001473 (2013), https://doi.org/10.1371/journal.pbio.1001473.

70. V. F. Niculescu, *MOJ Tumor Res.* **1**(1), 00004 (2018), doi: 10.15406/mojtr.2018.01.00004.

71. E. Zuckerkandl, Programs of Gene Action and Progressive Evolution, *Molecular Anthropology*, 387–447 (1976), doi.org/10.1007/978-1-4615-8783-5_20.

72. J. Austen, *Pride and Prejudice*, (Egerton, Whitehall, 1813).

73. C. Darwin, *Origin of Species* (1859).

74. F. Supek, B. Minana, J. Valcarcel, T. Gabaldon and B. Lehner, *Cell* **156**, 6, 1324–35 (2014).

75. A. Davis, R. Gao and N. Navin, *Biochim. Biophys. Acta* **1867** 151–61 (2017).

76. D. Chu and L. Wei, *BMC Cancer* **19**, Article number: 359 (2019).

77. D. Rokhsar, https://slate.com/technology/2017/04/cancer-has-been-with-us-since-the-origins-of-multicellularity.html (2010).

78. C. A. Aktipis and R. M. Nesse, *Evolutionary Applications* **6**, 144–159 (2012), doi: 10.1111/eva.12034.

79. C. A. Aktipis, A. M. Boddy, G. Jansen, U. Hibner, M. E. Hochberg, C. C. Maley and G. S. Wilkinson, *Philos. Trans. R. Soc. Lond. B Biol. Sci.* **370**, 1673 (2015).

80. A. P. Moczek, K. E. Sears, A. Stollewerk, P. J. Wittkopp, P. Diggle, I. Dworkin, C. Ledon-Rettig, D. Q. Matus, S. Roth, E. Abouheif, F. D. Brown, C.-H. Chiu, C. S. Cohen, A. W. De Tomaso, S. F. Gilbert, B. Hall, A. C. Love, D. C. Lyons, T. J. Sanger, J. Smith, C. Specht, M. Vallejo-Marin and C. G. Extavour, *Evol. Dev.* **17**, 198–219 (2015).

81. R. Dawkins and Y. Wong, *The Ancestor's Tale: A Pilgrimage to the Dawn of Evolution*, 2nd Edt. (Houghton Mifflin Harcourt, NY, 2016).

82. C. H. Lineweaver and A. Chopra, *Journal of Big History*, **III**(3), 69–82 (2019).

83. S. F. Altschul, W. Gish, W. Miller, E. W. Myers and D. J. Lipman. *J. Mol. Biol.* **215**, 403–410 (1990).

84. T. Domazet-Lošo and D. Tautz, *Nature* **468**(7325), 815–818 (2010a), doi: 10.1038/nature09632.

85. J. A. Capra, M. Stolzer, D. Durand and K. S. Pollard, *Trends in Genetics* Nov. 2013, **29**(11), 659–668 (2013).

86. T. Domazet-Lošo, A.-R. Carvunis, M. Mar Albà, M. S. Šestak, R. Bakarić, R. Neme and D. Tautz, (2017-01-12). *Molecular Biology and Evolution.* **34**(4), 843–856 (2017), doi: 10.1093/molbev/msw284.

87. B. A. Moyers and J. Zhang, (2018-08-01). Martin, Bill (ed.). *Genome Biology and Evolution.* **10**(8), 2037–2048 (2018), doi:10.1093/gbe/evy161.

88. T. Domazet-Lošo and D. Tautz, *BMC Biol.* **8**, 66 (2010b).

89. H. Chen, F. Lin, K. Xing and X. He, *Nature Communications* **6**, 6367 (2015).

90. A. Wu, Q. Zhang, G. Lambert, Z. Khin, R. A. Gatenby, H. J. Kim, N. Pourmand, K. Bussey, P. C. W. Davies, J. C. Sturm and R. H. Austin, *Proc. Natl Acad. Sci.* **112**(33), 10467–10472 (2015).

91. L. Cisneros, K. J. Bussey, A. J. Orr, M. Miočević, C. H. Lineweaver and P. Davies, *PLoS One* **12**(4), e0176258 (2017).

92. A. S. Trigos, R. B. Pearson, A. T. Papenfuss and D. L. Goode, *Proc. Natl Acad. Sci.* **114**, 6406–11 (2017).

93. A. S. Trigos, R. B. Pearson, A. T. Papenfuss and D. L. Goode, *Br. J. Cancer* **118**, 145–152 (2018).

94. A. S. Trigos, R. B. Pearson, A. T. Papenfuss and D. L. Goode, *eLife* **8**, e40947 (2019), doi: 10.7554/eLife.40947.

95. W. Chen, Y. Li and Z. Wang, *Scientific Reports*, (2018) **8**, 8256 (2018), doi: 10.1038/s41598-018-26653-5.

96. J. X. Zhou, L. Cisneros, T. Knijnenburg, K. Trachana, P. Davies and S. Huang, *Convergent Science Physical Oncology* **4**, 2 (2018).

97. C. Barras, New Scientist, 9 March, (2011).

Index

www.ingramcontent.com/pod-product-compliance
Lightning Source LLC
Chambersburg PA
CBHW060232120726
48009CB00004B/238